Plants

of

Southern
Ontario

Trees, Shrubs, Wildflowers,
Aquatic Plants, Grasses & Ferns

Richard Dickinson
France Royer

Lone Pine Publishing

Lone Pine Publishing
87 Pender Street East
Vancouver, BC
V6A 1S9

Website: www.lonepinepublishing.com

Library and Archives Canada Cataloguing in Publication

Dickinson, Richard, 1960–, author
      Plants of Southern Ontario / Richard Dickinson, author ; France Royer, photographer.

Includes bibliographical references and index.
ISBN 978-1-55105-906-8 (pbk.)

      1. Plants—Ontario—Identification. 2. Plants—Ontario—Pictorial works. I. Royer, France, 1951–, illustrator II. Title.

QK203.O5D53 2014   581.9713   C2013-906430-3

Editorial Director: Nancy Foulds
Project Editor: Nicholle Carrière
Editorial: Nicholle Carrière
Production Manager: Leslie Hung
Layout & Production: Volker Bodegom, Greg Brown, Alesha Braitenbach-Cartledge
Map: Greg Brown
Cover Design: Gerry Dotto

All photographs by France Royer.
Plant structure (p. 10) and glossary illustrations by Lana Kelly and the University of Alberta Press.
Illustrations for family descriptions are from *An Illustrated Flora of the Northern United States and Canada*, 2nd ed., by N.L. Britton and A. Brown. 3 vols. Charles Scribner's Sons: New York, 1913.

Disclaimer: This guide is not meant to be a "how-to" reference for using plants. We do not recommend experimentation by readers, and we caution that many plants, including some traditional medicines, may be poisonous or otherwise harmful. Self-medication with herbs is unwise, and wild plant foods should be used with caution and only with expert advice.

We acknowledge the financial support of the Government of Canada through the Canada Book Fund for our publishing activities.

 Canadian   Patrimoin
Heritage   canadien

PC: 20

TABLE OF CONTENTS

Acknowledgements

We would like to thank the following people for their support and encouragement during the production of this guide:

Brenda Beath
Chris Doran
Elemee Royer
Jean Frank
Ken Delorme and Bruce Hovey
Dee Macpherson
Bonnie McMillan
Jean Saindon and Kathie Conlin-Saindon
Marlene and Ritchie Armstrong
Pam and Ross Richardson
Nadia Talent and Graeme Hirst
Sonja Marshall
Madeleine Harding

We would also like to thank Nicholle Carrière for her tireless effort in editing this field guide.

Mossy stonecrop (*Sedum acre*)

Introduction

Plants are found on every continent and in every ocean in the world. From tropical rainforests to harsh climates such as the Kalahari Desert or the Canadian Arctic, plants thrive in a variety of habitats, and it is this diversity that has lead to the evolution of over 300,000 species of vascular plants. Vascular plants are distinguished from non-vascular plants by the presence of conductive tissue that transports food and water within the plant. They include pteridophytes, gymnosperms and angiosperms. Non-vascular plants are represented by mosses, liverworts and algae. Pteridophytes, commonly referred to as ferns and fern allies, include species that are among the oldest living vascular plants. These plants do not produce seeds, a feature that distinguishes them from gymnosperms and angiosperms. Gymnosperms, which are primarily coniferous trees and shrubs, produce seeds but lack flowers. Angiosperms, what we typically think of as "flowering plants," represent the largest component of the world's flora. Flowering plants can be divided further into monocotyledons, or monocots, and dicotyledons, or dicots. In general, monocots have parallel-veined leaves, flower parts in multiples of 3 and seeds with a single cotyledon. In Ontario, monocots are herbaceous plants and include lilies, orchids, grasses and sedges. Dicots usually have netted veins, flower parts in 4s or 5s, and 2 seed leaves. Well-known dicots include walnut trees, roses, asters and buttercups.

The area covered by this guide includes two large ecoregions, the Southern Deciduous Forest and the Great Lakes–St. Lawrence Forest.

The Southern Deciduous Forest is located along the north shores of Lake Erie and part of Lake Ontario. A small area covering less than 1% of Canada, it contains more than 25% of Canada's vascular plant species. This region boasts numerous southern plant species, including tulip-tree, magnolia, sassafras and redbud.

The Great Lakes–St. Lawrence Forest is a transition area between the Southern Deciduous Forest and the Boreal Forest. This mixedwood forest region contains numerous deciduous and coniferous tree species.

Over 2000 species of plants thrive in these two zones. Many of the plants in these regions are also found in the Boreal Forest farther north, as well as in the southern United States. Over 400 species are non-native plants that have been introduced into the natural environment.

Ecoregions of Ontario

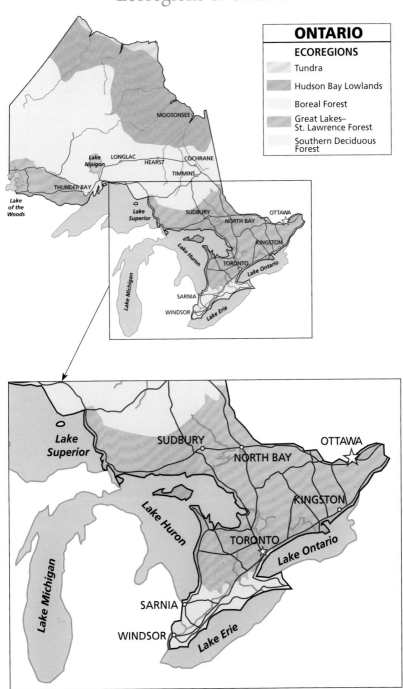

ONTARIO

ECOREGIONS

- Tundra
- Hudson Bay Lowlands
- Boreal Forest
- Great Lakes–St. Lawrence Forest
- Southern Deciduous Forest

MOOSONEE

Lake Nipigon

LONGLAC

HEARST

COCHRANE

TIMMINS

THUNDER BAY

Lake of the Woods

Lake Superior

SUDBURY

NORTH BAY

OTTAWA

KINGSTON

Lake Huron

TORONTO

Lake Ontario

Lake Michigan

SARNIA

WINDSOR

Lake Erie

Lake Superior

SUDBURY

NORTH BAY

OTTAWA

KINGSTON

Lake Huron

TORONTO

Lake Ontario

Lake Michigan

SARNIA

WINDSOR

Lake Erie

About This Book

Organization of Plant Families and Species

Species in this guide are arranged by scientific family and by species names within each plant form (i.e., trees, shrubs, wildflowers, grasses and grass-like plants, aquatic plants and ferns and fern allies). Each species appears alphabetically by the currently accepted scientific name and is also identified by locally accepted common names. It is important to remember that species belonging to the same family often share a number of common characteristics such as flower structure, fruit type and leaf arrangement. Family descriptions are usually found before the first species of each family encountered in the text and detail variations within the family. In cases where a family occurs in more than one section, the description precedes the largest number of species.

Plant Identification Keys

The identification keys in this book use thumbnail images of each species. Within each plant form, species are grouped by flower colour, leaf arrangement and other useful characteristics. The illustrated glossary and index also assist the reader in identifying a species.

Plant Names

Many plants have several common names but only one scientific name. The scientific name, written in Latin, is usually derived from Latin, Greek or Arabic words. Occasionally words from other languages are used. The scientific name is understood throughout the world's scientific community. For example, the common dandelion is known as *pissenlit* in French, but the scientific name, *Taraxacum officinale*, is not only understood by someone speaking French, but also by individuals who speak English, Russian, Japanese or any other language. A plant's scientific name is composed of a generic, or genus, name and a specific epithet or species name. *Taraxacum* is the generic name and *officinale* is the specific epithet or species name.

Occasionally, a plant species may have more than one scientific name. In the early days of botanical exploration, many plants were collected and given a new scientific name, even though they had been previously discovered and named by someone else. Since then, the botanical community has compared and determined many of these "new" species to have been previously named. These new scientific names are treated as synonyms in accordance with the standards set by the Integrated Taxonomic Information System (ITIS), www.itis.gov/index.html.

Blue-eyed grass (*Sisyrinchium montanum*)

Scarlet pimpernel (*Anagallis arvensis*)

How to Identify Plants in This Field Guide (1 pg)

This field guide is divided into 6 sections based on plant form and/or habitat. Thumbnail image identification keys appear at the beginning of each section. Each section is further divided based on easily identifiable characteristics. Below each photograph is the corresponding page number for that species.

To identify your specimen, first answer this question:

Is your plant a …

Tree (woody plant with a single main stem)?
Go to page 12.

Shrub or woody vine (woody plant often with several stems rising from the base)?
Go to page 77.

Wildflower (non-woody plant with showy flowers)?
Go to page 129.

Aquatic plant (submerged or floating on the surface of the water)?
Go to page 411.

Grass, sedge or rush (plants with narrow leaves and non-showy flowers)?
Go to page 436.

Fern or fern ally (plants reproducing by spores and without flowers)?
Go to page 479.

Toadflax (*Linaria vulgaris*)

Tatarian honeysuckle (*Lonicera tatarica*)

Plant Structures

Flower Structures

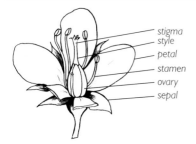

stigma
style
petal
stamen
ovary
sepal

regular flower

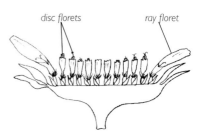

disc florets   ray floret

cross-section of a radiate head

Leaf Types

simple

*pinnately
compound*

*bipinnately
compound*

*palmately
compound*

Leaf Arrangement

basal

alternate

opposite

whorled

Trees

White spruce (*Picea glauca*)

TREES

Trees are single-stemmed, woody plants with evergreen or deciduous leaves.

Key to the Trees

Species in this section are identified by leaf type (i.e., needle-like or broadleaf).

LEAVES NEEDLE- OR SCALE-LIKE, USUALLY EVERGREEN

Leaves single

| *Juniperus virginiana* p. 17 | *Thuja occidentalis* p. 18 | *Abies balsamea* p. 19 | *Picea abies* p. 20 | *Picea glauca* p. 20 | *Picea mariana* p. 21 |

Leaves single

| *Pseudotsuga menziesii* p. 23 | *Tsuga canadensis* p. 24 |

Leaves in groups of 2 or more

| *Larix laricina* p. 19 | *Pinus banksiana* p. 21 | *Pinus resinosa* p. 22 | *Pinus strobus* p. 22 | *Pinus sylvestris* p. 23 |

LEAVES BROAD, NEVER EVERGREEN

Leaves alternate, simple, not lobed

| *Populus alba* p. 25 | *Populus balsamifera* p. 25 | *Populus deltoides* p. 26 | *Populus grandidentata* p. 26 | *Populus tremuloides* p. 27 | *Salix fragilis* p. 27 |

Leaves alternate, simple, not lobed

| *Salix nigra* p. 28 | *Alnus incana* ssp. *rugosa* p. 31 | *Betula alleghteniensis* p. 32 | *Betula papyrifera* p. 32 | *Carpinus caroliniana* ssp. *virginiana* p. 33 | *Ostrya virginiana* p. 33 |

Leaves alternate, simple, not lobed

| *Castanea dentata* p. 34 | *Fagus grandifolia* p. 35 | *Quercus muehlenbergii* p. 37 | *Celtis occidentalis* p. 39 | *Ulmus americana* p. 40 | *Ulmus pumila* p. 40 |

Leaves alternate, simple, not lobed

| *Ulmus rubra* p. 41 | *Maclura pomifera* p. 42 | *Morus rubra* p. 43 | *Magnolia acuminata* p. 44 | *Asimina triloba* p. 45 | *Amelanchier laevis* p. 49 |

Leaves alternate, simple, not lobed

| *Crataegus chryso- carpa* var. *chrysocarpa*, p. 50 | *Crataegus punctata* p. 51 | *Malus pumila* p. 51 | *Prunus americana* p. 52 | *Prunus nigra* p. 52 | *Prunus serotina* p. 53 |

Leaves alternate, simple, not lobed

| *Cercis canadensis* p. 55 | *Euonymus europaeus* p. 59 | *Tilia americana* p. 68 | *Tilia cordata* p. 68 | *Elaeagnus angustifolia* p. 69 | *Hippophae rhamnoides* p. 70 |

LEAVES BROAD, NEVER EVERGREEN

Leaves alternate, simple, lobed

*Cornus
alternifolia*
p. 71

*Nyssa
sylvatica*
p. 72

*Ginkgo
biloba*
p. 16

*Quercus
alba*
p. 35

*Quercus
bicolor*
p. 36

*Quercus
macrocarpa*
p. 36

Leaves alternate, simple, lobed

*Quercus
palustris*
p. 37

*Quercus
rubra*
p. 38

*Quercus
velutina*
p. 38

*Morus
alba*
p. 42

*Liriodendron
tulipifera*
p. 44

*Sassafras
albidum*
p. 46

Leaves alternate, simple, lobed

*Liquidambar
styraciflua*
p. 47

*Platanus ×
acerifolia*
p. 48

*Crataegus
monogyna*
p. 50

Leaves alternate, compound

*Carya
cordiformis*
p. 29

*Carya
ovata*
p. 29

*Juglans
cinerea*
p. 30

*Juglans
nigra*
p. 30

*Sorbus
americana*
p. 53

*Sorbus
aucuparia*
p. 54

Leaves alternate, compound

*Gleditsia
triacanthos*
p. 55

*Gymnocladus
dioica*
p. 56

*Robinia
pseudoacacia*
p. 56

*Ptelea
trifoliata*
p. 57

*Ailanthus
altissima*
p. 58

*Staphylea
trifolia*
p. 60

Leaves opposite, simple

| *Acer ginnala* p. 61 | *Acer pensylvanicum* p. 62 | *Acer platanoides* p. 63 | *Acer rubrum* p. 63 | *Acer saccharinum* p. 64 | *Acer saccharum* p. 64 |

Leaves opposite, simple

| *Acer spicatum* p. 65 | *Rhamnus cathartica* p. 67 | *Catalpa speciosa* p. 75 |

Leaves opposite, compound

| *Acer negundo* p. 62 | *Aesculus hippocastanum* p. 66 | *Fraxinus americana* p. 73 | *Fraxinus nigra* p. 73 | *Fraxinus pensylvanica* p. 74 | *Fraxinus quadrangulata* p. 74 |

MAIDENHAIR TREE FAMILY
Ginkgoaceae

All members of the maidenhair tree family are extinct except for a single species, *Ginkgo biloba*. Fossil records indicate that the species disappeared from North America about 7 million years ago when the climate changed from warm tropical to cool temperate. This living fossil is not known to occur in the wild even in its native range of China. It has been grown on monastery grounds since 1100 AD, and Buddhist priests are credited for saving the species from extinction. *Ginkgo biloba* is now cultivated as an ornamental throughout the world. It is disease and pollution resistant, making it an ideal specimen for street lines.

Maidenhair Tree
Ginkgo biloba

Habitat: Cultivated ornamental; introduced from China.
General: Tree 10–30 m tall; male or female; bark grey, becoming furrowed with age; twigs light green to brown; spurs knob-like, to 3 cm long.
Leaves: Deciduous, alternate or in clusters of 3–5, on short shoots, fan-shaped, 2-lobed, 2–9.5 cm long, 2–12 cm wide, dull light green; stalks 2.5–8.5 cm long; turning yellow in autumn. **Reproductive Structures:** Pollen cones 1.8–2 cm long; seed cones berry-like, 2.3–2.7 cm long, 1.9–2.3 cm wide, green turning yellow or orange, outer layer fleshy and ill-scented; stalks orange, 3–9.5 cm long.

Notes: The species name *biloba* refers to the 2-lobed leaves.

CYPRESS FAMILY
Cupressaceae

The cypress family, also known as the juniper family, is a family of cone-bearing, evergreen trees and shrubs with opposite or whorled, scale- or needle-like leaves. The pollen-producing male cones are small and appear at the ends of branches. In the genus *Thuja*, the female cones are small and dry, whereas in *Juniperus*, the female cones are somewhat fleshy and berry-like. The term "juniper berries" refers to these fleshy female cones.

Several members of this family are grown for ornamental purposes.

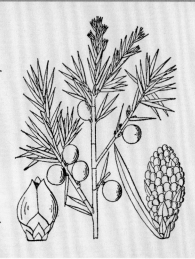

Eastern Red-cedar
Juniperus virginiana

Habitat: Dry, sandy or rocky fields. **General:** Tree to 10 m tall; male or female; crown narrow to cone-shaped; bark light reddish brown, shredding into long, narrow strips. **Leaves:** Evergreen, opposite, in 4 rows, bluish green; older branches with scale-like leaves 1–3 mm long; young branches with needle-like leaves 5–7 mm long. **Reproductive Structures:** Pollen cones 2–3 mm long, yellowish; seed cones berry-like, 3–6 mm across, bluish black to brownish blue; seeds 1–2.

Notes: *Juniperus* is the Latin name for juniper. The species name *virginiana* means "of Virginia."

17

Eastern White Cedar
Thuja occidentalis

Habitat: Moist to wet ground; swamps to limestone ridges.
General: Tree to 15 m tall; crown cone-shaped; older bark greyish brown to reddish brown, 6–9 mm thick, shredding into long, narrow strips. **Leaves:** Evergreen, oppo-site, in 4 rows, dull yellowish green,

scale-like, 2–5 mm long, overlapping those above; young leaves glandular-dotted. **Reproductive Structures:** Pollen cones red-dish, 1–2 mm long; seed cones dry, 7–14 mm long, brown, scales 8–12, greenish yellow when young, reddish brown at maturity; seeds 4–7 mm long.

Notes: "Arborvitae," another common name for this species, is Latin for "tree-of-life." • The leaves of this tree are rich in vita-min C and are credited with saving the lives of Jacques Cartier's crew from scurvy in the winter of 1535–36.

PINE FAMILY · Pinaceae

Members of this family are mostly evergreen trees and shrubs with resinous sap. The needle-shaped leaves are spirally arranged and appear singly or in groups called fascicles. The cones are of 2 types, male and female. The small male cones produce pollen, whereas the female cones produce seeds. At maturity, the female cones are woody and produce 2 seeds per cone scale.

Well-known members of this family include the Great Basin bristlecone pine (*Pinus longaeva*), with one specimen known to be over 5000 years old, and pinyon pine (*Pinus edulis*), which produces edible seeds that are often used in cooking.

Balsam Fir
Abies balsamea

Habitat: Swampy ground to well-drained, forested slopes. **General:** Tree 10–25 m tall; crown spire-like; young bark smooth, grey and blistered, mature bark brownish grey and scaly. **Leaves:** Evergreen, needle-like, the bases twisted and appearing opposite, 1–2.5 cm long, 1.5–2 mm wide, flattened, dark green above, 2 white bands on the underside. **Reproductive Structures:** Pollen cones red, purple, bluish, greenish or orange, 3–5 mm long; seed cones erect, cylindric, 4–10 cm long, 1.5–3 cm wide, dark purple when young; scales 1–1.5 cm long, shed with seeds; seeds 2–3 mm long, winged.

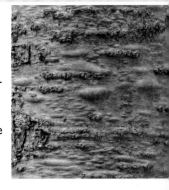

Notes: *Abies* is the Latin name for fir. The species name *balsamea* means "balsam-producing," a reference to the resin blisters on the bark.

Tamarack
Larix laricina

Habitat: Muskeg and swampy areas. **General:** Tree to 20 m tall; crown narrow; older bark reddish brown with small, flaky scales. **Leaves:** Deciduous, needle-like, 3-angled, soft, pale green, 1–2.5 cm long, 0.5–0.8 mm wide, in clusters of 10–20, turning pale yellow in autumn. **Reproductive Structures:** Pollen cones less than 1 cm long, yellow; seed cones erect, 1–2.5 cm long, 0.5–1 cm wide, brown, remaining on branches until the following summer; scales 10–30; bracts shorter than scales, awned; seeds 2–3 mm long, winged.

Notes: The species name *laricina* means "resembling *Larix*."

Norway Spruce
Picea abies

Habitat: Moist to well-drained soils; introduced ornamental from northern Europe. **General:** Tree to 40 m tall; crown cone-shaped; young bark reddish brown, mature bark purplish brown and scaly; twigs reddish brown. **Leaves:** Evergreen, needle-like, 4-sided, dark green, 1.2–2.5 cm long, blunt-tipped. **Reproductive Structures:** Pollen cones yellow, small; seed cones greyish to reddish brown, 10–18 cm long, borne near branch tips; scales thin, flat, diamond-shaped, 1.8–3 cm long; seeds winged.

Notes: The genus name *Picea* comes from the Latin word *pix*, meaning "pitch," a reference to the resinous sap.

White Spruce
Picea glauca

Habitat: Variety of soils and moisture regimes. **General:** Tree to 40 m tall; crown cone-shaped to spire-like; bark thin, greyish brown, scaly; twigs hairless, pinkish brown. **Leaves:** Evergreen, needle-like, 4-sided, 1–2.5 cm long, bluish green, sharp-pointed, retained on branches for several years. **Reproductive Structures:** Pollen cones yellow to purplish, borne in axils; seed cones cylindric, 2.5–6 cm long; scales light brown, fan-shaped, 1–1.6 cm long, tips rounded, usually shed in winter; seeds winged.

Notes: The species name means "glaucous" or "whitish," a reference to the colour of the leaves.

Black Spruce
Picea mariana

Habitat: Muskeg and moist, sandy soils.
General: Tree to 10 m tall; crown club-shaped to narrowly cone-shaped, easily identified at a distance; bark thin, greyish brown, scaly; twigs yellowish brown.
Leaves: Evergreen, needle-like, 4-sided, 0.6–1.5 cm long, pale bluish green, blunt-tipped. **Reproductive Structures:** Pollen cones yellowish brown, borne in axils; seed cones egg-shaped, 1.5–2.5 cm long, purplish to dark brown, remaining on branches for several years; scales fan-shaped, 0.8–1.2 cm long, margins irregularly toothed.

Notes: The species name *mariana* means "of Maryland."

Jack Pine
Pinus banksiana

Habitat: Well-drained, sandy soils.
General: Tree to 10 m tall; crown irregular; bark dark orange to reddish brown with scaly, irregular ridges and furrows; twigs orange-red to reddish brown.
Leaves: Evergreen, needle-like, in pairs, twisted, spreading, 2–5 cm long, 1–2 mm wide, yellowish green.
Reproductive Structures: Pollen cones cylindric, 1–1.5 cm long, yellow to orangey brown; seed cones 3–5 cm long, pointed toward branch tip, remaining on branches for several years; scales thick; seeds 4–5 mm long, brown to black, winged.

Notes: *Pinus* is the Latin name for pine. • The species name *banksiana* commemorates Sir Joseph Banks (1743–1820), a British explorer and naturalist.

Red Pine
Pinus resinosa

Habitat: Dry, rocky or sandy areas. **General:** Tree to 25 m tall; crown irregular with spreading branches; young bark reddish brown and scaly, mature bark scaly and plate-like; twigs orange to reddish brown. **Leaves:** Evergreen, needle-like, 10–16 cm long, in pairs, dark yellowish green, often clustered near branch tips. **Reproductive Structures:** Pollen cones elliptic, 1.2–1.7 cm long, dark purple, at the base of new growth; seed cones 4–7 cm long, light reddish brown; scales slightly thickened; seeds 3–5 mm long, brown, winged.

Notes: The species name means "resin-bearing," a reference to the resinous sap.

Eastern White Pine
Pinus strobus

Habitat: Dry, rocky soil to peat bogs. **General:** Tree to 30 m tall; crown cone-shaped to rounded; young bark greyish green, mature bark dark greyish green with purple-tinged scales; branches whorled. **Leaves:** Evergreen, needle-like, 5–15 cm long, 0.7–1 mm wide, in groups of 5, bluish green, soft, flexible, slightly twisted. **Reproductive Structures:** Pollen cones elliptic, 1–1.5 cm long, yellow, at the base of new growth; seed cones greyish brown, elliptic to lance-shaped, slightly curved, 8–20 cm long, on stalks 2–3 cm long; scales 50–80; seeds 5–6 mm long, reddish brown, winged.

Notes: The species name *strobus* means "cone." • Eastern white pine is the provincial tree of Ontario.

Scots Pine
Pinus sylvestris

Habitat: Escaped from cultivation; introduced from Europe. **General:** Tree to 35 m tall; crown cone-shaped to rounded or irregular; young bark orangey red, papery and peeling into strips, older bark greyish brown and breaking into irregular plates. **Leaves:** Evergreen, needle-like, 3.5–8 cm long, 1–2 mm wide, in pairs, bluish green, twisted, sharp-pointed.

Reproductive Structures: Pollen cones 0.8–1.2 cm long, yellowish or occasionally pink; seed cones curved, 2.5–8 cm long, yellowish to purplish brown when young, turning brown with age; scales flattened, bumpy at tip, falling off 2 years after pollination; seeds 2–4 mm long, brown, winged.

Notes: The species name *sylvestris* means "of the woods."

Douglas-fir
Pseudotsuga menziesii

Habitat: Escaped from cultivation; introduced from western Canada. **General:** Tree to 50 m tall; crown columnar; young bark grey and smooth, old bark very thick, becoming dark brown and deeply fur-rowed; lower branches drooping. **Leaves:** Ever-green, needle-like, flat, 2–3 cm long, blunt-tipped, yellowish green, bases twisted.

Reproductive Structures: Pollen cones small, reddish brown; seed cones pendent, 5–10 cm long, shed soon after seeds are released; scales broad, rounded; bracts exceeding cone scales, prominently 3-lobed.

Notes: Douglas-fir is the largest tree species in Canada.

Eastern Hemlock
Tsuga canadensis

Habitat: Moist, shady forests. **General:** Tree to 30 m tall; crown cone-shaped; young bark reddish brown and scaly, mature bark reddish brown and furrowed; twigs yellowish brown, densely haired. **Leaves:** Evergreen, needle-like, the bases twisted and appearing opposite, 1–2 cm long, flattened, blunt-tipped, dark yellowish green above, pale below with 2 white stripes. **Reproductive Structures:** Pollen cones yellow, round; seed cones oval, light brown, 1.2–2.5 cm long, 1–1.5 cm wide, on hairy stalks 2–3 mm long; scales few, 8–12 mm long; seeds light brown, 1–2 mm long, winged.

Notes: *Tsuga* is the Japanese name for hemlock. • The species name *canadensis* means "of Canada."

WILLOW FAMILY
Salicaceae

The willow family consists of 2 genera: willows (*Salix* spp.) and poplars (*Populus* spp.). Members are deciduous trees and shrubs with male and female flowers on separate plants. The alternate leaves have papery structures called stipules near the base. Flowers are produced in spike-like clusters called catkins. Male flowers are composed of 1 to many stamens with 1 small bract at the base. Female flowers have a single pistil and a small basal bract. The fruit, a capsule, contains numerous seeds with tufts of silky hairs.

Willows are easily distinguished from poplars. The winter buds of willows have a single scale, whereas poplar buds have several scales. Also, poplar catkins droop, while those of willows are usually erect. Several willow and poplar species are grown for their ornamental value.

White Poplar
Populus alba

Habitat: Roadsides and waste areas; introduced from Europe. **General:** Tree to 25 m tall, spreading by suckers; bark whitish grey; twigs and buds covered in white, woolly hairs. **Leaves:** Alternate, deciduous, oval, 5–10 cm long and wide, 3–7-lobed, often resembling maple leaves, upper surface dark green, underside covered in white, woolly hairs; stalks 2.5–5.5 cm long; turning reddish in autumn. **Flower Cluster:** Male catkins 4–10.5 cm long; female catkins 3–5 cm long; catkins covered in white, woolly hairs, appearing before leaves. **Flowers:** Green; male flowers with 5–10 stamens; female flowers with 1 pistil with 2–4 styles. **Fruit:** Capsule egg-shaped, 5–6.5 mm long; seeds covered in cottony hairs.

Notes: *Populus* is the Latin name for poplar. • The species name *alba* means "white," a reference to the underside of the leaves.

Balsam Poplar
Populus balsamifera

Habitat: Moist forests and riverbanks. **General:** Tree to 25 m tall; bark deeply furrowed; young branches greyish brown; buds covered in sticky gum. **Leaves:** Alternate, deciduous, oval to lance-shaped, 6–15 cm long, 4–7.5 cm wide, base round to wedge or heart-shaped, upper surface dark green and shiny, underside silvery white to yellowish green or rusty; margins smooth to toothed; stalks 7–10 cm long; turning yellow in autumn. **Flower Cluster:** Male catkins reddish, 7–10 cm long; female catkins greyish white, 8–13 cm long; appearing before leaves. **Flowers:** Green; male flowers with 12–30 stamens; female flowers with 1 pistil. **Fruit:** Capsule 6–10 mm long; seeds numerous, covered with white, cottony hairs.

Notes: The species name *balsamifera* means "balsam-bearing," a reference to the resinous sap.

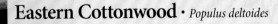

Eastern Cottonwood · *Populus deltoides*

Habitat: Riverbanks and moist, sandy areas.
General: Tree to 30 m tall; bark grey, deeply
furrowed; branches greyish brown. **Leaves:**
Alternate, deciduous, leathery, triangular to
heart-shaped, 5–18 cm long, 4.5–13 cm across,
base heart-shaped; margins coarsely toothed;
stalks flattened; turning yellow in autumn.
Flower Cluster: Male catkins 5–7 cm long;
female catkins 15–25 cm long; appearing before
leaves. **Flowers:** Green; male flowers with
about 60 stamens; female flowers with 1 pistil.
Fruit: Capsule 6–10 mm long; seeds numerous,
covered with white, cottony hairs.

Notes: The species name *deltoides* means
"deltoid" or "triangular,"
a reference to the leaf
shape. • The common
name "cottonwood"
refers to the large
number of cottony-
haired seeds shed in
early summer.

Large-toothed Aspen
Populus grandidentata

Habitat: Dry to moist sites; scrub-like on dry
sites with poor soil. **General:** Tree to 25 m tall;
young bark greenish white, becoming dark and furrowed with age; twigs and buds covered in greyish,
woolly hairs. **Leaves:** Alternate, deciduous, oval
to round, 5–12 cm long, 4.5–9 cm wide; margins
with 5–15 large teeth per side; stalks flattened;
turning yellow in autumn. **Flower Cluster:** Catkins
5–12 cm long; appearing before leaves. **Flowers:** Green;
male flowers with 5–12 stamens; female flowers with
1 pistil. **Fruit:** Capsule 3–7 mm long, on stalk 1–2 mm
long; seeds numerous, covered with white, cottony
hairs.

Notes: The species name *grandidentata* means "large-
toothed," a reference to the leaf margins.

Trembling Aspen
Populus tremuloides

Habitat: Forests and edges of meadows.
General: Tree to 30 m tall; young bark greenish white, becoming dark and furrowed; twigs yellowish green. **Leaves:** Alternate, deciduous, 4–7 cm long, 3–6 cm across, oval with an abrupt point, dark green above, pale below; margins finely toothed; stalks flattened. **Flower Cluster:** Male catkins greyish white, 2–4 cm long; female catkins 4–10 cm long; appearing before leaves. **Flowers:** Green; male flowers with 5–12 stamens; female flowers with 1 pistil. **Fruit:** Capsule 3–6 mm long; seeds numerous, covered with silky hairs.

Notes: The species name *tremuloides* means "like *Populus tremula*," the European aspen that this species resembles. • The flattened leaf stalks allow the leaves to move even in the slightest breeze, hence the common name "trembling aspen."

Crack Willow
Salix fragilis

Habitat: Wet woods and roadsides; introduced from Europe as a source of charcoal for gunpowder.
General: Tree to 30 m tall; older bark grey and deeply furrowed; branches spreading; twigs yellowish green or dark reddish brown. **Leaves:** Alternate, deciduous, lance-shaped, 7–15 cm long, 1–4 cm wide, green above, pale beneath; margins with 4–7 teeth per cm; stalks 7–15 mm long; stipules small or absent. **Flower Cluster:** Catkins 2.5–8 cm long, 0.8–1 cm wide; appearing with leaves. **Flowers:** Green; male flowers with 2–4 stamens; female flowers with 1 pistil. **Fruit:** Capsule lance-shaped, 4–6 mm long, in catkins 5–8 cm long; seeds covered in silky hairs.

Notes: *Salix* is the Latin name for willow. The species name *fragilis* refers to the brittle branches.

Black Willow
Salix nigra

Habitat: Wet forests, riverbanks and swamps.
General: Tree or large shrub to 20 m tall; bark black, deeply ridged; twigs brittle, yellow to reddish brown, becoming hairless with age.
Leaves: Alternate, deciduous, linear to lance-shaped, 5–15 cm long, 0.5–2 cm wide, dark green above, slightly paler below; margins finely toothed; stalks 4–10 mm long; stipules somewhat heart-shaped, to 12 mm long, margins minutely toothed; turning yellow in autumn. **Flower Cluster:** Catkins 2.5–7.5 cm long, 0.8–1.2 cm wide, on stalks 1–3 cm long; appearing with leaves.
Flowers: Greenish yellow; male flowers with 1 yellow bract and 3–7 stamens; female flowers with 1 yellow bract and 1 pistil. **Fruit:** Capsule 3–5 mm long, brown, hairless, on stalk 1–2 mm long; seeds covered in silky hairs.

Notes: The species name *nigra* means "black."
• Black willow is the largest native willow in Canada.

WALNUT FAMILY
Juglandaceae

Members of the walnut family are deciduous trees with alternate, pinnately compound leaves. Stipules are absent. Flowers are produced in male or female catkins on the same plant and appear on the previous year's branches, at the base of the current year's growth. Male flowers have a 2–6-lobed calyx and 3–50 stamens. Female flowers have a 4-lobed calyx and 1 pistil. A single bract is found below each flower. The fruit, a nut, is enclosed by a fleshy or hard husk.

Well-known members of this family include black walnut (*Juglans nigra*) and pecan (*Carya illinoinensis*), both which produce edible nuts.

Bitternut Hickory
Carya cordiformis

Habitat: Rich, moist woods to hillsides.
General: Tree 15–25 m tall; bark grey to
brown, smooth to ridged or plate-like; twigs
tan-coloured. **Leaves:** Alternate, deciduous, 20–40 cm long, pinnately compound
with 5–13 (commonly 7–9) leaflets, stalk
3–7 cm long; leaflets oval to lance-shaped,
3–19 cm long, 1–7 cm wide, yellowish green
above, green and slightly hairy below, uppermost leaflet
largest; margins finely to coarsely toothed; turning yellow in autumn. **Flower Cluster:** Male catkins 7–16 cm

long, bracts scaly; female spikes 1–2-flowered; appearing
with leaves. **Flowers:** Green; male flowers with 2–3-
lobed calyx and 3–10 (commonly 4–5) stamens; female
flowers with cup-like bract, 4-lobed calyx and 1 pistil.
Fruit: Nut greenish brown, round, 2–3 cm long,
2–3.2 cm wide; husk smooth, yellowish green, 2–3 mm
thick, breaking into 4 pieces at maturity; seeds bitter.

Notes: The genus name comes from *karya*, the Greek name for walnut. The
species name *cordiformis* means "heart-shaped," a reference to shape of the fruit.

Shagbark Hickory
Carya ovata

Habitat: Rich, moist woods. **General:** Tree
19–25 m tall; bark light grey, separating into
long strips or plates and curling away from the
trunk; twigs green, reddish or orangey brown.
Leaves: Alternate, deciduous, 15–60 cm long,
stalks 4–13 cm long, pinnately compound with
3–7 (commonly 5) leaflets; leaflets oval to
oblong or elliptic, 4–26 cm long, 1–14 cm
wide, yellowish green above; margins finely to coarsely
toothed; turning golden brown in autumn. **Flower
Cluster:** Male catkins 10–13 cm long; female spikes 1- or
5-flowered; appearing with leaves. **Flowers:** Green; male
flowers with 2–3-lobed calyx and 3–10 (commonly 4)
stamens; female flowers with 1 cup-like bract, 4-lobed
calyx and 1 pistil. **Fruit:** Nut brown to reddish brown,
round, 2.5–4 cm long and wide; husk rough, 4–15 mm
thick, splitting into 4 pieces at maturity; seeds sweet.

Notes: The species name *ovata* means "ovate," a reference to the leaflet shape. • The common name refers to the shaggy appearance
of the bark.

Butternut · *Juglans cinerea*

Habitat: Rich woods and dry, rocky slopes. **General:** Tree 12–30 m tall; crown irregular; bark light grey to brown, smooth to scaly plated; twigs tan to grey. **Leaves:** Alternate, deciduous, 30–60 cm long, stalks 3.5–12 cm long, pinnately compound with 11–17 leaflets; leaflets oval to oblong or lance-shaped, 5–12 cm long, 1.5–6.5 cm wide, yellowish green above,

pale and hairy below, 3 terminal leaflets largest; margins toothed; turning yellowish brown in autumn. **Flower Cluster:** Male catkins 6–14 cm long; female spikes 4–8-flowered; appearing with leaves. **Flowers:** Green; male flowers with 3–6-lobed calyx and 7–15 stamens; female flowers with 3-lobed, cup-like bract, small, 4-lobed calyx and 1 pistil. **Fruit:** Nut 3–6 cm long, sweet, edible; husk green, elliptic to oval or cylindric (lemon-shaped), 4–8 cm long, densely sticky-haired; in clusters of 3–7.

Notes: The species name *cinerea* means "ashy grey," a reference to the colour of the leaf underside.

Black Walnut · *Juglans nigra*

Habitat: Rich woods. **General:** Tree 20–50 m tall; bark grey to brown or black, narrowly ridged; twigs orangey brown. **Leaves:** Alternate, deciduous, 20–60 cm long, stalks 6.5–14 cm long, pinnately compound with 9–23 (commonly 15–19) leaflets; leaflets oval to lance-shaped, 5–15 cm long, 1.5–5.5 cm wide, middle leaflets largest; margins toothed; turning yellow in autumn. **Flower Cluster:** Male catkins 5–10 cm long; female spikes 1–5-flowered;

appearing with leaves. **Flowers:** Green; male flowers with 3–6-lobed calyx and 17–50 (commonly 20–30) stamens; female flowers with 3-lobed, cup-like bract, small, 4-lobed calyx and 1 pistil. **Fruit:** Nut 3–4 cm long; husk yellowish green, globe-shaped, 3.5–8 cm across, surface warty and glandular-haired; in clusters of 1–3.

Notes: *Juglans* comes from the Latin words *jovis*, "Jupiter," and *glans*, " acorn" or "nut." The species name *nigra* means "black," a reference to the colour of the bark.

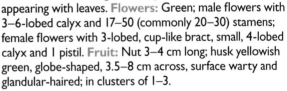

BIRCH FAMILY
Betulaceae

The birch family contains deciduous trees and shrubs with simple, alternate leaves. The branches do not have terminal buds but elongate from lateral buds. Flowers are borne in catkins, which appear before the leaves. Male and female catkins appear on the same plant. Male catkins have 1–3 flowers composed of 1–10 stamens and are borne in the axil of each bract. Female flowers have 1 pistil with 2 styles. The fruit is a 1-seeded, winged nutlet, often referred to as a samara.

Well-known members of this family include filbert (*Corylus maxima*), an edible nut, and weeping birch (*Betula pendula*), which is grown as an ornamental.

Speckled Alder
Alnus incana ssp. *rugosa*

Habitat: Streambanks, lakeshores, bogs and roadside ditches. **General:** Small tree to large shrub 3–9 m tall; bark dark grey to reddish brown, smooth; lenticels horizontal, white to pale orange. **Leaves:** Alternate, deciduous, oval to elliptic, 4–11 cm long, 3–8 cm wide, base wedge-shaped to round; veins 6–12 per side; margins coarsely toothed; remaining green in autumn. **Flower Cluster:** Male catkins 2–8 cm long, in clusters of 2–4, each scale 3-flowered;

female catkins in clusters of 2–6, each scale 2-flowered; appearing before leaves. **Flowers:** Green to brown; male flowers with 4-lobed calyx and 4 stamens; female flowers with 4-lobed calyx and 1 pistil. **Fruit:** Samara 2–3.5 mm long; in woody, persistent, cone-like catkins 1–1.7 cm long, 0.8–1.2 cm wide; scales 5-lobed.

Notes: The species name *incana* means "hoary." The subspecies name *rugosa* means "wrinkled."

Yellow Birch
Betula alleghaniensis

Habitat: Moist woods, swamps and streambanks. **General:** Tree 15–30 m tall; young bark dark reddish brown, older bark yellowish or greyish, somewhat shiny, irregularly peeling; lenticels horizontal. **Leaves:** Alternate, deciduous, oval to oblong, 6–10 cm long, 3–5.5 cm wide; veins 9–18 pairs, hairy on leaf underside; margins coarsely toothed; turning yellow in autumn. **Flower Cluster:** Male catkins 2–8 cm long; female catkins erect, 1–2.5 cm long; appearing after leaves. **Flowers:** Green; male flowers with 2- or 4-lobed calyx and 2 stamens, 3 per scale; female flowers with 4-lobed calyx and 1 pistil. **Fruit:** Samara 2.5–4.5 mm long; in oval, cone-like structures 1.5–3 cm long, 1–2.5 cm wide, remaining intact after fruit is shed.

Notes: *Betula* is the Latin name for birch. The species name means "of or from the Allegheny Mountains."

Paper Birch
Betula papyrifera

Habitat: Dry, sandy hillsides to moist woods. **General:** Tree 15–30 m tall; bark white, papery, peeling easily from trunk; young twigs and branches reddish brown; lenticels horizontal. **Leaves:** Alternate, deciduous, oval to diamond-shaped, 2–8 cm long, 3–6 cm across, stalked; veins 5–9 per side; margins coarsely toothed; turning yellow in autumn. **Flower Cluster:** Male catkins 7–9 cm long; female catkins 1–3 cm long, erect; appearing before leaves. **Flowers:** Green or brown; male flowers with 2–3 bracts and 3 stamens; female flowers with 1 pistil. **Fruit:** Samara 1.5–2 mm long; in catkins 3–5 cm long; maturing in fall and shed in winter.

Notes: The species name *papyrifera* means "paper-bearing," a reference to the bark.

Blue-beech, Muscle-wood

Carpinus caroliniana ssp. *virginiana*

Habitat: Rich, moist forests, streambanks and wet hillsides. **General:** Tree 4–12 m tall; crown spreading; trunk short, crooked, shallow to deeply ridged, resembling muscles; bark bluish to ash grey, smooth; buds 2–4 mm long, reddish brown. **Leaves:** Alternate, deciduous, oval to elliptic, 6–12 cm long, 3.5–6 cm wide, stalked; lower veins hairy and covered with tiny, dark brown glands; margins toothed; turning red to golden yellow in autumn. **Flower Cluster:** Male catkins 2–6 cm long; female catkins 1–3 cm long; appearing with leaves. **Flowers:** Green; male flowers with 1 scale and several stamens; female flowers with 2 small bracts (enlarging in fruit) and 1 pistil. **Fruit:** Nutlet 6–9 mm long; in catkins 4.5–15 cm long; bracts 2–3.5 cm long, 1.4–2.8 cm wide, 3-lobed, pointed.

Notes: *Carpinus* is the Latin name for hornbeam. The species name means "of Carolina."

Hop-hornbeam, Ironwood

Ostrya virginiana

Habitat: Moist woods to dry, upland slopes. **General:** Tree 7–12 m tall; bark greyish brown to steel grey, shredding into narrow, ragged strips; twigs densely haired. **Leaves:** Alternate, deciduous, oval to elliptic, oblong or lance-shaped, 6–13 cm long, 4–5 cm wide; lower veins often covered in woolly hairs; margins coarsely toothed; turning yellow in autumn. **Flower Cluster:** Male catkins 1.5–5 cm long, 3 flowers per bract; female catkins 0.5–1.5 cm long, 2 flowers per bract; appearing with leaves. **Flowers:** Green; male flowers with 1 scale and 3 stamens; female flowers with oval, hairy bracts (falling off as flowers appear), the 2 inner bracts fused and surrounding the pistil. **Fruit:** Nutlet 4–6 mm long, enclosed by an inflated sac; in catkins 3.5–6.5 cm long, 2–2.5 cm wide, resembling the fruit of common hops (*Humulus lupulus*), p. 91; bracts 1–1.8 cm long, 0.8–1 cm wide.

Notes: *Ostrya* comes from *ostrys*, the Greek name for hornbeam. The species name means "of Virginia." • The common name "ironwood" refers to this tree's hard, heavy wood.

BEECH FAMILY
Fagaceae

The beech family includes evergreen or deciduous trees and shrubs. The alternate, simple leaves are often lobed. Stipules are present but fall off early in the season. Male flowers, borne in catkins, have 4–6 sepals and 6–12 stamens. Female flowers composed of 4–6 sepals and 1 pistil are borne singly or in small clusters. Petals are absent. The fruit is a nut. An important feature of this family is the cupule, a structure formed by the fusion of the stem and floral bracts.

Well-known members of this family include American chestnut (*Castanea dentata*), beech (*Fagus grandifolia*) and numerous species of oak (*Quercus* spp.).

American Chestnut
Castanea dentata

Habitat: Rich deciduous and mixedwood forests. **General:** Tree 3–10 m tall (to 35 m prior to the introduction of the blight); young bark dark brown and smooth, older bark with large, flat-topped ridges; twigs reddish brown, shiny. **Leaves:** Alternate, deciduous, oblong to oval or lance-shaped, 9–30 cm long, 3–10 cm wide; margins sharply toothed, the teeth incurved and bristle-tipped; stalks 1–3 cm long. **Flower Cluster:** Catkins 5–20 cm long, completely male or with a few female flowers at the base; cupules 2–4-valved, with 1–3 female flowers; appearing after leaves. **Flowers:** Creamy white, fragrant; male flowers with 6 sepals and 12–18 stamens; female flowers with 6 sepals and 1 pistil. **Fruit:** Nut 1.8–2.5 mm long, edible, surrounded by 4-valved cupule 5–6 cm wide; cupule spiny, the spines to 10 mm long.

Notes: *Castanea* comes from the Latin name for Castania, a town in northern Greece known for its sweet chestnut trees. The species name *dentata* means "toothed," a reference to the leaves. • In about 1904, chestnut blight was introduced into North America. It is estimated that by 1937, 99% of North American chestnuts had been infected or destroyed (about 3.5 billion trees).

Beech
Fagus grandifolia

Habitat: Rich, deciduous woods.
General: Tree 18–25 m tall; bark grey, smooth; twigs hairless; buds reddish brown, 1.5–2 cm long. **Leaves:** Alternate, deciduous, oval to oblong, 6–12 cm long, 2.5–7.5 mm wide, leathery; veins 9–14 per side, each ending in a tooth; margins coarsely toothed; stalks 4–12 mm long; turning brown to bronze in autumn, often remaining attached to lower branches throughout winter. **Flower Cluster:** Male flowers in dense, globe-shaped catkins, 2.2–2.8 cm long; female flowers in catkins, 2–4 flowers per cupule; appearing with leaves. **Flowers:** Yellowish green; male flowers with 4–8-lobed calyx and 8–16 stamens; female flowers with 6-lobed calyx and 1 pistil. **Fruit:** Nut 1.5–2 cm long, 1–1.8 cm wide, 3-angled; cupule brown to reddish brown, 1.5–2 cm long, with prickles 4–10 mm long, opening at maturity; edible.

Notes: *Fagus* is the Latin name for beech. The species name *grandifolia* means "large-leaved," a reference to the leaf size.

White Oak · *Quercus alba*

Habitat: Moist to dry, deciduous forests. **General:** Tree 15–35 m tall; bark light grey, scaly; twigs green to reddish, becoming grey with age. **Leaves:** Alternate, deciduous, oblong to elliptic or oval, 10–22 cm long, 7–11 cm wide, bright green above, pale below; margins deeply 5–9-lobed, the lobes extending ⅓–⅞ the distance to the midrib; stalks 1–2.5 cm long; turning reddish purple in autumn. **Flower Cluster:** Male flowers in catkins; female flowers solitary or in short spikes, 2–4-flowered; appearing with leaves. **Flowers:** Green; male flowers with 6-lobed calyx and 6 stamens; female flowers with 1 cupule, 6-lobed calyx and 1 pistil. **Fruit:** Nut (acorn) light brown at maturity, oval to elliptic or oblong, 1.2–

2.1 cm long, 0.9–1.8 cm wide, tip pointed, on stalks to 2.5 cm long; cupule enclosing ¼–⅓ of nut; scales overlapping, giving a knobby appearance; maturing the first season.

Notes: *Quercus* is the Latin name for oak. The species name *alba* means "white."

Swamp White Oak
Quercus bicolor

Habitat: Low-lying forests, swamps and poorly drained soils. **General:** Tree 12–30 m tall; bark dark grey, scaly; twigs pale brown, hairless. **Leaves:** Alternate, deciduous, oblong to oval or elliptic, 12–20 cm long, 0.7–1.1 cm wide, widest above the middle, dark green above, pale below; margins irregularly toothed to deeply 3–7-lobed; stalks 1–2.5 cm long; turning reddish brown in autumn. **Flower Cluster:** Male flowers in catkins 7.5–10 cm long; female flowers solitary or in short spikes, 2–5-flowered (commonly 2-flowered); appearing with leaves. **Flowers:** Green; male flowers with 6-lobed calyx and 6 stamens; female flowers with 1 cupule, 6-lobed calyx and 1 pistil. **Fruit:** Nut (acorn) light brown, oval to elliptic or oblong, 1.5–2.5 cm long, 0.9–1.8 cm wide; in groups of 1–3, on stalks 4–7 cm long; cupule enclosing ½–¾ of nut; scales overlapping, covered in grey, woolly hairs, those near the rim somewhat awn-like; maturing the first year.

Notes: The species name *bicolor* means "2-coloured," a reference to the colour contrast between the upper and lower leaf surfaces.

Burr Oak · *Quercus macrocarpa*

Habitat: Valley bottom, often on poorly drained soils. **General:** Tree 12–30 m tall; bark grey, scaly to flat-ridged; twigs grey to reddish, often with cork-like ridges. **Leaves:** Alternate, deciduous, oblong to oval or fiddle-shaped, 7–25 cm long, 5–13 cm wide, dark green above, pale below; margins moderately to deeply 8–14-lobed, the lobes extending ⅔ the distance to the midrib; stalks 1.5–2.5 cm long; turning yellowish brown in fall. **Flower Cluster:** Male flowers in catkins 10–13 cm long; female flowers solitary or in short spikes, 2–5-flowered; appearing with leaves. **Flowers:** Yellowish green to reddish; male flowers with 6-lobed calyx and 6 stamens; female flowers with 1 cupule, 6-lobed calyx and 1 pistil. **Fruit:** Nut (acorn) light brown to greyish, 2.5–5 cm long, 2–4 cm wide, on stalks 0.6–2 cm long; cupule enclosing ½–⅞ of nut; scales triangular, covered in grey-woolly hairs, those near the margins with soft awns 0.5–1 cm long, forming a fringe around the nut; maturing the first year.

Notes: The species name *macrocarpa* means "large-fruited."

Chinquapin Oak
Quercus muehlenbergii

Habitat: Mixed deciduous forests.
General: Tree 12–30 m tall; bark grey,
flaky; twigs brownish, becoming hairless
with age; buds reddish brown, 4–6 mm
long. **Leaves:** Alternate, deciduous,
oblong to oval or lance-shaped, 5–18 cm
long, 4–8 cm wide, leathery, dark green
and shiny above, light green below;

margins toothed to wavy or shallowly lobed, often with
8–15 sharp, incurved teeth per side; stalks 1–3 cm long;
turning reddish brown in autumn. **Flower Cluster:** Male
flowers in catkins 7.5–10 cm long; female flowers solitary
or in short spikes, 2–5-flowered; appearing with leaves.
Flowers: Green; male flowers with 6-lobed calyx and 6
stamens; female flowers with 1 cupule, 6-lobed calyx and
1 pistil. **Fruit:** Nut (acorn) light brown at maturity, oblong
to oval, 1.5–2 cm long, 1–1.3 cm wide, on stalks to 8 mm
long; cupule enclosing ¼–½ of nut; scales overlapping; maturing the first year.

Notes: The species name commemorates Henry Ernst Muehlenberg (1735–1815),
a prominent Pennsylvanian botanist.

Northern Pin Oak
Quercus palustris

Habitat: Valley bottoms and poorly
drained, clay soils. **General:** Tree
15–25 m tall; bark greyish brown, fissures
shallow; twigs reddish brown. **Leaves:**
Alternate, deciduous, elliptic to oblong,
5–16 cm long, 5–12 cm wide, green and
shiny above, pale below; margins deeply
5–7-lobed, 10–30-awned, notches
U-shaped; stalks 2–6 cm long; turning red
in autumn. **Flower Cluster:** Male flowers
in catkins 5–7.5 cm long, hairy; female
flowers solitary or in short spikes,

2–4-flowered; appearing with leaves. **Flowers:** Green;
male flowers with 6-lobed calyx and 6 stamens; female
flowers with 1 cupule, 6-lobed calyx and 1 pistil.
Fruit: Nut (acorn) globe-shaped to oval, 1–1.6 cm
long, 0.9–1.5 cm wide; cupule enclosing about ¼ of nut;
scales pointed; maturing the second year.

Notes: The species name means "marsh-loving,"
a reference to this tree's habitat.

Red Oak · *Quercus rubra*

Habitat: Moist to well-drained soils in deciduous forests. **General:** Tree 18–25 m tall; bark grey, ridged; twigs reddish brown. **Leaves:** Alternate, deciduous, oval to elliptic or oblong, 10–20 cm long, 6–12 cm wide, dull green above, pale below; margins 7–11-lobed, 12–50-awned, lobes somewhat triangular, extending less than ½ the distance to the midrib; stalks 2.5–5 cm long, often reddish; turning red in autumn. **Flower Cluster:** Male flowers in catkins 10–13 cm long; female flowers solitary or in short spikes, 2–5-flowered; appearing with leaves. **Flowers:** Green; male flowers with 6-lobed calyx and 6 stamens; female flowers with 1 cupule, 6-lobed calyx and 1 pistil. **Fruit:** Nut (acorn) oval to oblong, 1.5–3 cm long, 1–2.1 cm wide; cupule enclosing ¼–⅓ of nut; scales reddish brown; maturing in the second year of growth.

Notes: The species name *rubra* means "red," a reference to the leaves' autumn colour.

Black Oak · *Quercus velutina*

Habitat: Dry hillsides and sandy lowlands. **General:** Tree 15–25 m tall; bark dark brown to black, deeply furrowed; twigs reddish brown. **Leaves:** Alternate, deciduous, oval to oblong, 10–30 cm long, 8–15 cm wide, dark green and glossy above, pale green below; margins deeply 5–9-lobed, 15–50-awned; stalks 2.5–7 cm long. **Flower Cluster:** Male flowers in catkins 10–15 cm long; female flowers solitary or in short spikes, 2–5-flowered; appearing with leaves. **Flowers:** Green; male flowers with 6-lobed calyx and 6 stamens; female flowers with 1 cupule, 6-lobed calyx and 1 pistil. **Fruit:** Nut (acorn) globe-shaped to oval, 1–2 cm long, 1–1.8 cm wide; cupule covering about ½ of nut; scale tips loose; maturing in the second year.

Notes: The species name *velutina* means "velvety," a reference to the buds.

ELM FAMILY · Ulmaceae

The elm family consists of deciduous trees and shrubs. The simple, alternate leaves each have a pair of stipules at the base. The flowers can be perfect or unisexual and are borne singly or in cymes, racemes or fascicles. The perfect flowers have 4–8 (commonly 5) sepals, 4–8 stamens and 1 pistil. Petals are absent. The fruit is a samara, nut or drupe.

A well-known species in this family is the American elm (*Ulmus americana*), a native tree susceptible to Dutch elm disease. First observed in Ohio in 1930, this disease has spread throughout North America, destroying over 50% of the American elm population. The fungal disease is transmitted from tree to tree by 2 species of bark beetles.

Hackberry

Celtis occidentalis

Habitat: Moist stream-sides, hillsides and woods. **General:** Tree or large shrub 1–15 m tall; bark grey, becoming deeply furrowed with age. **Leaves:** Alternate, deciduous, oval to lance-shaped or triangular, 5–12 cm long, 3–6 cm wide, leathery, bluish green and rough above, paler and hairy below; margins toothed, 15–40 teeth per side; stalks 5–12 mm long; stipules deciduous; leaves turning yellow in autumn. **Flower Cluster:** Male flowers in cymes; female flowers solitary or in 2–3-flowered cymes; a few perfect flowers may be present; appearing with leaves. **Flowers:** Green, 2–4 mm across; male flowers with 5-lobed calyx and 4–5 stamens; female flowers with 4–5-lobed calyx and 1 pistil. **Fruit:** Drupe dark orange or red to purplish black, 6–13 mm across; stalk 5–15 mm long; edible.

Notes: *Celtis* is the Greek name for a similar tree. The species name *occidentalis* means "western," a reference to this tree's New World range.

American Elm
Ulmus americana

Habitat: Moist, deciduous woods, swampy forests and pastures. **General:** Tree 18–35 m tall; bark light brown to grey, deeply furrowed; twigs brown. **Leaves:** Alternate, deciduous, oval to oblong, 7–15 cm long, 3–7 cm wide, prominently 30–40-veined; margins double-toothed; stalks 4–6 mm long; turning yellow in autumn. **Flower Cluster:** Fascicle raceme-like, less than 2.5 cm long; bracts 2; appearing with leaves. **Flowers:** Green, 2–4 mm wide; mostly perfect with a few unisexual flowers; sepals 6–9, margins hairy; stamens 7–9, anthers red; pistil 1. **Fruit:** Samara oval, 8–10 mm long, narrowly winged, deeply notched at tip, margins hairy, green turning creamy yellow at maturity; stalk 1–2 cm long; shed before leaves expand.

Notes: *Ulmus* is the Latin name for elm. The species name means "of America."

Siberian Elm
Ulmus pumila

Habitat: Waste places, roadsides and fence lines; introduced from Siberia as a quick-growing windbreak tree. **General:** Tree 15–30 m tall; bark grey to brown, deeply furrowed. **Leaves:** Alternate, deciduous, elliptic to lance-shaped, 2–6.5 cm long, 1.2–3.5 cm wide, dark green above, pale below; margins single-toothed; stalks 2–4 mm long; turning yellow in autumn. **Flower Cluster:** Fascicle raceme-like, 6–15-flowered, 4–6 mm long; appearing with leaves. **Flowers:** Green, 2–4 mm wide; calyx 4–5-lobed; stamens 4–8, anthers brownish red; pistil 1. **Fruit:** Samara creamy yellow, 1–1.4 cm across, broadly winged, deeply notched at tip; very short-stalked.

Notes: The species name *pumila* means "dwarf," a reference to the stature of this tree. • Siberian elm has been widely planted for its resistance to Dutch elm disease.

Slippery Elm, Red Elm · *Ulmus rubra*

Habitat: Floodplains, streambanks and valley bottoms. **General:** Tree 15–35 m tall; bark reddish brown, deeply furrowed; inner bark fragrant and slimy; twigs grey, becoming hairless with age. **Leaves:** Alternate, deciduous, oblong to oval, 8–20 cm long, 5–7.5 cm wide, bristly haired above, woolly haired below; margins double-toothed; stalks 5–7 mm long, hairy; turning yellow in autumn. **Flower Cluster:** Fascicle raceme-like, 8–20-flowered, less than 2.5 cm long; appearing before leaves. **Flowers:** Green to reddish, 2–4 mm across; sepals 5–9, hairs reddish; stamens 5–9, anthers dark red; pistil 1, stigmas pink or reddish. **Fruit:** Samara creamy yellow, 1.2–1.8 cm across, broadly winged, slightly notched at tip, rusty-haired on body of fruit; stalk 1–2 mm long; shed before leaves expand.

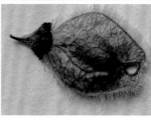

Notes: The species name refers to the reddish hairs on the buds and the red inner bark. • The common names refer to the sticky to slimy, red inner bark of this species.

Mulberry Family · Moraceae

Members of the mulberry family are trees and shrubs with milky latex juice. The alternate, simple leaves have 2 stipules at the base. The unisexual flowers are borne in spikes or heads. Male flowers have 4 sepals (occasionally absent) and 4 stamens opposite the sepals. Female flowers have a similar calyx and a single pistil. Petals are absent. The fruit is an achene, nut, drupe or aggregate.

A well-known member of this family is white mulberry (*Morus alba*), the primary food source for silkworms.

Osage-orange
Maclura pomifera

Habitat: Lowlands and thickets; introduced ornamental from the southern U.S. **General:** Tree 3–12 m tall; bark dark orangey brown, irregularly ridged; young branches greenish yellow; thorns 1–2 cm long. **Leaves:** Alternate, deciduous, oval to lance-shaped, 4–12 cm long, 2–6 cm wide, shiny above, dull green below; margins smooth; stalks 1–5 cm long; stipules lance-shaped, 1.5–2 mm long; turning yellow in autumn. **Flower Cluster:** Male flowers in globe-shaped to cylindric heads, 1.3–3.8 cm across, on stalks 2–10 mm long; female flowers in globe-shaped heads, 1.8–2.5 cm across, on stalks 2–2.5 mm long. **Flowers:** Male flowers yellowish green, with 4 sepals and 4 stamens; female flowers green, 2–5 mm across, with 4 sepals and 1 pistil. **Fruit:** Syncarp or aggregate of achenes 6–14 cm across, resembling a grapefruit, yellowish green, dimpled, containing milky juice; achenes embedded in the enlarged, fleshy calyx; receptacle becoming enlarged and fleshy.

Notes: The genus name commemorates William Maclure (1763–1840), an American geologist. • The species name *pomifera* means "pome-bearing," a reference to the fruit's resemblance to an apple.

White Mulberry · *Morus alba*

Habitat: Waste areas, forest margins and fence lines; introduced from China in about 1750. **General:** Tree 5–15 m tall; bark brown, yellow- or red-tinged, narrowly ridged; young branches orangey brown to reddish green; lenticels reddish brown. **Leaves:** Alternate, deciduous, oval, 8–10 cm long, 3–6 cm wide; margins deeply to irregularly 3–7-lobed, coarsely toothed; stalks 2.5–5 cm long, hairy; stipules oval to lance-shaped, 5–9 mm long, hairy; turning yellow in autumn. **Flower Cluster:** Male catkins 2.5–4 cm long; female catkins 5–8 mm long; appearing with leaves. **Flowers:** Green; male flowers with 4–5-lobed calyx, 1–2 mm long, green with red tip and 4 stamens; female flowers with 4-lobed calyx (2 large and 2 small lobes) and 1 pistil. **Fruit:** Syncarp or aggregate of achenes, 1.5–2.5 cm long, resembling a blackberry; sepals red, becoming fleshy and purplish black or white at maturity; edible.

Notes: *Morus* is the Latin name for black mulberry. The species name *alba* means "white," a reference to the fruit, which may be white. • White mulberry was introduced into North America in an attempt to create a silkworm industry.

Red Mulberry · *Morus rubra*

Habitat: Moist forests and woods. **General:** Tree 5–10 m tall; bark reddish brown with scaly strips; branches reddish brown to greenish brown; lenticels pale. **Leaves:** Alternate, deciduous, broadly oval, 10–18 cm long, 8–12 cm wide; margins irregularly 2–5-lobed, wavy or toothed; stalks 2–2.5 cm long; stipules linear, 1–1.3 cm long, hairy; turning yellow in autumn. **Flower Cluster:** Male catkins 3–5 cm long; female catkins 8–12 mm long, 5–7 mm wide; appearing before or with leaves. **Flowers:** Male flowers reddish green, with 4 sepals, 2–2.5 mm long, and 4 stamens; female flowers green, with 4 sepals and 1 pistil. **Fruit:** Syncarp or aggregate of achenes, 2.5–4 cm long, 1 cm wide, resembling a blackberry; sepals becoming fleshy and red to purple or black at maturity; edible.

Notes: The species name *rubra* means "red," a reference to the fruit colour. • Red mulberry is rare in Canada and often produces hybrids with white mulberry (*M. alba*). This hybridization is a threat to native stands of red mulberry.

Magnolia Family
Magnoliaceae

The magnolia family consists of deciduous and evergreen trees and shrubs. Members of this family are among the most ancient and primitive of the flowering plants. The simple, alternate leaves have smooth margins. The stipules, which often surround the stem, fall off early or late in the season. The solitary terminal flower has 1–2 bracts, 6–18 tepals appearing in 3 or more whorls, and numerous stamens and pistils. The cone-like fruit is an aggregate of follicles or samaras.

Several species of magnolias and their hybrids are grown for their ornamental value. Saucer magnolia (*Magnolia × soulangiana*), widely cultivated for its pink, early spring flowers, is a hybrid between lily-flowered magnolia (*M. liliiflora*) and Yulan magnolia (*M. denudata*). Another popular ornamental, star magnolia (*M. stellata*), has bright white flowers that bloom in early spring.

Tulip-tree
Liriodendron tulipifera

Habitat: Rich woods.
General: Tree 10–45 m tall; bark light grey, deeply furrowed.
Leaves: Alternate, deciduous, somewhat square, 4–6-lobed (commonly 4-lobed), 7.5–15 cm long, 12.5–18.5 cm wide; margins smooth; stalks 5–11.5 cm long; stipules elliptic to oblong or lance-shaped, 2–4.5 cm long. **Flower Cluster:** Solitary; bract 1, brown; appearing with leaves. **Flowers:** Greenish yellow with orange base, bell-shaped, resembling a tulip, 4–5 cm across; petals 6, 4–6 cm long, 2–3 cm wide; sepals 3, green, reflexed; stamens 20–50, 4–5 cm long; pistils 60–100. **Fruit:** Aggregate of samaras, cone-like, 4.5–8.5 cm long; samaras 3–5.5 cm long, 0.5–1 cm wide, separating at maturity.

Notes: *Liriodendron* is from the Greek words *leiron*, "a lily," and *dendron*, "a tree," a reference to this tree's lily-like flowers. The species name *tulipifera* means "tulip-bearing," another reference to the flowers.

Cucumber-tree
Magnolia acuminata

Habitat: Moist to wet, open forests. **General:** Tree 5–25 m tall; bark dark grey, furrowed. **Leaves:** Alternate, deciduous, oval to elliptic to oblong, 10–25 cm long, 4–15 cm wide, pale green; stipules 3.2–4.3 cm long, 1.4–1.6 cm wide. **Flower Cluster:** Solitary, terminal; bracts 2; appearing before or with leaves. **Flowers:** Greenish to orangey yellow, cup-shaped, 6–9 cm across; petals 6, 3–8 cm long, 1.8–3 cm wide; sepals 3, petal-like, reflexed; stamens 60–122, 0.5–1.3 cm long; pistils 40–45. **Fruit:** Aggregate of follicles, cone-like, oblong to cylindric, 2–7 cm long, 0.8–2.7 cm wide; follicles leathery, dark red, fleshy.

Notes: The genus name honours French botanist Pierre Magnol (1638–1715). The species name *acuminata* means "long-pointed," a reference to the leaves. • Cucumber-tree, rare in Ontario, is the only magnolia native to Canada.

CUSTARD-APPLE FAMILY
Annonaceae

The custard-apple family is a small family of deciduous or evergreen trees and shrubs. The bark, leaves and flowers are aromatic. The alternate leaves are simple. Stipules are absent. Flowers are borne singly or in few-flowered fascicles. The perfect flowers have 6 petals in 2 whorls of 3 (the outer larger, the inner fleshy), 3 sepals, 10 to many stamens and 1 to many pistils. The fruit is a syncarp or an aggregate of 1–12 berries.

Well-known tropical members of this family include custard-apple (*Annona cherimola*), a Peruvian tree widely cultivated throughout the tropics, and ylang-ylang (*Cananga odorata*), whose flower oils are used to add fragrance to soaps and shampoos.

Pawpaw
Asimina triloba

Habitat: Floodplains and wet forests. **General:** Tree or tall shrub 1.5–11 m tall; bark shallowly furrowed; young branches with rust-coloured hairs. **Leaves:** Alternate, deciduous, oblong to oval or lance-shaped, 15–30 cm long, 7.5–13 cm wide; margins smooth; stalks 0.5–1 cm long. **Flower Cluster:** Solitary or a 2–4-flowered fascicle borne on previous year's wood; stalk 1.5–2 cm long, nodding; bracts 1–2, 2–3 mm long, hairy; appearing with leaves. **Flowers:** Reddish purple to maroon, 3–4 cm wide, bell-shaped, ill-scented; petals 6 (outer 3 larger), 1.5–2.5 cm long, veiny; sepals 3, 0.8–1.2 cm long, deciduous; stamens numerous; pistils 3–12. **Fruit:** Berry yellowish green to brownish, 5–15 cm long, 3–4 cm wide, often in groups of 1–3; edible.

Notes: The genus name *Asimina* comes from the First Nations words *assimin* or *rassimin*, the name for pawpaw. The species name *triloba* means "3-lobed," a reference to the 3-parted flowers.

LAUREL FAMILY
Lauraceae

The laurel family consists of ever-green (rarely deciduous) trees and shrubs with aromatic leaves. The simple, alternate leaves have smooth to lobed margins. Stipules are absent. Flowers are borne in panicles, racemes or cymes. The flowers are perfect or unisexual and have 6 tepals in 2 whorls of 3, 9 stamens in 3 whorls and 1 pistil. A well-developed hypanthium is usually present. The fruit is a drupe that is often embedded in a cup-shaped receptacle (cupule).

Well-known members of this family include cinnamon (*Cinnamomum zeylanicum*), cam-phor (*Cimmamomum camphora*), avocado (*Persea gratissima*) and bayleaf (*Laurus nobilis*).

Sassafras · *Sassafras albidum*

Habitat: Forests, woods and old fields. **General:** Tree 5–35 m tall; bark dark brown, cork-like ridges, fragrant; twigs pale green. **Leaves:** Alternate, deciduous, oval to ellip-tic, entire or 2–3-lobed, 10–16 cm long, 5–10 cm wide, bright green above, pale below; stalks long; turn-ing yellow to red in autumn. **Flower Cluster:** Raceme 1–5 cm long; bracts 0.8–1.3 cm long; appearing with leaves. **Flowers:** Greenish yellow, fragrant (lemon-scented); male flowers with 6 tepals and 9 stamens in 3 whorls (inner 3 with glands at the base); female flowers with 6 tepals, 6 sterile sta-mens and 1 pistil; flower stalks elongating in fruit. **Fruit:** Drupe berry-like, blue, 1–1.5 cm across, on a fleshy, red, club-shaped base; stalk 3.5–4 cm long.

Notes: The genus name *Sassafras* comes from the Spanish common name *salsafras*. The species name *albidum* means "whitish," a reference to the leaf underside.

WITCH-HAZEL FAMILY

Hamamelidaceae

The witch-hazel family is a small family of deciduous trees and shrubs. The alternate leaves are simple to deeply 5–7-lobed. Stipules are present, but drop early in the season. Flowers are borne in heads, spikes or racemes. The perfect or unisexual flowers have 4–5 petals, 4–5 sepals, 4–34 stamens and 1 pistil. Petals and sepals are occasionally absent. The fruit is a capsule with a leathery exterior.

A well-known member of this family is witch-hazel (*Hamamelis virginiana*), renowned for its medicinal properties.

Sweetgum

Liquidambar styraciflua

Habitat: Old fields, woods and floodplains. **General:** Tree 10–30 m tall; bark grey, deeply furrowed; young branches with cork-like ridges. **Leaves:** Alternate, deciduous, palmately 5–7-lobed (star- or maple-leaf-shaped), 7–19 cm long, 4–16 cm wide, resinous-scented when crushed; margins toothed; stalks 6–10 cm long; stipules linear to lance-shaped, 3–4 mm long, deciduous; turning red or crimson in autumn. **Flower Cluster:** Male flowers in racemes 3–10 cm long; female flowers in globe-shaped heads; both types of clusters on long, drooping stalks. **Flowers:** Green; sepals and petals absent; male flowers with 4–8 stamens; female flowers with 5–8 sterile stamens and 1 pistil. **Fruit:** Aggregate of woody capsules, brown, globe-shaped, 2.5–4 cm across; each capsule ending in 2 curved, prickly points; seeds winged.

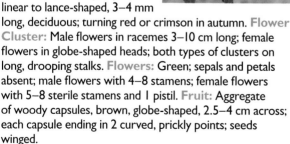

Notes: The genus name comes from the Latin word *liquidus*, "liquid," and the Arabic word *ambar*, "amber," a reference to the fragrant resin. The species name *styraciflua* means "flowing with styrax," an aromatic balsam.

PLANE-TREE FAMILY
Platanaceae

The plane-tree family is a small family of deciduous trees. A distinguishing feature of this family is the bark. Young bark is smooth at first, then exfoliates into thin plates, leaving the trunk and branches mottled. The simple, alternate leaves are palmately lobed. Stipules often form a sheath around the stem. The unisexual male and female flowers are borne in globe-shaped heads. The inconspicuous flowers have 3–4 petals, 3–4 sepals, 3–4 stamens and 3–4 pistils. The fruit is an achene, borne in globe-shaped clusters.

London Plane-tree
Platanus × acerifolia

Habitat: Along streets and in ornamental landscapes; introduced from Europe. **General:** Tree 15–21 m tall; bark flaky, with patches of brown, green and grey; twigs greenish. **Leaves:** Alternate, deciduous, palmately 3–5-lobed, 13–25 cm long and wide, shiny and green above, pale below; margins smooth or with a few large teeth; stalks long. **Flower Cluster:** Heads globe-shaped, appearing with leaves. **Flowers:** Green; male flowers with 3–6 petals (very small or absent), 3–6 sepals and numerous stamens; female flowers with no petals, 3–4 sepals, 3–4 sterile stamens and 3–9 pistils. **Fruit:** Aggregate of achenes, 2–3 cm across, usually in pairs on long stalks.

Notes: The species name *acerifolia* means "leaves like *Acer* (maples)."
• London plane-tree is believed to be a hybrid between our native sycamore (*P. occidentalis*) and the European plane-tree (*P. orientalis*).

ROSE FAMILY
Rosaceae

The diverse rose family is often separated into 4 subfamilies: Maloideae, Prunoideae, Rosoideae and Spiraeoideae. Species may be herbs, shrubs or trees with alternate, simple or compound leaves. Stipules are present. The flowers have 5 petals, 5 sepals and numerous stamens and pistils. Often 5 small bracts alternate with the sepals. Members of the rose family have a well-developed hypanthium, a structure formed by the fusion of the lower parts of the sepals, petals and stamens. Its shape varies from disc- to cup- or flask-shaped. The fruit is an achene, aggregate, berry, follicle, drupe or pome.

The subfamilies are distinguished by the following characteristics:

Spiraeoideae	Rosoideae	Prunoideae	Maloideae
Woody plants Pistils usually free Ovary superior Fruit an aggregate of follicles	Herbs and shrubs Pistils free Ovary superior Fruit an aggregate of achenes or drupelets	Woody plants Pistil 1 Ovary superior Fruit a drupe	Woody plants Pistils 2–5, fused Ovary inferior Fruit a pome

Smooth Serviceberry
Amelanchier laevis

Habitat: Forest edges, fence lines and roadsides. **General:** Tree or tall shrub 2–10 m tall; bark grey with dark vertical lines; twigs purplish. **Leaves:** Alternate, deciduous, oval to elliptic, 3–8 cm long, 2–4 cm wide, red to copper when young, dark green when fully expanded, both surfaces hairless; veins 10 or fewer per side; margins with about 25 teeth per side; stalks 1–3 cm long. **Flower Cluster:** Raceme many-flowered; stalks 1–3 cm long; bracts soon deciduous; appearing when leaves are half expanded. **Flowers:** White; petals 5, 1–2 cm long; sepals 5, 3–4 mm long; stamens about 20; pistil 1. **Fruit:** Pome berry-like, dark reddish purple to black, 0.6–1 cm wide, fleshy, juicy; stalk 2.5–4.5 cm long; seeds 5–10.

Notes: The species name *laevis* means "smooth," a reference to the hairless leaves.

Scarlet Hawthorn

Crataegus chrysocarpa var. *chrysocarpa*

Habitat: Dry river valleys and open woods.
General: Tree or tall shrub 1–6 m tall; bark scaly; thorns stout, 2–8 cm long, shiny, black, straight or curved; buds round. **Leaves:** Alternate, deciduous, round to oval, 2–7 cm long and wide; margins toothed to shallowly 7–13-lobed; stalks with reddish glands. **Flower Cluster:** Corymb 6–15-flowered; appearing after leaves. **Flowers:** White, 1–1.5 cm across, ill-scented; petals 5; sepals 5; stamens about 10, anthers white or pale yellow; pistil 1. **Fruit:** Pome berry-like, red or yellowish orange, 0.8–1 cm across; seeds hard, bony.

Notes: The species name *chrysocarpa* means "golden-fruited," though the fruit of this variety is often red.

English Hawthorn

Crataegus monogyna

Habitat: Waste areas, roadsides and pastures; introduced from Europe.
General: Tree or tall shrub 8–10 m tall; bark dark grey to brownish, scaly; thorns 1–2 cm long, in leaf axils and at twig tips. **Leaves:** Alternate, deciduous, oval to triangular, 1.5–5 cm long, 1–5 cm wide, deeply 3–7-lobed; margins coarsely toothed. **Flower Cluster:** Corymb many-flowered, 2–8 cm wide; appearing after leaves. **Flowers:** White, 0.8–1.5 cm across, ill-scented; petals 5; sepals 5; stamens about 20, anthers pink, pistil 1. **Fruit:** Pome berry-like, bright red to purple, 5–8 mm across; seed 1.

Notes: The species name *monogyna* refers to the single-seeded fruit.

Dotted Hawthorn
Crataegus punctata

Habitat: Rocky, open ground, clearings and floodplains. **General:** Tree or tall shrub 8–10 m tall; bark brownish grey, furrowed; thorns 2–8 cm long, straight, slightly curved or branched. **Leaves:** Alternate, deciduous, elliptic to oblong or oval, 2–8 cm long, 2–5 cm wide, widest above the middle; margins toothed. **Flower Cluster:** Corymb 2–9-flowered; appearing after leaves. **Flowers:** White, 1–2 cm across, ill-scented; petals 5; sepals 5; stamens about 20, anthers pink, red or yellow; pistil 1. **Fruit:** Pome berry-like, dull red to orange, 1–1.5 cm across, round to pear-shaped; seeds 3–5.

Notes: The species name *punctata* refers to the dotted surface of the fruit.

Common Apple
Malus pumila

Habitat: Open forests, roadsides and waste areas; introduced from Europe. **General:** Tree 5–15 m tall; bark brownish grey, flaky; branches with stout twigs (not thorny). **Leaves:** Alternate, deciduous, elliptic to oval, 4–10 cm long, 3–6 cm wide, lower surface and stalks hairy; margins finely toothed; turning yellow in autumn. **Flower Cluster:** Umbel or corymb 3–7-flowered, 4–6 cm across; appearing after leaves. **Flowers:** Pink fading to white, 2.5–3.5 cm across; petals 5; sepals 5; stamens numerous, anthers yellow; pistil 1. **Fruit:** Pome apple-like, 6–12 cm across, green to yellow or red at maturity, juicy, edible.

Notes: *Malus* is the Latin name for apple. The species name *pumila* means "dwarf." • The ancestral range of all apples is believed to be southwestern Asia.

Wild Plum
Prunus americana

Habitat: Moist woods, roadsides and fence lines. **General:** Tree or tall shrub 6–10 m tall; bark smooth, reddish brown to grey or black; branches with thorn-like twigs; twigs reddish brown to grey. **Leaves:** Alternate, deciduous, oval to oblong or lance-shaped, 4–12 cm long, 2–5 cm wide, dull green above; margins finely toothed; stalks 0.8–1.2 cm long. **Flower Cluster:** Umbel 2–5-flowered; flower stalks 1–2.5 cm long; appearing before or with leaves. **Flowers:** White, fragrant, 1.5–2.5 cm across; petals 5; sepals 5; stamens about 20; pistil 1. **Fruit:** Drupe plum-like, yellow to red, 2–3 cm long, flesh yellow, sour but edible; seed 1.

Notes: *Prunus* is the Latin name for plum. The species name means "of America."

Canada Plum · *Prunus nigra*

Habitat: Moist, open woods, pastures and fence lines. **General:** Tree or tall shrub 6–10 m tall; young bark reddish brown to blackish, older bark greyish brown, scaly; twigs short, thorny or spine-like. **Leaves:** Alternate, deciduous, oblong to oval, 6–14 cm long, 2.5–7 cm wide, dark green above, pale and somewhat hairy below; margins toothed; stalks with 1–2 glands at the blade. **Flower Cluster:** Umbel 2–4-flowered; flower stalks 1–2 cm long, red; appearing before or with leaves. **Flowers:** White, turning pink, 1.5–3 cm across, fragrant; petals 5; sepals 5; stamens about 20; pistil 1. **Fruit:** Drupe plum-like, red to yellow, 2.5–3 cm long, flesh yellow; seed 1.

Notes: The species name *nigra* means "black."

Black Cherry · *Prunus serotina*

Habitat: Open woods, roadsides and burned-over areas. **General:** Tree 20–30 m tall; young bark reddish brown to black, older bark dark grey to black and scaly. **Leaves:** Alternate, deciduous, oblong to lance-shaped, 5–15 cm long, 3–5 cm wide, dark green and waxy above, pale below; lower midvein with rust-coloured hairs on each side; margins finely toothed; turning yellow or reddish in autumn. **Flower Cluster:** Raceme 10–15 cm long; flower stalks 4–6 mm long; appearing with leaves. **Flowers:** White, 0.9–1.2 cm wide; petals 5; sepals 5; stamens about 20; pistil 1. **Fruit:** Drupe cherry-like, red to black, 0.8–1 cm across, juicy, flesh purple, in clusters of 6–12; seed 1.

Notes: The species name *serotina* means "late-flowering or ripening," a reference to the late-maturing fruit of this tree. • Black cherry is the largest native cherry in Ontario.

American Mountain-ash

Sorbus americana

Habitat: Moist, shady woods, swamp edges and rocky hillsides. **General:** Tree or tall shrub 4–10 m tall; bark reddish brown and scaly, young bark grey with lenticels; twigs reddish brown to grey. **Leaves:** Alternate, deciduous, 15–20 cm long, pinnately compound with 11–17 leaflets (commonly 13–15), stalks 5–10 cm long; leaflets oblong to lance-shaped, 4–10 cm long, 1–2.5 cm wide; margins coarsely toothed; turning yellow in autumn. **Flower Cluster:** Corymb 5–15 cm wide; appearing after leaves. **Flowers:** White, 5–7 mm wide; petals 5; sepals 5; stamens 15–20; pistils 2–4. **Fruit:** Pome berry-like, bright orange to red, 4–6 mm across; seeds 1–2.

Notes: *Sorbus* is the ancient Latin name for mountain-ash. The species name means "of America."

European Mountain-ash
Sorbus aucuparia

Habitat: Forest edges, roadsides and fields; introduced from Europe. **General:** Tree 5–10 m tall; bark greyish brown, scaly; twigs greyish, shiny; buds dark purple, covered in white, woolly hairs. **Leaves:** Alternate, deciduous, 10–20 cm long, pinnately compound with 9–17 leaflets; leaflets oblong to lance-shaped, 2.5–5 cm long, 1.2–1.9 cm wide, dull green above, white-woolly beneath; margins coarsely toothed. **Flower Cluster:** Corymb 7.5–15 cm wide, 75–100-flowered; appearing after leaves. **Flowers:** White, 0.9–1.1 cm wide; petals 5; sepals 5; stamens 15–20; pistils 2–4. **Fruit:** Pome berry-like, orangey red, 1–1.2 cm across; seeds 1–2.

Notes: The species name *aucuparia* means "bird-catching," a reference to the brightly coloured fruit, which is attractive to birds.

LEGUME OR PEA FAMILY · Fabaceae

Formerly called Leguminosae, the legume family is the third-largest plant family in the world with over 20,000 species of trees, shrubs, vines and herbaceous plants. All species have alternate, compound leaves. The flowers are composed of 5 petals, 5 sepals, 5 to many stamens and 1 pistil. There are 2 types of fruit, legumes and loments.

This large family is often divided into 3 subfamilies: Mimosoideae, Caesalpinioideae and Papilionoideae. The subfamilies are distinguished by the following characteristics:

Mimosoideae	Caesalpinioideae	Papilionoideae
Leaves bipinnate Flowers regular; petals free or fused Stamens 10 to many	Leaves pinnate or bipinnate, occasionally simple Flowers regular to irregular; petals free Stamens 10 or fewer	Leaves pinnate or palmate Flowers irregular (1 upper petal or "standard"; 2 lateral petals; lower 2 petals or "keel") Stamens 10 (9 fused, 1 free)

Redbud
Cercis canadensis

Habitat: Moist woods; native to southwestern Ontario. **General:** Tree 4–8 m tall; bark reddish brown, scaly; twigs reddish brown. **Leaves:** Alternate, deciduous, simple, heart-shaped, 7–12 cm long and wide, prominently 5–9-veined, pale bluish green above, pale below; margins smooth; stalks long; turning yellow in autumn. **Flower Cluster:** Raceme 4–8-flowered; stalks 6–12 cm long; appearing before leaves. **Flowers:** Pink, 0.9–1.4 cm long, somewhat pea-like; petals 5; sepals 5; stamens 10; pistil 1. **Fruit:** Legume pod-like, 6–10 cm long, 1–1.5 cm wide, reddish brown, pointed at both ends; seeds 10–12.

Notes: The genus name comes from the Greek word *kerkis,* meaning "a weaver's shuttle," a reference to the shape of the woody fruit. • This tree is rare in Ontario. The northern limit of its native range is believed to have been Pelee Island.

Honey Locust
Gleditsia triacanthos

Habitat: Rich, moist woods; native to southwestern Ontario. **General:** Tree 15–30 m tall; trunk with branched thorns to 40 cm long; bark furrowed; branches reddish with forked thorns. **Leaves:** Alternate, deciduous, 15–30 cm long, pinnately to bipinnately compound with 18–30 leaflets; leaflets oblong to lance-shaped, 2.5–5 cm long, 0.5–2 cm wide, dark green above, pale below; margins smooth; turning yellow in autumn. **Flower Cluster:** Male racemes 4–10 cm long; female racemes 7–9 cm long; appearing after leaves. **Flowers:** Greenish white, 0.4–1 cm across, somewhat bell-shaped, fragrant; male or female; petals 3–5; sepals 3–5; stamens 3–10; pistil 1. **Fruit:** Legume pod-like, dark brown, leathery, spirally twisted, 15–40 cm long, 3–4 cm wide; seeds separated by fleshy, edible pulp.

Notes: The genus name commemorates Johann Gottlieb Gleditsch (1714–1786), a German botanist. The species name *triacanthos* means "3-spined."

Kentucky Coffee-tree
Gymnocladus dioicus

Habitat: Moist woods. **General:** Tree 15–25 m tall; bark dark grey, thinly ridged; branches greyish brown. **Leaves:** Alternate, deciduous, 30–90 cm long, bipinnately compound with 3–7 pairs of branches and about 70 leaflets; leaflets oval, 4–6 cm long, 2–5 cm wide, bluish green; turning yellow in autumn. **Flower Cluster:** Raceme 6–20 cm long; appearing after leaves. **Flowers:** Greenish white, 1.5–2 cm across; petals 5; sepals 5; stamens 10, free; pistil 1. **Fruit:** Legume pod-like, leathery, thick, 8–25 cm long, 3–5 cm wide, stalks 2–3 cm long; seeds few, embedded in sticky pulp, poisonous.

Notes: The species name *dioicus* means "dioecious" or "2 houses," a reference to the trees being either male or female. • Kentucky coffee-tree has the largest leaf of any tree in Canada.

Black Locust
Robinia pseudoacacia

Habitat: Roadsides, waste areas and open woods; introduced from the eastern U.S. **General:** Tree 9–15 m tall; bark dark brown, furrowed; branches reddish brown with a pair of spines at each leaf base; spines (modified stipules) 8–13 cm long. **Leaves:** Alternate, deciduous, 25–30 cm long, compound with 7–19 leaflets; leaflets oval, 2–5 cm long, 1.2–1.9 cm wide, dull green; margins smooth; leaves turning yellow in fall. **Flower Cluster:** Raceme 10–20 cm long; appearing after leaves. **Flowers:** White, 1.5–2.5 cm long, fragrant, pea-like; petals 5; sepals 5; stamens 10; pistil 1. **Fruit:** Legume pod-like, reddish brown to black, 5–10 cm long, thin-walled; seeds 3–14.

Notes: The genus name commemorates Jean Robin (1550–1629), herbalist to King Henry IV of France. The species name *pseudoacacia* means "false acacia," a reference to the thorny, acacia-like branches.

RUE FAMILY
Rutaceae

The rue or citrus family is a large family of trees and shrubs with alternate, simple or compound leaves. The leaves are often dotted with glands. The perfect or unisexual flowers, usually borne in cymes, have 4–5 petals, 4–5 sepals, 5 or 10 stamens and 1 pistil. The fruit is a capsule, drupe or berry.

The rue family is an economically important one. The best-known genus is *Citrus*, which gives us orange, lemon, lime and grapefruit.

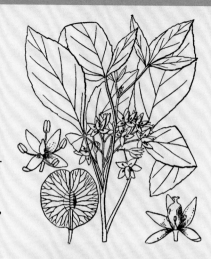

Hoptree, Wafer-ash
Ptelea trifoliata

Habitat: Moist, sandy areas, usually along shorelines; native to the north shore of Lake Erie. **General:** Tree 4–5 m tall; male or female; bark grey, rough, scaly; twigs yellowish to reddish brown. **Leaves:** Alternate, deciduous, 10–18 cm long, stalks 7–15 cm long, ill-scented (somewhat lemony) when crushed, compound with 3 leaflets; leaflets oval to elliptic, 5–10 cm long, 2–5 cm wide, dark green above, pale below; margins smooth; turning yellow in autumn. **Flower Cluster:** Cyme 4–8 cm wide; appearing after leaves. **Flowers:** Greenish white, ill-scented, 9–11 mm wide; petals 4–5, 4–7 mm long; sepals 4–5; stamens 4–5; pistil 1. **Fruit:** Samara 2–2.5 cm across, tan-coloured; wing flat and circular, prominently veined; seeds 2.

Notes: The genus name *Ptelea* is the Greek name given to a species of elm. The fruit of hoptree resembles that of elms (*Ulmus* spp.). The species name *trifoliata* means "3-leaved."

QUASSIA FAMILY
Simaroubaceae

The quassia family is a small family of trees and shrubs. Plants are either male or female. The alternate leaves are pinnately compound. The unisexual flowers are borne in racemes or panicles. Male flowers have 3–8 petals, 3–8 sepals and 6–16 stamens, whereas female flowers have similar petals and sepals and a single pistil. The fruit is a schizocarp, capsule or aggregate of drupes or samaras.

Tree-of-heaven
Ailanthus altissima

Habitat: Waste areas and road-sides; ornamental introduced from China in 1784. **General:** Tree 10–30 m tall; bark grey, smooth; twigs light brown, covered in hairs. **Leaves:** Alternate, deciduous, ill-scented when crushed, 30–100 cm long, compound with 11–41 leaflets; leaflets lance-shaped, 5–15 cm long, 2.5–5 cm wide, both surfaces with minute hairs and glands; base with 1–2 teeth, gland-bearing; turning yellow in autumn. **Flower Cluster:** Panicle 10–30 cm long; appearing after leaves. **Flowers:** Greenish yellow, 6–8 mm long, ill-scented; male flowers with 5–6 petals, 5–6 sepals and 6–16 stamens; female flowers with similar petals and sepals and 1 pistil. **Fruit:** Schizocarp yellow-green to reddish brown, breaking into 2–5 winged mericarps (samara-like); mericarps 3–5 cm long, 0.7–1.3 cm wide, twisted; fruiting cluster to 30 cm wide.

Notes: The genus name comes from the Chinese common name *ailanthos,* meaning "to reach for the sky." The species name *altissima* means "tallest."

STAFF-TREE FAMILY
Celastraceae

The staff-tree family is a small family of trees, shrubs and woody vines. The simple leaves may be alternate or opposite. The perfect flowers are borne in cymes or fascicles. The flowers consist of 3–5 petals (occasionally absent), 3–5 sepals, 3–5 stamens and 1 pistil. The fruit may be a berry, drupe, samara or capsule. An important diagnostic feature of this family is the seeds, which have a brightly coloured aril.

A few members of this family are grown for their ornamental value.

Spindle-tree
Euonymus europaeus

Habitat: Thickets and edges of woods; introduced from Europe. **General:** Tree or tall shrub 2–6 m tall; bark greyish green with red streaks; branches somewhat 4-sided. **Leaves:** Opposite, deciduous, elliptic to lance-shaped, 5–12 cm long, dull green above, pale below; margins finely toothed; stalks to 2.5 cm long; turning red in autumn. **Flower Cluster:** Cyme 3–7-flowered; stalks 2–5 cm long; appearing after leaves. **Flowers:** Green to yellowish white, 5–7 mm across; petals 4; sepals 4; stamens 4–5; pistil 1. **Fruit:** Capsule 4-lobed, red to purplish, 1–1.4 cm across; seeds 4, surrounded by fleshy, orange aril.

Notes: The species name *europaeus* refers to Europe, this tree's native range.

BLADDERNUT FAMILY
Staphyleaceae

Members of the bladdernut family are trees and shrubs with alternate or opposite, compound leaves. Stipules are present. Flowers are borne in racemes or panicles. The perfect flowers are composed of 5 petals, 5 sepals, 5 stamens and 1 pistil. The fruit is a capsule.

Bladdernut
Staphylea trifolia

Habitat: Moist, rocky woods and hillsides. **General:** Tree or tall shrub 3–5 m tall; bark green with mottled stripes, smooth; twigs green. **Leaves:** Opposite, deciduous, 15–25 cm long, compound with 3 leaflets; leaflets oval to lance-shaped, dark green above, pale below; terminal leaflet 5–10 cm long, stalked; lateral leaflets 3–10 cm long, 1–5 cm wide, stalkless; margins finely toothed; stipules 0.5–2.5 cm long, deciduous; turning greenish yellow in autumn. **Flower Cluster:** Panicle drooping, 4–10 cm long; appearing after leaves. **Flowers:** White to cream or greenish, 0.8–1.2 cm long, bell-shaped; petals 5; sepals 5; stamens 5; pistil 1. **Fruit:** Capsule inflated (bladder-like), veiny, 3–6 cm long, 3-lobed, pointed, green to yellowish brown; seeds 3–12.

Notes: The genus name comes from the Greek word *staphule,* meaning "cluster," a reference to the panicle of flowers. The species name *trifolia* means "3-leaved."

MAPLE FAMILY
Aceraceae

Members of the maple family are trees or shrubs with opposite, simple or compound leaves. The small flowers are unisexual. The male flowers are composed of 5 small petals, 5 separate or united sepals and 4–10 stamens. The female flowers are composed of 5 petals, 5 sepals and 1 pistil with 2 styles. The fruit, a samara, is composed of 2 winged seeds connected at the base.

A well-known member of this family, the sugar maple (*Acer saccharum*), is tapped for its sap, which is made into maple syrup. It is also a sugar maple leaf that appears on the Canadian flag. Other maple species are grown for their hard wood, which is used in furniture and flooring.

Amur Maple
Acer ginnala

Habitat: Roadsides, fence lines and waste areas; introduced from northeastern Asia and Japan in about 1860. **General:** Tree or tall shrub 4–6 m tall; bark dark grey, scaly; twigs yellowish brown. **Leaves:** Opposite, deciduous, 3–10 cm long, 1–4 cm wide, palmately 3-lobed (2 small lateral lobes, 1 large terminal lobe), notches wedge-shaped, dark and glossy green above, pale below; margins irregularly toothed; stalks 3–4 cm long; turning yellow to red in autumn. **Flower Cluster:** Corymb 10–15 cm long; appearing with leaves. **Flowers:** Cream to pale yellow, fragrant, 4–6 mm long; petals 5, very small; sepals 5; stamens 5; pistil 1.

Fruit: Samara 2–2.5 cm long, red to pinkish brown; in pairs.

Notes: *Acer* is the Latin name for maple. The species name *ginnala* is the common name for this plant in its native range.

Manitoba Maple
Acer negundo

Habitat: Riverbanks, roadsides and waste areas. **General:** Tree 8–20 m tall; male or female; bark dark greyish brown; twigs brown to purplish green; buds covered in fine, white hairs. **Leaves:** Opposite, deciduous, long-stalked, compound with 3–5 leaflets; leaflets oval to lance-shaped, 5–12 cm long, 2.5–4 cm wide, short-stalked; margins irregularly toothed or lobed; turning yellow in autumn. **Flower Cluster:** Raceme drooping, 2–14 cm long, male or female; appearing before leaves. **Flowers:** Yellowish green, 3–5 mm long; male flowers with no petals, 4–5 small sepals and 4–10 stamens; female flowers with no petals, 5 sepals and 1 pistil. **Fruit:** Samara 3–5 cm long; in pairs at an angle of less than 45°.

Notes: The species name *negundo* comes from the resemblance of the leaves to those of the Chinese chastetree (*Vitex negundo*). • This species is the only maple with compound leaves.

Striped Maple, Moosewood
Acer pensylvanicum

Habitat: Cool, moist and shady woods. **General:** Tree or tall shrub 4–10 m tall; bark greenish brown with darkened stripes; twigs reddish brown, shiny. **Leaves:** Opposite, deciduous, 10–18 cm long and wide, palmately 3-lobed, yellowish green, base rounded to heart-shaped; margins finely toothed; stalks 2.5–8 cm long; turning yellow in autumn. **Flower Cluster:** Raceme 3–14 cm long; appearing after leaves. **Flowers:** Greenish yellow, 3–7 mm across, bell-shaped; male flowers with 5 petals, 5 sepals and 5 stamens; female flowers with similar petals and sepals, 1 pistil. **Fruit:** Samara 2.5–3 cm long, green, in pairs at an angle of 90–120°.

Notes: The species name means "of Pennsylvania."

Norway Maple
Acer platanoides

Habitat: Forests, waste areas, roadsides and open woods; introduced from Europe in 1756. **General:** Tree 15–20 m tall; bark dark grey, ridged; twigs purplish to reddish green. **Leaves:** Opposite, deciduous, 8–16 cm long, 10–16 cm wide, palmately 5–7-lobed, dark green above, long-stalked; turning yellow in autumn. **Flower Cluster:** Corymb 15–30-flowered; appearing with leaves. **Flowers:** Greenish yellow, 8–11 mm across; perfect or male; petals 5; sepals 5; stamens 5; pistil 1. **Fruit:** Samara 3.5–5 cm long; in pairs at an angle of almost 180°.

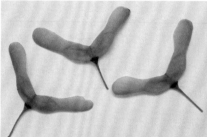

Notes: Norway maple is often confused with sugar maple (*A. saccharum*). The leaf stalks of Norway exude milky juice when broken, whereas those of sugar maple do not. The species name means "like *Platanus*," referring to the leaves, which resemble those of sycamore (*Platanus occidentalis*).

Red Maple
Acer rubrum

Habitat: Moist sites, swamps, and streambanks. **General:** Tree 20–25 m tall; bark dark greyish brown, scaly; twigs reddish, shiny. **Leaves:** Opposite, deciduous, 5–15 cm long, 5–12 cm wide, palmately 3–5-lobed, pale green above, whitish below; margins irregular, coarsely toothed; stalks 5–15 cm long, red; turning bright red in autumn. **Flower Cluster:** Umbel 2–12-flowered; appearing long before leaves. **Flowers:** Red, 2–4 mm across; male flowers with 5 petals, 5 sepals and 5 stamens; female flowers with similar petals and sepals, 1 pistil. **Fruit:** Samara 1.5–2.5 cm long, red to reddish brown; in pairs at an angle of 50–60°.

Notes: The species name *rubrum*, "red," refers to the colour of the flowers and leaf stalks.

Silver Maple
Acer saccharinum

Habitat: Moist to wet sites along streams and rivers. **General:** Tree 20–30 m tall; bark silvery grey; twigs shiny, ill-scented when broken. **Leaves:** Opposite, deciduous, 8–15 cm long and wide, palmately 5–7-lobed (middle lobe often 3-lobed), pale green above, silvery white below; margins irregularly to coarsely toothed; stalks 7.5–10 cm long; turning yellow to pale brown in autumn. **Flower Cluster:** Panicle several-flowered; appearing long before leaves. **Flowers:** Greenish yellow to reddish, 5–7 mm long; male or female; male flowers with no petals, 5 sepals and 5 stamens; female flowers with no petals, 5 sepals and 1 pistil. **Fruit:** Samara 4–7 cm long, yellowish green to brownish; in pairs at an angle of 90°; often only 1 fruit reaching maturity in late May.

Notes: The species name *saccharinum* means "sugary," a reference to the sweetish sap that can be collected to make maple syrup. The sap yield of silver maple is much lower than that of sugar maple (*A. saccharum*).

Sugar Maple
Acer saccharum

Habitat: Rich, dry woods. **General:** Tree 20–35 m tall; bark grey, irregularly ridged; twigs reddish brown. **Leaves:** Opposite, deciduous, 8–20 cm long and wide, palmately 5-lobed, dark yellowish green above, pale below, notches rounded; margins with a few irregular, blunt teeth; stalks 4–8 cm long; turning yellow to red in autumn. **Flower Cluster:** Umbel few-flowered; stalks 3–7 cm long; appearing before leaves. **Flowers:** Greenish yellow, 4–6 mm long; male or female; petals absent; sepals 5; stamens 5; pistil 1. **Fruit:** Samara 2–4 cm long; in pairs forming a U-shaped angle; seeds plump.

Notes: The species name *saccharum* is the Latin name for sugar cane. • Each tree may produce 23–270 L of sap. It takes about 157 L to produce 4.5 L of syrup or 2 kg of sugar.

Mountain Maple
Acer spicatum

Habitat: Moist woods, swamps and rocky hillsides.
General: Tree or tall shrub 3–5 m tall; bark greenish grey to black; branches purplish grey. **Leaves:** Opposite, deciduous, 5–12 cm long, 5–10 cm wide, prominently 3-lobed, base heart-shaped, yellowish green above, lower surface whitish with hairs; margins irregularly coarsely toothed; stalks 6–13 cm long; turning red, yellow or brown in autumn. **Flower Cluster:** Panicle erect, 6–10 cm long; branches 2–4-flowered; appearing after leaves. **Flowers:** Yellowish green, 5–7 mm across, perfect or unisexual; petals 5; sepals 5; stamens 5; pistil 1. **Fruit:** Samara 1.8–2 cm long, red to yellow or pinkish brown; in pairs at an angle of 90° or less.

Notes: The species name *spicatum*, "spike," refers to the spike-like flower cluster.

HORSECHESTNUT FAMILY
Hippocastanaceae

The horsechestnut family is a small family of trees and shrubs with opposite, palmately compound leaves. The irregular, perfect flowers, borne in panicles, are composed of 4–5 petals, 5 fused sepals, 5–8 stamens and 1 pistil. The fruit is a leathery capsule with 1 seed.

Horsechestnut
Aesculus hippocastanum

Habitat: Edges of forests and along roadsides; introduced from Europe. **General:** Tree 12–25 m tall; bark grey, fissured; twigs stout; buds sticky, 2–4 cm long. **Leaves:** Opposite, deciduous, palmately compound with 5–9 (commonly 7) leaflets; leaflets oblong to oval or wedge-shaped, 10–25 cm long, 2.5–9 cm wide; margins irregularly toothed; turning yellow in autumn. **Flower Cluster:** Panicle cone-shaped, 20–30 cm long; appearing after leaves. **Flowers:** White with red or yellow markings, bell-shaped, 2–3 cm long; perfect or unisexual; petals 4–5, unequal; sepals 5; stamens 5–8; pistil 1. **Fruit:** Capsule green, spiny, leathery, 4–6 cm across; seeds 1–2, brown, shiny, poisonous.

Notes: *Aesculus* is the Latin name for an oak with edible acorns. The species name *hippocastanum* means "horsechestnut."

BUCKTHORN FAMILY
Rhamnaceae

The buckthorn family consists of trees, shrubs and woody vines with alternate or opposite, simple leaves. Stipules are present. Flowers are borne in cymes, umbels, spikes or heads. The perfect or unisexual flowers have 4–5 petals (occasionally absent), 5 sepals, 5 stamens opposite the petals and 1 pistil. The fruit is a capsule or drupe-like berry.

Common Buckthorn
Rhamnus cathartica

Habitat: Open forests and pastures, waste areas and roadsides; introduced from Europe in the early 1800s. **General:** Tree or tall shrub 4–6 m tall; male or female; bark light grey to brownish, slightly scaly; twigs spine-tipped. **Leaves:** Opposite, deciduous, elliptic to oblong, 3–8 cm long, 2–5 cm wide, dark green above, yellowish green below; veins 2–4 pairs curving toward the tip; margins with 6–8 teeth per cm; stipules 3–5 mm long, toothed, often absent; remaining green into late autumn. **Flower Cluster:** Male cymes 2–7-flowered; female cymes 2–15 flowered; appearing after leaves. **Flowers:** Greenish yellow, fragrant, 4–6 mm wide; male flowers with 4 petals, 4 sepals and 4 stamens; female flowers with 4 petals, 4 sepals and 1 pistil; petals lance-shaped, 1–1.5 mm long. **Fruit:** Drupe berry-like, purplish black, 5–9 mm across, juicy; seeds 4.

Notes: The species name *cathartica* means "purging," a reference to this plant's medicinal use. • Consumption of about 20 berries may lead to abdominal pain and vomiting.

LINDEN FAMILY
Tiliaceae

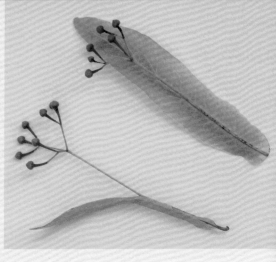

Members of the linden family are trees, shrubs and herbs with simple, alternate leaves. Stipules are present. The flowers, borne in cymes, are perfect and consist of 5 petals, 5 sepals, numerous stamens and 1 pistil. The fruit is a capsule or schizocarp. A distinguishing characteristic of this family is the large, leaf-like bract.

Important members of this family include 2 jute species (*Corchorus capsularis* and *C. olitorius*), both cultivated for their fibres.

American Basswood
Tilia americana

Habitat: Cool, moist, deciduous forests, often near water. **General:** Tree 18–22 m tall; bark dark greyish brown, ridges flat-topped; twigs yellowish brown. **Leaves:** Alternate, deciduous, heart-shaped, 12–15 cm long and wide, dark green and shiny above, pale below, base unequal; margins coarsely toothed; stalks long; turning yellow or brownish in autumn. **Flower Cluster:** Cyme 6–20-flowered, originating from the centre of a leaf-like bract 7–10 cm long; appearing after leaves. **Flowers:** Cream to yellowish, 1.1–1.5 cm across, fragrant; petals 5, 7–12 mm long; sepals 5; stamens numerous; pistil 1. **Fruit:** Capsule round, nut-like, 6–8 mm across, covered in brown, woolly hairs; seeds 1–2.

Notes: The genus name *Tilia* is believed to originate from the Greek word *ptilon*, "wing," a reference to the large, wing-like floral bract. The species name means "of America."

Littleleaf Linden
Tilia cordata

Habitat: Landscapes and roadsides; introduced from Europe. **General:** Tree 12–15 m tall; bark greyish brown, ridges block-like; twigs reddish brown. **Leaves:** Alternate, deciduous, heart-shaped to round, 3.5–8 cm long and wide, dark green and shiny above, pale below with rust-coloured hairs in axils; margins coarsely toothed; stalks long; turning yellow to tan in autumn. **Flower Cluster:** Cyme 3–4.5 cm wide, 5–8-flowered, originating from the centre of a leaf-like bract; bract 3.5–8 cm long; appearing after leaves. **Flowers:** Greenish yellow, fragrant, 1–1.2 cm across; petals 5; sepals 5; stamens numerous; pistil 1. **Fruit:** Capsule round, nut-like, 4–6 mm across, covered in rusty brown hairs; seed 1.

Notes: The species name means "heart-shaped," a reference to the leaf shape.

OLEASTER FAMILY
Elaeagnaceae

Members of the oleaster family are shrubs and trees with alternate or opposite leaves. The simple leaves are often covered in star-shaped or scale-like hairs. Stipules are absent. Flowers are borne in racemes or appear solitary. The perfect flowers have 2 or 4 fused or partly fused sepals, 4 or 8 stamens and 1 pistil. Petals are absent. The fruit, an achene surrounded by a fleshy calyx tube, often resembles a drupe-like berry.

Russian-olive
Elaeagnus angustifolia

Habitat: Roadsides, waste areas and open fields; introduced from Russia in the early 1900s. **General:** Tree 5–12 m tall; bark dark grey, smooth; branches reddish brown, ending in a spine. **Leaves:** Alternate, deciduous, oblong to elliptic or lance-shaped, 2–10 cm long, 1–4 cm wide, dull greyish green to silver-coloured, scaly and brown-dotted below; margins smooth; stalks 0.6–1.3 cm long. **Flower Cluster:** Solitary or raceme of 2–3 flowers; appearing after leaves. **Flowers:** Yellow, fragrant, 3–12 mm long, bell-shaped; perfect or unisexual; sepals 4, silvery grey outside, yellow inside; stamens 4; pistil 1. **Fruit:** Achene 7–8 mm long, surrounded by silvery grey, mealy calyx, berry-like, 1–2 cm long.

Notes: The genus name comes from the Greek words *helodes*, "growing in marshes," and *hagnos*, "pure," a reference to the whitish fruiting clusters. The species name *angustifolia* means "narrow-leaved."

Sea-buckthorn
Hippophae rhamnoides

Habitat: Roadsides, waste areas and open fields; introduced from northern Europe and Asia. **General:** Tree or tall shrub 2–6 m tall; bark brown to black, rough-textured; branches grey, spiny; young growth covered with star-shaped hairs and silver scales. **Leaves:** Alternate, deciduous, linear to lance-shaped, 2–8 cm long, 2–8 mm wide, silvery. **Flower Cluster:** Male racemes 4–6-flowered; female flowers solitary; appearing before leaves. **Flowers:** Yellowish; male flowers with 2 sepals and 4 stamens; female flowers with 2 sepals and I pistil. **Fruit:** Achene dark brown to black, surrounded by yellowish orange, fleshy calyx, berry-like, 4–9 mm long; remaining on the stem through winter.

Notes: The species name *rhamnoides* refers the resemblance of this plant to the buckthorns (*Rhamnus* spp.).

DOGWOOD FAMILY
Cornaceae

The dogwood family has over 100 species worldwide. Members of this family are herbs, shrubs and trees with simple, alternate or opposite leaves. Stipules are absent. Flowers are borne in cymes, umbels or heads, and often resemble a single flower. The perfect flowers are composed of 4 petals, 4 sepals, 4 stamens and I pistil. The fruit is a drupe or berry.

Several species of dogwood are grown for their ornamental value.

Alternate-leaved Dogwood, Pagoda Dogwood
Cornus alternifolia

Habitat: Rich, deciduous forests. **General:** Tree or tall shrub 4–6 m tall; bark greenish to reddish brown; branches horizontal; twigs reddish green to purplish. **Leaves:** Alternate, deciduous, oval, 4–13 cm long, 2–7.5 cm wide, green above, greyish and slightly hairy below; margins smooth to wavy; stalks 0.8–6 cm long; turning red in autumn. **Flower Cluster:** Cyme flat-topped, 5–10 cm across; appearing after leaves. **Flowers:** White; petals 4; sepals 4, small; stamens 4; pistil 1. **Fruit:** Drupe berry-like, blue to bluish black, 5–7 mm across, on short, red stalks; seed 1.

Notes: The species name *alternifolia* means "alternate-leaved," a reference to the leaf arrangement.

SOURGUM FAMILY
Nyssaceae

The sourgum family, a small group of trees and shrubs, is often included with the dogwoods (Cornaceae). Members have simple, alternate leaves and no stipules. Plants are often male or female. Flowers are borne singly or in racemes, heads or umbels. Plants with perfect flowers have 5 petals, 5 sepals, 5–12 stamens and 1 pistil. The fruit is a fleshy drupe or samara.

Black Tupelo
Nyssa sylvatica

Habitat: Wet lowlands; native to the north shore of Lake Erie. **General:** Tree 10–20 m tall; male or female; bark dark grey with thick, irregular segments; twigs reddish brown. **Leaves:** Alternate, deciduous, oval to oblong or wedge-shaped, 5–15 cm long, 3–5 cm wide, dark green and shiny above, whitish below; margins wavy, occasionally toothed; turning golden yellow or scarlet in autumn. **Flower Cluster:** Male umbels on stalks 1–3 cm long; female umbels 2–8-flowered, on stalks 3–5 cm long; appearing after leaves. **Flowers:** Greenish white, inconspicuous; petals 5, small or absent; sepals 5, very small; stamens 10; pistil 1. **Fruit:** Drupe plum-like, dark blue to black, 1–1.5 cm long, fleshy; in clusters of 1–3 on stalks 3–6 cm long; seed 1.

Notes: The genus name commemorates Nysa, a mythical water nymph, alluding to this tree's preference for wetter sites. The species name *sylvatica* means "of the woods."

OLIVE FAMILY
Oleaceae

The olive family is a family of trees and shrubs with opposite, simple or compound leaves. Stipules are absent. The flowers, borne in racemes and panicles, may be perfect or unisexual. Perfect flowers have 4 united petals, 4 united sepals, 2 stamens, and 1 pistil. The fruit is a capsule, samara or drupe.

Well-known members of this family include olive (*Olea europaea*), lilac (*Syringa vulgaris*) and jasmine (*Jasminum* spp.).

White Ash · *Fraxinus americana*

Habitat: Upland forests with well-drained soils. **General:** Tree 5–20 m tall; male or female; bark grey with intersecting ridges; twigs greyish, hairless; terminal bud reddish brown, 0.5–1.4 cm long, flanked by smaller buds. **Leaves:** Opposite, deciduous, 20–40 cm long, compound with 5–9 (commonly 7) leaflets; leaflets oval to oblong, 6–15 cm long, 3–6 cm wide, dark green above, whitish below, stalks 0.5–1.5 cm long; margins smooth to finely toothed; turning yellow to bronze or purple in autumn. **Flower Cluster:** Panicle densely flowered; appearing before leaves. **Flowers:** Purplish to yellowish green, inconspicuous, unisexual, 4–7 mm long; petals absent; sepals 4, very small; stamens 2; pistil 1. **Fruit:** Samara pale green to yellowish, 2.5–5 cm long; wing enclosing upper ⅓ of seed, tip rounded.

Notes: The genus name *Fraxinus* is the Latin name for ash. The species name means "of America," a reference to the native range of this tree. • White ash is the most common ash species in Ontario.

Black Ash · *Fraxinus nigra*

Habitat: Cool, wet sites, often in swampy woods. **General:** Tree 10–20 m tall; male or female; bark grey, scaly; twigs stout, dull grey; terminal bud 4–10 mm long, slightly above the 2 lateral buds. **Leaves:** Opposite, deciduous, 22–45 cm long, compound with 7–11 leaflets; leaflets oblong to lance-shaped, 7–14 cm long, 2.5–4 cm wide, dark green, lower surface with rust-coloured hairs on the veins, stalkless; margins fine-toothed; turning reddish brown in autumn. **Flower Cluster:** Panicle compact; appearing before leaves. **Flowers:** Purplish, 2–3 mm long; perfect or unisexual; petals absent; sepals 4, very small; stamens 2; pistil 1. **Fruit:** Samara pale green to yellowish, 2.5–4 cm long, 0.6–1 cm wide; winged to the base of the seed, the wing round or notched at tip.

Notes: The species name *nigra* means "black."

Red Ash, Green Ash
Fraxinus pensylvanica

Habitat: Moist to wet floodplains and swamps.
General: Tree or tall shrub 10–15 m tall; male
or female; bark greyish brown, often reddish,
flaky with irregular ridges; twigs often hairy;
terminal bud 3–8 mm long with small buds at
the side. **Leaves:** Opposite, deciduous,
23–35 cm long, compound with 5–9 (usually 7)
leaflets; leaflets oval to lance-shaped, 8–15 cm
long, 2.5–4 cm wide; margins toothed above
the middle; turning yellow in autumn. **Flower
Cluster:** Panicle many-flowered, compact;
appearing before leaves. **Flowers:** Purplish to
yellowish green, 2–5 mm long, perfect or unisexual; petals absent; sepals 4, very small; stamens 2; pistil 1. **Fruit:** Samara pale green to
yellowish, 3–6 cm long, 4–8 mm wide; wing
enclosing upper ½ of seed, notched at tip.

Notes: The species name means "of Pennsylvania," a reference to the native range of this tree.

Blue Ash · *Fraxinus quadrangulata*

Habitat: Wet woods along beaches.
General: Tree 10–20 m tall; male or
female; bark grey with scaly plates; inner
bark turning blue when exposed to air;
twigs 4-sided with cork-like ridges;
buds dark brown, 6–8 mm long, slightly
flattened. **Leaves:** Opposite, deciduous,
20–40 cm long, compound with
5–11 leaflets; leaflets lance-shaped,
8–14 cm long, 2.5–4 cm wide, dark yellowish
green above, pale below, short-stalked; margins coarsely toothed; turning yellow in
autumn. **Flower Cluster:** Panicle many-
flowered, compact; appearing before leaves.
Flowers: Purplish, 2–3 mm long; perfect;
petals absent; sepals 5, very small;
stamens 2; pistil 1. **Fruit:** Samara green
to yellowish green, 2.5–4 cm long,
0.6–1.2 cm wide; winged to the base of the seed, the wing rounded to notched
at tip.

Notes: The species name *quadrangulata* means "4-angled," a reference to the
twigs. • The common name refers to the inner bark, which turns blue when
exposed to air.

TRUMPET-CREEPER FAMILY
Bignoniaceae

Members of this family are trees, shrubs and woody vines. The opposite leaves may be simple or compound. Stipules are absent. The perfect, irregular flowers are borne in cymes and consist of 5 petals, 5 sepals, 2 or 4 stamens and 1 pistil. The fruit, a capsule, produces seeds that are winged or tufted with hairs.

A well-known member of this family is the African sausage-tree (*Kigelia africana*), whose sausage-like fruit can weigh as much as 9 kg.

Northern Catalpa
Catalpa speciosa

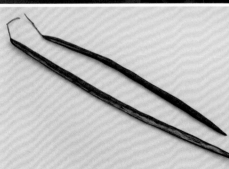

Habitat: Landscapes and occasionally roadsides; introduced from the southern U.S. **General:** Tree 10–30 m tall; bark dark reddish brown, scaly. **Leaves:** Opposite or in whorls of 3, simple, heart-shaped, 10–30 cm long, 10–20 cm wide, yellowish green above, pale and soft-haired below; margins smooth; stalks 10–16 cm long; turning black with the first frost. **Flower Cluster:** Panicle 10–20 cm long; appearing after leaves. **Flowers:** White with 2 yellow stripes and pinkish purple spots, 2-lipped, bell-shaped, 5–7 cm long and wide; petals 5; sepals 5; stamens 2; pistil 1. **Fruit:** Capsule 25–60 cm long, 1–1.5 cm wide, green to purplish brown; seeds 2–3 cm wide, numerous.

Notes: *Catalpa* is a First Nations name for this tree. The species name *speciosa* means "showy." • Another common name, cigar-tree, refers to the long fruit.

Shrubs & Woody Vines

Purple clematis (*Clematis occidentalis*)

SHRUBS & WOODY VINES

Shrubs and woody vines often have several stems rising from the same crown or rootstock. Several small shrub species are often mistaken for herbaceous plants and have been included in the **Wildflowers** section.

Key to the Shrubs & Woody Vines

Species in this section are identified by leaf arrangement (i.e., alternate, opposite or whorled) and leaf margin and complexity (i.e., simple entire, simple lobed or compound).

Leaves needle- or scale-like, often evergreen

| *Juniperus communis* p. 81 | *Juniperus horizontalis* p. 81 | *Taxus baccata* p. 82 | *Taxus canadensis* p. 83 | *Tamarix ramosissima* p. 113 |

Leaves alternate, simple, not lobed

| *Smilax tamnoides* p. 84 | *Salix bebbiana* p. 84 | *Salix candida* p. 85 | *Salix cordata* p. 85 | *Salix discolor* p. 86 | *Salix eriocephala* p. 86 |

Leaves alternate, simple, not lobed

| *Salix exigua* p. 87 | *Salix petiolaris* p. 87 | *Myrica gale* p. 89 | *Betula pumila* p. 89 | *Corylus cornuta* p. 90 | *Berberis vulgaris* p. 93 |

Leaves alternate, simple, not lobed

| *Lindera benzoin* p. 94 | *Hamamelis virginiana* p. 97 | *Aronia melanocarpa* p. 97 | *Prunus pensylvanica* p. 99 | *Prunus pumila* p. 99 | *Prunus virginiana* p. 100 |

LEAVES BROAD

Leaves alternate, simple, not lobed

| *Spiraea alba* p. 103 | *Spiraea tomentosa* p. 104 | *Ilex verticillata* p. 108 | *Nemopanthus mucronatus* p. 108 | *Celastrus orbiculatus* p. 109 | *Ceanothus americanus* p. 109 |

Leaves alternate, simple, not lobed

| *Rhamnus alnifolia* p. 110 | *Elaeagnus umbellata* p. 113 | *Arctostaphylos uva-ursi* p. 116 | *Chamaedaphne calyculata* p. 117 | *Gaultheria procumbens* p. 117 | *Ledum groenlandicum* p. 118 |

Leaves alternate, simple, not lobed

Vaccinium myrtilloides p. 119

Leaves alternate, simple, lobed

| *Comptonia peregrina* p. 88 | *Menispermum canadense* p. 94 | *Ribes aureum* p. 95 | *Ribes glandulosum* p. 96 | *Ribes triste* p. 96 | *Physocarpus opulifolius* p. 98 |

Leaves alternate, simple, lobed

| *Rubus odoratus* p. 103 | *Vitis riparia* p. 112 | *Solanum dulcamara* p. 120 |

Leaves alternate, compound

Dasiphora fruticosa
p. 98

Rosa acicularis
p. 100

Rosa multiflora
p. 101

Rosa rubiginosa
p. 101

Rubus allegheniensis
p. 102

Rubus idaeus
p. 102

Leaves alternate, compound

Caragana arborescens
p. 104

Zanthoxylum americanum
p. 105

Rhus aromatica
p. 106

Rhus typhina
p. 106

Toxicodendron radicans
p. 107

Parthenocissus quinquefolia
p. 111

Leaves alternate, compound

Parthenocissus vitacea
p. 111

Leaves opposite, simple, not lobed

Shepherdia canadensis
p. 114

Cornus amomum
p. 114

Cornus racemosa
p. 115

Cornus rugosa
p. 115

Cornus sericea ssp. *sericea*
p. 116

Kalmia polifolia
p. 118

Leaves opposite, simple, not lobed

Syringa vulgaris
p. 119

Cephalanthus occidentalis
p. 120

Diervilla lonicera
p. 121

Lonicera canadensis
p. 122

Lonicera dioica
p. 122

Lonicera tatarica
p. 123

LEAVES BROAD

Leaves opposite, simple, not lobed

| *Symphoricarpos albus* p. 124 | *Symphoricarpos occidentalis* p. 125 | *Viburnum lantana* p. 125 | *Viburnum lantanoides* p. 126 | *Viburnum nudum* var. *cassinoides* p. 126 |

Leaves opposite, simple, lobed

| *Humulus lupulus* p. 91 | *Viburnum opulus* var. *americanum* p. 127 | *Viburnum opulus* var. *opulus* p. 127 |

Leaves opposite, compound

| *Clematis occidentalis* p. 91 | *Clematis virginiana* p. 92 | *Sambucus nigra* ssp. *canadensis* p. 123 | *Sambucus racemosa* var. *racemosa*, p. 124 |

CYPRESS FAMILY · Cupressaceae

Family description on p. 17.

Ground Juniper
Juniperus communis

Habitat: Open, wooded areas. **General:** Shrub to 1.5 m tall; often forming clumps or mats; male or female. **Leaves:** Whorls of 3, evergreen, needle-like, 0.5–1.5 cm long, less than 2 mm wide, sharp-pointed, upper surface grooved and whitish green. **Reproductive Structures:** Male cones small, appearing near branch ends; female cones pale blue, berry-like, 0.6–1.3 cm across, appearing in leaf axils, 1–3-seeded.

Notes: *Juniperus* is the Latin name for juniper. The species name *communis* means "common."

Creeping Juniper
Juniperus horizontalis

Habitat: Sandy to rocky, open or wooded areas. **General:** Shrub; stems horizontal, to 5 m long, often forming mats; branches ascending, 10–30 cm high; male or female. **Leaves:** Opposite, evergreen; young stem leaves needle-like, 5–7 mm long; older branch leaves scale-like, 1–2 mm long, overlapping. **Reproductive Structures:** Male cones small, appearing near branch ends; female cones dark blue, berry-like, 5–7 mm across, 1–6-seeded.

Notes: The species name means "horizontal," a reference to this species' prostrate growth habit.

YEW FAMILY
Taxaceae

Members of this family are evergreen trees and shrubs with alternate, needle-like leaves. The 2-sided needles give the branches a flat-tened appearance, and the leaf bases extend down the stem to the leaf node below. Male and female cones are produced on separate plants. The male (pollen) cones have 4–32 spirally arranged scales. The female (seed) cones are berry-like and contain a single seed. These fleshy cones are usually red or green.

English Yew
Taxus baccata

Habitat: Waste areas and roadsides; introduced from Europe as an ornamental shrub. **General:** Tree or shrub 3–20 m tall; bark thin, scaly, brown. **Leaves:** Alternate, evergreen, needle-like, flat, 1–4 cm long, 2–3 mm wide, dark green; bases twisted, making leaves appear to be in 2 flat rows. **Reproductive Structures:** Males cones 3–6 mm across, producing pollen in early spring, then falling off; female cones berry-like, red, 8–15 mm across; aril fleshy, edible, sweet-tasting; seeds and cones poisonous.

Notes: The species name *baccata* means "bearing berries."

Canada Yew
Taxus canadensis

Habitat: Moist woods and ravine slopes. **General:** Shrub to 2 m tall; male or female; young branches green, older branches brown and scaly. **Leaves:** Alternate, evergreen, needle-like, flat, 1–2.5 cm long, 1–3 mm wide, dark green above, pale below, tip sharp-pointed; stalks twisted, making leaves appear to lay flat along stem. **Reproductive Structures:** Male cones small, readily visible in spring when pollen is shed; female cones red, berry-like, 0.5–1 cm across; aril fleshy, edible; seeds reportedly poisonous.

Notes: *Taxus* is the Latin name for yew. The species name means "of Canada."

CATBRIER FAMILY
Smilacaceae

Members of this family are shrubs, herbs or vines, and are often included in the lily family (Liliaceae). The erect or climbing stems are usually prickly. Tendrils are present. The opposite or alternate leaves are prominently 3-veined, with netted venation between the primary veins. Flowers are usually borne in umbels, but occasionally occur in racemes or spikes. Flowers are male or female, and are borne on separate plants. Male flowers have 6 tepals in 2–3 whorls. Female flowers have 6 tepals and 1 pistil. The fruit is a 1–6-seeded berry.

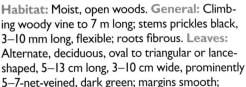

Bristly Catbrier, Carrion-flower
Smilax tamnoides

Habitat: Moist, open woods. **General:** Climbing woody vine to 7 m long; stems prickles black, 3–10 mm long, flexible; roots fibrous. **Leaves:** Alternate, deciduous, oval to triangular or lance-shaped, 5–13 cm long, 3–10 cm wide, prominently 5–7-net-veined, dark green; margins smooth; stalks 6–13 mm long; tendrils (modified stipules) in pairs. **Flower Cluster:** Umbel 5–25-flowered, on drooping stalks 1.5–6 cm long. **Flowers:** Greenish yellow to bronze, 4–5 mm long, ill-scented; male flowers with 6 tepals and 6 stamens; female flowers with 6 tepals and 1 pistil; flower stalks 4–12 mm long. **Fruit:** Berry black, 0.6–1 cm wide; seeds 1–2.

Notes: *Smilax* is the Greek name for catbrier. The species name *tamnoides* means "like *Tamnus*," referring to a climbing plant in the yam family (Dioscoreaceae). • The ill-scented flowers attract carrion flies, hence the common name "carrion-flower."

WILLOW FAMILY · Salicaceae
Family description on p. 24.

Beaked Willow · *Salix bebbiana*

Habitat: Wet meadows, swamps, riverbanks and moist, open forests. **General:** Shrub or small tree to 5 m tall; branches reddish brown, hairy or hairless. **Leaves:** Alternate, deciduous, oval to lance-shaped, 2–6 cm long, 1–2.5 cm wide, upper surface dull grey, underside grey and prominently veined; margins smooth to wavy; stalks 0.3–1 cm long. **Flower Cluster:** Male catkins 1–4 cm long; female catkins 2–5 cm long; appearing before or with leaves. **Flowers:** Male flowers with 2 stamens and 1 bract; female flowers with 1 pistil and 1 bract. **Fruit:** Capsule 6–9 mm long; seeds with fine, white hairs.

Notes: *Salix* is the Latin name for willow. The species name commemorates Michael Schuck Bebb (1833–95), a willow specialist.

Hoary Willow
Salix candida

Habitat: Sandy, peaty or marshy ground. **General:** Shrub 2–3 m tall; young branches yellowish brown, surface with white, felt-like hairs; old bark smooth, reddish brown to grey. **Leaves:** Alternate, deciduous, linear to lance-shaped, 3–10 cm long, 0.5–2 cm wide, upper surface dark green, underside with white, felt-like hairs, midrib often yellow; margins smooth, often rolled under; stalks 3–10 mm long. **Flower Cluster:** Male catkins 1–2.5 cm long; female catkins 2–4 cm long; appearing with leaves. **Flowers:** Male flowers with 2 stamens, anthers reddish purple, bract 1; female flowers with 1 pistil and 1 bract. **Fruit:** Capsule 5–8 mm long; seeds with fine, white hairs.

Notes: The species name means "white" or "woolly," referring to the leaves.

Heart-leaved Willow
Salix cordata

Habitat: Riverbanks and lakeshores, usually on sandy soil. **General:** Shrub 1–3 m tall; branches reddish, covered with white to greyish hairs when young, older branches smooth. **Leaves:** Alternate, deciduous, oval to lance-shaped, 3–10 cm wide, 1–5 cm wide, both surfaces covered in soft, shiny hairs; margins glandular-toothed; stalks 2–10 mm long, hairy; stipules oval to heart-shaped, 0.6–1.5 cm long; young leaves often red-tinged. **Flower Cluster:** Male catkins 2–4.5 cm long; female catkins 3–8 cm long; appearing with leaves. **Flowers:** Male flowers with 2 stamens, barely exceeding the brownish black scales; female flowers with 1 pistil. **Fruit:** Capsule 5–8 mm long, green to brown, hairless; seeds with white hairs.

Notes: The species name means "heart-shaped."

Pussy Willow · *Salix discolor*

Habitat: Wet meadows, lakeshores, riverbanks, swamps and roadside ditches. **General:** Shrub 2–3 m tall; bark greyish brown; older branches dark reddish brown and hairless. **Leaves:** Alternate, deciduous, oblong to elliptic or lance-shaped, 3–10 cm long, 1–3 cm wide, bright green above, pale below; margins irregularly toothed to wavy; stalks 0.5–1.5 cm long; young leaves often reddish. **Flower Cluster:** Male catkins 2–4 cm long; female catkins 2–6 cm long, densely flowered; appearing before leaves. **Flowers:** Male flowers with 2 stamens; female flowers with 1 pistil. **Fruit:** Capsule 7–12 mm long, on stalks 2–2.5 mm long; seeds with fine, white hairs.

Notes: The species name *discolor* refers to the contrasting colours of the leaf surfaces.

Missouri River Willow, Heartleaf Willow

Salix eriocephala

Habitat: Streambanks, ditches, swamps and wet woods. **General:** Shrub 3–4 m tall; branches reddish brown to yellowish green, becoming hairless with age; stems often with silvery insect galls. **Leaves:** Alternate, deciduous, oblong to lance-shaped, 5–15 cm long, 1–4 cm wide, dark green above, pale below; margins toothed; stalks 0.5–1.5 cm long; stipules oval to lance- or heart-shaped, 0.5–2 cm long, becoming leathery with age and turning reddish purple in autumn; young leaves often tinged red or purple. **Flower Cluster:** Male catkins 1–3 cm long; female catkins 2–8 cm long; appearing before or with leaves. **Flowers:** Male flowers with 2 stamens; female flowers with 1 reddish green pistil. **Fruit:** Capsule 4–6 mm long, reddish turning brown with age; seeds with white, cottony hairs.

Notes: The species name *eriocephala* means "woolly headed."

Sandbar Willow
Salix exigua

Habitat: Riverbanks and sandy shorelines. **General:** Shrub or small tree, to 4 m tall; bark grey to reddish brown. **Leaves:** Alternate, deciduous, linear, 3–15 cm long, 0.3–1.5 cm wide, green above, paler below; margins smooth to finely toothed; stalks 0.5–5 mm long. **Flower Cluster:** Male catkins 3–4 cm long; female catkins 1–7 cm long; on young branches 3–6 cm long; appearing before leaves (occasionally throughout summer). **Flowers:** Male flowers with 2 stamens and 1 bract; female flowers with 1 pistil and 1 bract. **Fruit:** Capsule 5–8 mm long; seeds with fine, white hairs.

Notes: The species name *exigua* means "little" or "weak," a reference to the plant's stature and branches.

Slender Willow
Salix petiolaris

Habitat: Wet meadows, lakeshores and streambanks. **General:** Shrub 1–3 m tall; young branches yellow to reddish brown, older branches reddish brown. **Leaves:** Alternate, deciduous, elliptic, 4–12 cm long, 0.8–2.5 cm wide, velvety-haired when young, becoming hairless with age; margins finely toothed to wavy; stalks 0.3–1 cm long, often yellow. **Flower Cluster:** Male catkins 1–2 cm long; female catkins 1.5–2.5 cm long; appearing with leaves. **Flowers:** Male flowers with 2 stamens and 1 bract; female flowers with 1 pistil and 1 bract. **Fruit:** Capsule 6–8 mm long; seeds with fine, white hairs.

Notes: The species name *petiolaris* refers to the yellowish leaf stalks (petioles).

BAYBERRY FAMILY
Myricaceae

Members of the bayberry family are trees and shrubs with simple, alternate leaves. Male and female flower clusters are found on different plants. Male flowers are composed of 1 bract with 2–20 stamens in the bract axil. Female flowers have 1 pistil subtended by 2 thickened, opposite bracts. Petals and sepals are absent in both flower types. The fruit may be a drupe or nutlet.

Sweet-fern
Comptonia peregrina

Habitat: Dry, sandy to rocky soils in open pine forests. **General:** Shrub to 1.5 m tall, fragrant; branches reddish brown to grey, hairy; rhizomes creeping. **Leaves:** Alternate, deciduous, fernlike, aromatic when crushed, 3–15.5 cm long, 0.3–2.9 cm wide, dark green above, pale greyish green and glandular-dotted below; margins 2–10-lobed, the lobes alternate to nearly opposite; stipules heart-shaped. **Flower Cluster:** Male catkins 1–5 cm long, at ends of branches, bracts oval with hairy margins; female catkins bristly, 3–5 mm long at flowering, elongating to 2 cm long in fruit, bracts to 1.3 cm long, hairy and glandular-dotted. **Flowers:** Green; male flowers with 1 bract and 3–8 stamens; female flowers with 3 bracts and 1 pistil. **Fruit:** Drupe 1.2–5.5 mm long; fruiting cluster bristly, oblong to oval, about 2.5 cm across.

Notes: The genus name commemorates Henry Compton (1632–1713), Bishop of London and an amateur botanist. The species name *peregrina* means "foreign," a name given by Linnaeus, who was not familiar with the species. • The common name refers to the aromatic, fern-like leaves.

Sweet Gale · *Myrica gale*

Habitat: Bogs, swamps and streambanks. **General:** Shrub 50–150 cm tall; branches dark grey to reddish brown, glandular-dotted, fragrant when broken. **Leaves:** Alternate, deciduous, wedge-shaped, 2.5–10 cm long, 0.5–2.5 cm wide, surfaces with small, yellow, resinous glands; margins coarsely toothed; stalks 1–6 mm long. **Flower Cluster:** Male catkins 1–2 cm long; female catkins 0.8–1 cm long; resembling small cones at maturity; appearing before leave;. **Flowers:** Male flowers with 4–16 stamens; female flowers with 2 small bracts and 1 pistil. **Fruit:** Nutlet oval, 2.5–3 mm long; in cone-like clusters.

Notes: The genus name comes from the Greek word *myrike*, meaning "tamarisk." The species name *gale* comes from the Old English word *gagel*, meaning "native."

BIRCH FAMILY · Betulaceae

Family description on p. 31.

Dwarf Birch · *Betula pumila*

Habitat: Swamps and bogs. **General:** Shrub 1–3 m tall; branches glandular-dotted; older stems smooth, dark brown; young branches often densely haired. **Leaves:** Alternate, deciduous, round to wedge-shaped, 1–4 cm long, 1–2 cm wide, shiny, base round to wedge-shaped; lateral veins 3–5 per side; margins with 10–15 rounded teeth per side; stalks 3–6 mm long. **Flower Cluster:** Male catkins erect, 1.2–2 cm long; female catkins 1.2–2.5 cm long; appearing with leaves. **Flowers:** Male flowers with 2–3 sepals and 2 stamens; female flowers with 1 pistil. **Fruit:** Nutlet 1.5–2 mm long, surrounded by a thin, papery membrane, shed from August–September.

Notes: The species name *pumila* means "dwarf," a reference to the plant's stature.

Beaked Hazelnut
Corylus cornuta

Habitat: Moist, deciduous forests.
General: Shrub to 3 m tall; branches numerous; bark mottled brown and grey. **Leaves:** Alternate, deciduous, oval, 5–10 cm long, 2.5–7 cm wide, base heart-shaped; margins coarsely toothed; stalks 0.8–1.8 cm long. **Flower Cluster:** Male catkins yellowish brown, 1–2 cm long; female catkins resembling a scaly bud with 2–5 reddish pink pistils; appearing before leaves, often when snow is still on the ground. **Flowers:** Male flowers with 2 small bracts and 4 stamens; female flowers with 3 bracts (1 large, 2 small, the smaller becoming enlarged in fruit) and 1 pistil. **Fruit:** Nut; in clusters of 2–6, edible; bract enlarging and becoming flask-shaped, bristly, 1–5 cm long.

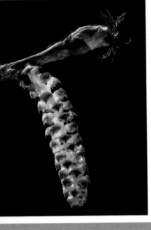

Notes: *Corylus* comes from *korylos*, the Greek name for hazelnut. The species name *cornuta* means "horned," a reference to the long beak of the fruit.

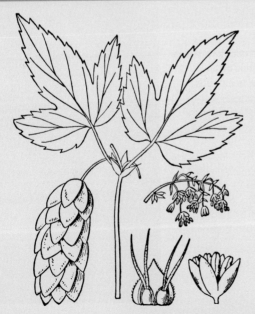

HEMP FAMILY
Cannabaceae

Members of this family are herbs with lobed, opposite or alternate leaves. The flowers, male or female, are borne in leaf axils. The male flowers are stalked and composed of 5 sepals and 5 stamens. The stalkless female flowers consist of 5 sepals and 1 pistil. Petals are absent. The fruit, an achene, is surrounded by the calyx.

This family includes common hops (*Humulus lupulus*), which are used in beer production, as well as *Cannabis sativa*, a source of hemp fibre.

Common Hops · *Humulus lupulus*

Habitat: Thickets, abandoned farmyards and waste areas; introduced from Europe. **General:** Twining vine; stems 2–4 m long, prickly; male or female. **Leaves:** Opposite, palmately 3–7-lobed, 2–8 cm wide, upper surface rough, lower surface with small, white and yellow glands, base heart-shaped; stalks prickly; stipules present. **Flower Cluster:** Male flowers in raceme or panicle; female flowers in drooping spike. **Flowers:** Greenish; male flowers with 5 sepals and 5 stamens; female flowers with 1 pistil in axil of broad, greenish yellow bracts. **Fruit:** Achene enclosed by prominent bract; in cone-like clusters 2–5 cm long.

Notes: *Humulus* comes from the Old German word *humela*, meaning "hops." The species name *lupulus* means "small wolf," from an old common name for this plant, "willow wolf."

CROWFOOT OR BUTTERCUP FAMILY

Ranunculaceae · Family description on p. 215.

Purple Clematis · *Clematis occidentalis*

Habitat: Open woods. **General:** Woody vine; stems to 2 m long, trailing or climbing on adjacent vegetation. **Leaves:** Opposite or in whorls of 3, deciduous, compound with 3 leaflets; leaflets oval, 2–8 cm long, 2.5–5 cm wide; margins smooth; stalks 5–10 cm long, twisting. **Flower Cluster:** Solitary, on stalks 5–15 cm long. **Flowers:** Blue, 1–6 cm long; petals absent; sepals 4–6, petal-like; stamens and pistils numerous. **Fruit:** Achene 2–4 mm long; style feathery, 3–5 cm long.

Notes: *Clematis* is the Greek name for an unknown climbing plant. The species name *occidentalis* means "western."

Virgin's Bower
Clematis virginiana

Habitat: Open fields and edges of woods. **General:** Woody vine; male or female; stems 1–5 m long, purplish brown, trailing or climbing. **Leaves:** Opposite, deciduous, compound with 3 leaflets; leaflets oval to heart-shaped, 5–9 cm long, 2.5–6 cm wide; margins lobed to irregularly toothed; stalks 5–9 cm long, twining. **Flower Cluster:** Panicle of male or female flowers; stalks 1–8 cm long. **Flowers:** White, fragrant, 1.8–2.5 cm wide; petals absent; sepals 4–5, 0.6–1 cm long; stamens and pistils numerous. **Fruit:** Achene 3–5 mm long; style feathery, 2.5–3.8 cm long; in globe-shaped heads.

Notes: The species name means "of Virginia."

BARBERRY FAMILY
Berberidaceae

Members of the barberry family are primarily shrubs with alternate, simple or compound leaves. Stipules are absent. The flowers are composed of 6 petals, 6 sepals, 6 stamens and 1 pistil. The stamens, attached to the petals, spring free and dust insects with pollen when they land on the flower. The fruit is a berry or follicle.

Some plants in this family, including barrenwort (*Epimedium* spp.) and Oregon-grape (*Mahonia* spp.), are often cultivated for their ornamental value. A few species are noxious weeds.

Common Barberry
Berberis vulgaris

Habitat: Fence lines and waste areas; introduced from Europe as an ornamental shrub. **General:** Shrub 1–3 m tall, much-branched; bark grey to brown. **Leaves:** Alternate, deciduous, 2–5 cm long, 1–2.5 cm wide; margins finely toothed; stalks short; 3-branched spine at the base of each leaf, the spines 0.9–1.8 cm long. **Flower Cluster:** Raceme drooping, 3–6 cm long, 10–20-flowered. **Flowers:** Yellow, 0.5–1.5 cm across; petals 6; sepals 6; stamens 6; pistil 1. **Fruit:** Berry red, oblong, 6–9 mm long.

Notes: The species name *vulgaris* means "common." • In the 1800s, it was discovered that this plant is an alternate host for black rust, a fungus that devastates wheat, oat and barley crops. The fungus overwinters on the leaves of common barberry and spreads to cereal crops, causing severe reductions in crop yield. Although widely exterminated, a few specimens still exist in the wild.

MOONSEED FAMILY
Menispermaceae

The moonseed family is composed of deciduous, woody vines and occasionally trees or shrubs. The simple, alternate leaves are stalked. Flowers are borne in cymes, racemes or panicles, with male and female flowers on separate plants. Male flowers have 6 petals, 6 sepals and 6 to numerous stamens. Female flowers usually have 6 petals, 4–6 sepals and 1–6 pistils. The fruit is a drupe.

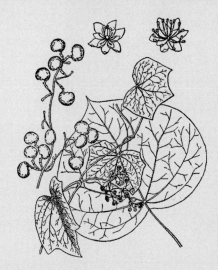

Well-known members of this family include strychnine (*Strychnos toxifera*) and pareira (*Chondrodendron tomentosum*), both which contain toxic alkaloids and are constituents of South American dart poison.

Moonseed Vine, Yellow Parilla · *Menispermum canadense*

Habitat: Moist woods and shady streambanks. **General:** Woody vine; stems 2–5 m long; rhizomes to 1 cm in diameter. **Leaves:** Alternate, deciduous, 5–23 cm long, 5–24 cm wide, oval, shield-like (leaf stalk inserted at least 1.1 cm from margin), 3–7-lobed, base heart-shaped; prominently 7–12-veined; margins smooth; stalks 5–20 cm long. **Flower Cluster:** Panicle 3–18 cm long; male or female. **Flowers:** Creamy white; male flowers with 4–12 petals, 5–8 sepals and 12–36 stamens; female flowers with 6 petals, 6 sepals, 6–9 sterile stamens and 2–4 pistils. **Fruit:** Drupe berry-like, globe-shaped, dark blue to black, 0.6–1.3 cm across, resembling a small, dusty grape; poisonous.

Notes: The genus name comes from the Greek words *mene*, "moon," and *sperma*, "seed," a reference to the crescent-shaped seeds. The species name means "of Canada."

LAUREL FAMILY · Lauraceae

Family description on p. 46.

Spicebush
Lindera benzoin

Habitat: Low-lying woods, streambanks and swamps. **General:** Shrub 2–5 m tall; male or female; bark greyish brown; twigs hairless. **Leaves:** Alternate, deciduous, oblong to oval, 6–15 cm long, 2–6 cm wide, somewhat leathery, base wedge-shaped; margins smooth; stalks 0.5–1.2 cm long; aromatic when crushed. **Flower Cluster:** Umbel-like, 4–6-flowered with 4 bracts at the base; appearing before leaves. **Flowers:** Yellow; male flowers with 6 tepals and 9 stamens; female flowers with 6 tepals and 1 pistil. **Fruit:** Drupe berry-like, bright red, oblong, 0.8–1.1 cm long, in clusters of 2–6 on stalks 3–5 mm long.

Notes: The genus name commemorates Swedish physician and botanist Johann Linder (1676–1724). The species name *benzoin* comes from the Arabic name for a similar plant.

CURRANT OR GOOSEBERRY FAMILY
Grossulariaceae

The currant or gooseberry family, composed of shrubs with alternate, palmately lobed leaves, was once included in the saxifrage family (Saxifragaceae). It is The flowers have 5 petals, 5 sepals, 5 stamens and 1 pistil with 2 styles. The fruit is a berry.

Currants and gooseberries are distinguished from one another by the presence or absence of prickles on the stem. Species without prickles are referred to as currants, whereas those with prickles are gooseberries.

Members of this family are often grown for their fruit.

Golden Currant
Ribes aureum

Habitat: Fields, roadsides and fence lines; native to western North America. **General:** Shrub 1–2 m tall; bark grey; branches yellowish brown to brownish grey; prickles absent. **Leaves:** Alternate, deciduous, fan-shaped, 3–5-lobed, 2–4 cm long, 1.5–4 cm wide, base wedge to heart-shaped; margins toothed; stalks 1–4.8 cm long. **Flower Cluster:** Raceme 3–7 cm long, 5–15-flowered, in leaf axils; appearing after leaves. **Flowers:** Yellow with red tips, clove-scented, 0.6–1 cm long; petals 5; sepals 5; stamens 5; pistil 1. **Fruit:** Berry black, 4–7 mm wide.

Notes: The species name *aureum* means "golden," a reference to the flower colour.

Skunk Currant
Ribes glandulosum

Habitat: Moist woods and streambanks. **General:** Shrub with stems to 1 m long, spreading; bark purplish brown to grey; skunk-like odour when bruised. **Leaves:** Alternate, deciduous, 2–8 cm long, 2–8 cm across, round with 3–7 lobes, underside with glandular hairs; margins coarsely toothed; stalks 3–5.5 cm long. **Flower Cluster:** Raceme erect to ascending, 2.5–6 cm long, 6–15-flowered. **Flowers:** White or reddish, 2–2.5 mm long; petals 5; sepals 5; stamens 5; pistil 1. **Fruit:** Berry red, bristly, 5–8 mm across; edible but disagreeable tasting.

Notes: *Ribes* comes from the Arabic word *ribas*, meaning "gooseberry." The species name *glandulosum* means "glandular," a reference to the sticky hairs throughout the plant.

Wild Red Currant
Ribes triste

Habitat: Moist woods. **General:** Shrub to 1 m tall; young bark grey to brown, old bark reddish purple to black. **Leaves:** Alternate, deciduous, round to kidney-shaped, 4–10 cm long, 5–10 cm across, 3–5-lobed; margins coarsely toothed; stalks 2.5–6 cm long. **Flower Cluster:** Raceme drooping, 2.5–10 cm long, 6–20-flowered; appearing with leaves. **Flowers:** Greenish pink to purplish red, 4–6 mm across; petals 5, about 1 mm long; sepals 5, greenish purple; stamens 5; pistil 1. **Fruit:** Berry bright red, 3–6 mm in diameter; edible.

Notes: The species name *triste* means "sad," a possible reference to the drooping flower cluster.

WITCH-HAZEL FAMILY · Hamamelidaceae

Family description on p. 47.

Witch-hazel · *Hamamelis virginiana*

Habitat: Moist, shady woods. **General:** Shrub or small tree 4–8 m tall; trunks crooked; bark light brown to grey; twigs zigzagged; buds 1–1.4 cm long, densely covered with yellowish brown hairs. **Leaves:** Alternate, deciduous, oval to oblong, 6–15 cm long, 3–10 cm wide, dark green above, pale below; veins 5–7 per side; margins wavy to coarsely toothed; stalks short; turning yellow in autumn. **Flower Cluster:** Raceme 3-flowered; appearing as leaves fall in autumn (September–October). **Flowers:** Yellow, fragrant; petals 4, twisted, ribbon-like, 1.5–2 cm long; sepals 4, orangey brown; stamens 4; pistil 1. **Fruit:** Capsule 1–1.6 cm long, woody, maturing the following year, exploding at maturity and ejecting seeds to 8 m from the parent; seeds shiny, black, 0.7–1 cm long.

Notes: The species name means "of Virginia." • The common name comes from the Old English word *wyche*, meaning "bendable" or "pliant," a reference to the use of forked witch-hazel twigs to divine underground water sources. • Early settlers thought that the shrub resembled hazelnut (*Corylus* spp.), hence the name "witch-hazel."

ROSE FAMILY · Rosaceae

Family description on p. 49.

Chokeberry · *Aronia melanocarpa*

Habitat: Peat bogs, wet woods and swamps. **General:** Shrub 1–2.5 m tall; branches greyish brown to purple. **Leaves:** Alternate, deciduous, oval to oblong or elliptic, 2–8 cm long, 1–4 cm wide, dark green above with a row of dark, hair-like glands along the midvein, pale below; margins finely toothed, the teeth gland-tipped; stalks 0.2–1 cm long; stipules falling off as leaves expand. **Flower Cluster:** Corymb 5–15-flowered, 2–5 cm wide. **Flowers:** White, 0.5–1 cm across; petals 5; sepals 5; stamens about 20; pistil 1. **Fruit:** Pome berry-like, purple to black, 0.6–1 cm across; edible but very tart and bitter.

Notes: The species name *melanocarpa* means "black-fruited."

Shrubby Cinquefoil
Dasiphora fruticosa

Habitat: Dry to wet, open woods, rock ledges and cliff crevices. **General:** Shrub 30–100 cm tall, profusely branched; bark brown, shredding. **Leaves:** Alternate, deciduous, stalks 1–2 cm long, pinnately compound with 5–7 leaflets; leaflets leathery, linear to oblong, 1–2 cm long, 2–7 mm wide, often with rolled margins; 3 terminal leaflets often fused at the base. **Flower Cluster:** Solitary or a few-flowered cyme. **Flowers:** Yellow, 1–2.5 cm across; petals 5; sepals 5; stamens and pistils numerous; bracts 5, small, alternating with sepals. **Fruit:** Achene densely haired; in compact clusters.

Notes: *Dasiphora* is from the Greek word *dasys,* "shaggy" or "rough," referring to the stems.

Ninebark
Physocarpus opulifolius

Habitat: Sandy to rocky soil along shorelines. **General:** Shrub 2–3 m tall; branches greenish, somewhat angled; older stems greyish brown with peeling bark. **Leaves:** Alternate, deciduous, oval to round, 3–7 cm long, 3–6 cm wide, 3-lobed, dark green above, pale below; margins toothed; stalks 1–2 cm long; stipules deciduous. **Flower Cluster:** Corymb 2–5 cm across, many-flowered; stalks 1–2 cm long. **Flowers:** White, 0.4–1 cm across; petals 5; sepals 5; stamens 20–40; pistil 1–5. **Fruit:** Follicle inflated, reddish brown, 0.5–1 cm long; seeds 3–4.

Notes: *Physocarpus* comes from the Greek words *phusa,* "bladder," and *karpos,* "fruit," a reference to the inflated follicles. The species name *opulifolius* means "leaves like *opulus,*" because the leaves resemble those of *Viburnum opulus* var. *americanum* (p. 127).

Pin Cherry
Prunus pensylvanica

Habitat: Dry, sandy woods and forest edges. **General:** Shrub or small tree to 8 m tall; bark smooth, reddish brown. **Leaves:** Alternate, deciduous, oblong to oval or lance-shaped, 5–12 cm long, 1–2 cm wide; margins finely toothed; stalks red, 1–3 cm long. **Flower Cluster:** Corymb umbel-like, 5–15-flowered, appearing with leaves. **Flowers:** White, 1.2–1.8 cm across; petals 5; sepals 5; stamens about 20; pistil 1. **Fruit:** Drupe cherry-like, bright red, 5–8 mm across; edible.

Notes: The species name means "of Pennsylvania."

Sandcherry
Prunus pumila

Habitat: Dry to wet, sandy or gravelly lakeshores and sand dunes. **General:** Shrub 30–200 cm tall, prostrate to trailing; branches diffuse. **Leaves:** Alternate, deciduous, elliptic to oblong, 4–10 cm long, 1–3 cm wide, somewhat leathery, glossy above, pale below; margins gland-tipped above midleaf; stalks 0.3–1 cm long; stipules deciduous. **Flower Cluster:** Corymb 2–4-flowered; stalks 0.4–1.5 cm long. **Flowers:** White, 0.7–2 cm across; petals 5; sepals 5; stamens about 20; pistil 1. **Fruit:** Drupe cherry-like, purple to black, 1–1.5 cm across; bitter but edible.

Notes: The species name *pumila* means "dwarf," a reference to the plant's stature.

Chokecherry
Prunus virginiana

Habitat: Dry, open areas, forest edges and riverbanks. **General:** Shrub or small tree to 10 m tall; young bark reddish brown, becoming dark greyish brown with age. **Leaves:** Alternate, deciduous, elliptic to oblong or oval, 2–10 cm long, 2–6 cm wide; margins finely toothed; stalks 0.5–2 cm long. **Flower Cluster:** Raceme cylindric, 5–15 cm long, 10–35-flowered. **Flowers:** White, 1–1.5 cm across; petals 5; sepals 5; stamens about 20; pistil 1. **Fruit:** Drupe cherry-like, reddish purple to purplish black, 6–8 mm across; edible.

Notes: The species name means "of Virginia."

Prickly Wild Rose
Rosa acicularis

Habitat: Edges of woods, roadsides, pastures and riverbanks. **General:** Shrub to 2.5 m tall; bark reddish brown; prickles straight, 3–4 mm long. **Leaves:** Alternate, deciduous, stalked, pinnately compound with 3–7 leaflets; leaflets oval, 1.2–5 cm long, 0.8–3.5 cm across; margins coarsely toothed; stipules with soft, glandular hairs. **Flower Cluster:** Solitary, terminal. **Flowers:** Pink, fragrant, 5–8 cm across; petals 5; sepals 5; stamens and pistils numerous; hypanthium pear- to globe-shaped. **Fruit:** Achenes numerous; in a red, pear-shaped hip 1.5–2.5 cm long.

Notes: The species name means "needle-like," a reference to the stiff prickles.

Sweetbrier, Eglantine
Rosa rubiginosa

Habitat: Pastures, fields, fence lines and roadsides; introduced from Europe. **General:** Shrub 1–3 m tall; branches brownish grey, covered in numerous downward-curved, hook-like prickles to 1 cm long and scattered straight prickles. **Leaves:** Alternate, deciduous, pinnately compound with 5–9 (commonly 7) leaflets; leaflets oblong to oval, 1–3 cm long, 0.5–2 cm wide, dull green above, scurfy and covered in gland-

tipped hairs below, apple-scented when crushed; margins toothed; stipules elongated, with sharp-pointed lobes. **Flower Cluster:** Solitary or a few-flowered corymb. **Flowers:** Pink to white, 2–5 cm across; petals 5; sepals 5, margins comb-like; stamens and pistils numerous. **Fruit:** Achenes numerous; in an orangey red, globe-shaped hip 1–1.5 cm long.

Notes: The species name means "rusty." • The common name "eglantine" is a translation from the Old French *aiglantine*, which is from the Latin *aquilentum*, meaning "prickly."

Multiflora Rose
Rosa multiflora

Habitat: Pastures, fence lines and roadsides; introduced from Japan, Korea and China in 1886 as a rootstalk for ornamental roses; currently spreading throughout southern Ontario. **General:** Shrub 1.5–4.6 m tall; bark greyish brown to brown, hairless; branches erect to arching or trailing on the ground, covered in stout, recurved prickles; twigs reddish green, hairless; often forming impenetrable clumps to 10 m across. **Leaves:** Alternate, deciduous, 8–11 cm long, stalks 1–1.3 cm long, pinnately compound with 5–11 (usually 7 or 9) leaflets; leaflets elliptic to oblong, 1–5 cm long, 0.8–2.8 cm wide, hairless above, slightly hairy below; margins with 5–8 teeth per cm; stipules feathery or comb-like, often glandular. **Flower Cluster:** Panicle pyramidal, 25–100-flowered; stalks 1–1.5 cm long, often glandular-hairy; bracts 1–2. **Flowers:** White to pinkish white, 1.3–2.5 cm across; petals 5; sepals 5; stamens and pistils numerous. **Fruit:** Achenes numerous; in a red, globe-shaped hip 5–8 mm across, surface leathery, remaining on stem through winter.

Notes: The species name *multiflora* means "many-flowered."

Common Blackberry
Rubus allegheniensis

Habitat: Old fields and pastures, roadsides and forest edges. **General:** Shrub 1–2.5 m tall; stems arching, covered in glandular hairs and straight bristles; flowering stems brownish to reddish purple, with scattered, broad-based prickles. **Leaves:** Alternate, deciduous, palmately compound with 3 or 5 leaflets (occasionally simple); leaflets oval to lance-shaped, 5–20 cm long, 3–10 cm wide, margins toothed, stalks glandular-hairy and prickly, the 2 basal leaflets stalkless; stipules linear to lance-shaped, glandular-hairy. **Flower Cluster:** Raceme 5–30-flowered; stalks long, covered in glandular hairs and scattered prickles; on second-year stems only. **Flowers:** White, 1.5–2.5 cm across; petals 5, 1–2 cm long, 5–8 mm wide; sepals 5; stamens and pistils numerous. **Fruit:** Aggregate of drupelets (raspberry), black, cylindric to thimble-shaped, 1–2.5 cm long; edible.

Notes: *Rubus* is the Latin name for blackberry. The species name means "of Allegheny," an upland area of southwestern Pennsylvania.

Red Raspberry · *Rubus idaeus*

Habitat: Open aspen forests, riverbanks and roadside ditches. **General:** Shrub to 2 m tall, bristly; first-year stems green to reddish brown; second-year stems brown with shredded bark, producing flowers and fruit, then dying that autumn. **Leaves:** Alternate, deciduous, pinnately compound with 3–5 leaflets; leaflets oval, 5–10 cm long, 1.5–6 cm wide; margins coarsely toothed; stalks long, hairy. **Flower Cluster:** Solitary or a 2–5-flowered raceme; on second-year stems only. **Flowers:** White, 0.8–1.5 cm across; petals 5; sepals 5; stamens and pistils numerous. **Fruit:** Aggregate of drupelets (raspberry), red, 0.8–1.4 cm across; edible.

Notes: The species name *idaeus* means "of Mt. Ida," a sacred mountain in Greek mythology.

Purple-flowering Raspberry
Rubus odoratus

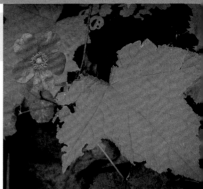

Habitat: Open woods, ravines, riverbanks and roadsides. **General:** Shrub 1–1.5 m tall; stems green to purplish, glandular-hairy when young, becoming brown to grey with peeling bark. **Leaves:** Alternate, deciduous, 10–20 cm long and wide, 3–5-lobed, base heart-shaped, lower surface hairy on veins; margins toothed; stalks 5–15 cm long, sticky-haired; stipules pale, lance-shaped, withering soon after leaves expand. **Flower Cluster:** Raceme 3–10-flowered; flower stalks covered in reddish purple, glandular hairs. **Flowers:** Pink to purplish, showy, 3–5 cm across; petals 5; sepals 5, long-tipped, covered in sticky, reddish hairs; stamens and pistils numerous. **Fruit:** Aggregate of drupelets (raspberry), pink to red, 1–2 cm across; edible but dry and tasteless.

Notes: *Rubus* is the Latin word for blackberry, a reference to the fruit colour of some species. The species name *odoratus* means "fragrant" or "sweet-smelling," referring to the lightly scented flowers.

Narrow-leaved Meadowsweet
Spiraea alba

Habitat: Moist meadows and shorelines. **General:** Shrub 1–2 m tall; twigs yellowish to reddish brown, ridged; old stems purplish grey with peeling bark. **Leaves:** Alternate, deciduous, oblong to lance-shaped, 3–6 cm long, 1–2 cm wide; margins finely sharp-toothed; stalks 2–6 mm long. **Flower Cluster:** Panicle pyramid-shaped, compact, 5–20 cm long; flowers numerous. **Flowers:** White, 5–8 mm across; petals 5; sepals 5; stamens 15 or more; pistils 5–8. **Fruit:** Follicle 3–9 mm long; in clusters of 5–8; seeds 2–5.

Notes: The species name *alba* means "white," a reference to the flower colour.

Steeplebush, Hardhack · *Spiraea tomentosa*

Habitat: Marshy edges of ponds and swamps. **General:** Shrub 0.5–1.5 m tall; branches greenish to brown with rust-coloured, woolly hairs; older stems purplish to reddish brown. **Leaves:** Alternate, deciduous, oval to elliptic or oblong, 1–5 cm long, 0.5–2 cm wide, dull green above, grey-ish green because of felt-like hairs and prominently veined below; margins coarsely toothed; stalks less than 4 mm long or absent. **Flower Cluster:** Panicle 5–15 cm long, 1–7 cm wide, branches with 11–20 flowers per cm and covered in woolly hairs. **Flowers:** Pink, 3–4 mm wide; petals 5; sepals 5; stamens 15 or more; pistils 5. **Fruit:** Follicle covered in woolly hairs; in clusters of 5.

Notes: *Spiraea* comes from the Greek word *speiraira*, a name given to plants used in garlands. The species name *tomentosa* means "densely woolly," a reference to the stem and leaves.

LEGUME OR PEA FAMILY · **Fabaceae**

Family description on p. 54.

Caragana · *Caragana arborescens*

Habitat: Roadsides and edges of forests; introduced from Siberia as a hedge or wind-break. **General:** Shrub or small tree 3–4 m tall; bark greenish brown. **Leaves:** Alternate, deciduous, pinnately compound with 8–12 leaf-lets; leaflets oval, spine-tipped, 1–2.5 cm long, 0.8–1.5 cm wide; margins smooth; stipules spine-tipped. **Flower Cluster:** Solitary; on long stalks originating from previous year's scaly bud. **Flowers:** Bright yellow, irregular, 1.5–2.5 cm long; petals 5; sepals 5; stamens 10; pistil 1. **Fruit:** Legume pod-like, narrow, 4–6 cm long, turning brown by late summer.

Notes: The genus name comes from the Mongolian common name *caragan*. The species name *arborescens* means "tree-like."

RUE FAMILY · Rutaceae

Family description on p. 57.

Prickly-ash · *Zanthoxylum americanum*

Habitat: Open pastures, thickets, fence lines and roadsides. **General:** Shrub 1–3 m tall, branches thorny; prickles sharp-pointed, flat-based, 0.5–1 cm long, in pairs at leaf bases; bark reddish to purplish, turning brownish grey with age. **Leaves:** Alternate, deciduous, 13–25 cm long, 5–10 cm wide, pinnately compound with 7–11 leaflets, stalks prickly; leaflets elliptic to oval, 2.5–5 cm long, 1–4 cm wide, lower surface glandular-dotted, margins smooth to wavy, stalkless; aromatic when crushed. **Flower Cluster:** Umbel-like; appearing before leaves. **Flowers:** Yellowish green, unisexual, 3–5 mm wide; male flowers with 5 petals and 4–5 stamens; female flowers with 5 petals and 2–5 pistils. **Fruit:** Follicle pod-like, somewhat fleshy, elliptic, reddish brown, 4–6 mm long.

Notes: The genus name comes from the Greek words *xanthos*, "yellow," and *xylon*, "wood," a reference to the colour of the wood in some species.
• Another common name, "toothache tree," comes from the use of the fresh bark by early settlers to treat toothaches.

SUMAC FAMILY
Anacardiaceae

The sumac family is composed of trees, shrubs and woody vines with alternate, simple or pinnately compound leaves. Plants may be male, female or both. Flowers are borne in a panicle and have 3–7 (commonly 5) separate petals, 3–7 (commonly 5) fused sepals, 5–10 stamens and 1 pistil. A nectary disc is found at the base of the stamens and petals. The fruit is a drupe or berry.

Well-known members of the sumac family include poison-ivy (*Toxicodendron radicans*), cashew (*Anacardium officinale*), mango (*Mangifera indica*) and pistachio (*Pistacia vera*). The toxic sap of wax-tree (*Rhus succedanea*) is often collected and used as a natural lacquer.

Fragrant Sumac · *Rhus aromatica*

Habitat: Sandy to rocky soils in pastures and open forests. **General:** Shrub 50–150 cm tall, often forming thickets; branches brownish to purplish grey. **Leaves:** Alternate, deciduous, 6–10 cm long, compound with 3 leaflets; leaflets elliptic to oval, 4–8 cm long, 2.5–4 cm wide, dark green above, pale below, hairy on both surfaces, margins with 3–6 coarse teeth per side, stalkless. **Flower Cluster:** Panicle 5–10 cm long; spikes compact, 1–2 cm long; appearing before leaves. **Flowers:** Yellowish, less than 5 mm across; petals 5; sepals 5; stamens 5; pistil 1. **Fruit:** Drupe berry-like, bright red, 6–9 mm across, densely haired.

Notes: The species name means "aromatic."

Staghorn Sumac

Rhus typhina

Habitat: Forest edges, open fields and rocky ridges. **General:** Shrub 3–9 m tall, often forming thickets; bark dark brown, becoming scaly with age; young twigs covered in soft hairs, exuding milky juice when broken. **Leaves:** Alternate, deciduous, 30–60 cm long, pinnately compound with 9–31 leaflets, leaf axis red; leaflets narrowly oblong to lance-shaped, 5–12 cm long, 1–3 cm wide, dark green above, white-hairy below, margins finely to coarsely toothed; turning bright red to purple in autumn. **Flower Cluster:** Panicle cone-shaped, 12–25 cm long, branches covered in soft, brown hairs; appearing after leaves. **Flowers:** Greenish, 3–5 mm across; petals 5; sepals 5; stamens 5; pistil 1. **Fruit:** Drupe berry-like, red, 3–6 mm across, densely haired; remaining attached through winter.

Notes: The common name comes from the densely haired, wide-spreading branches that resemble the velvet-covered antlers of a male deer.

Poison-ivy
Toxicodendron radicans

Habitat: Open woods, fields and roadsides.
General: Shrub 30–120 cm tall or climbing vine 3–15 m tall/long; bark brownish grey. **Leaves:** Alternate, deciduous, 6–40 cm long, compound with 3 leaflets; leaflets shiny, oval to diamond-shaped, 3–15 cm long, 2.5–6 cm wide; reddish purple in spring, turning glossy green by midsummer. **Flower Cluster:** Panicle 2.5–10 cm long; flowers numerous.
Flowers: Yellowish white, 2–5 mm across, unisexual or perfect; petals 5; sepals 5; stamens 5; pistil 1.
Fruit: Drupe berry-like, waxy, creamy white, 3–7 mm across, remaining on stem throughout winter.

Notes: The saying, "Leaves of three, let it be. Berries white, take flight," aids in the identification of this troublesome species. Poison-ivy contains the allergen urushiol, which initially causes severe itching, then 24–48 hours later, the skin becomes red and blistered. Urushiol is found in the sap of leaves, stems and roots. Burning the plants releases the toxin in the ash and smoke, so avoid inhalation. Contaminated clothing should be hot laundered to neutralize the urushiol. Poisoning can also occur when using contaminated tools.

HOLLY FAMILY
Aquifoliaceae

The holly family includes trees and shrubs, often with evergreen, alternate leaves. Both species found in southern Ontario are deciduous. The perfect flowers are borne singly in leaf axils or in cymes. The flowers have 4–8 sepals, 4–8 petals, 4–8 stamens and 1 pistil. The fruit is a drupe.

A well-known species, English holly (*Ilex aquifolium*), is widely used as a decoration at Christmastime.

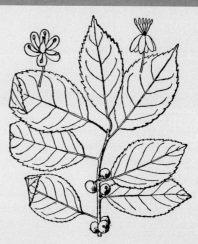

Winterberry, Black Alder
Ilex verticillata

Habitat: Swampy woods, peat bogs and roadside ditches. **General:** Shrub 3–4 m tall; bark brown turning grey to black with age; lenticels numerous, warty; branches stout, ridged. **Leaves:** Alternate, deciduous, 3–9 cm long, 1–4 cm wide, elliptic to oblong or oval, dull green above, pale and downy below; margins sharply toothed; stalks 0.7–1.2 cm long; stipules deciduous. **Flower Cluster:** Cyme, in leaf axils; male flowers numerous per cluster; female flowers 1–5 per axil; appearing with leaves. **Flowers:** Green to yellowish white, 2–5 mm across; male or female; petals 4–8; sepals 4–6; stamens 4–8; pistil 1. **Fruit:** Drupe orange to yellow or red, 5–7 mm across; seeds 3–5.

Notes: *Ilex* is from the Latin name for holly oak (*Quercus ilex*), an evergreen. The species name *verticillata* means "whorled." • The common name "winterberry" refers to the fruit, which remains attached to the branches throughout winter, an easily identifiable characteristic.

Mountain-holly
Nemopanthus mucronatus

Habitat: Swamps, bogs and low-lying woods. **General:** Shrub 1–3 m tall; bark grey; lenticels numerous, pale; branches purplish grey. **Leaves:** Alternate, deciduous, crowded, often appearing whorled, 2–7 cm long, 1–2.5 cm wide, elliptic to oblong or oval, bright green above, pale below, tip pointed; margins smooth with a few scattered teeth; stalks 5–12 mm long, purplish. **Flower Cluster:** Cyme (male flowers) or solitary in leaf axils (female flowers); stalks 1–2 cm long; appearing with leaves. **Flowers:** Greenish, 1–3 mm across; male or female; petals 4–5, 1–3 mm long; sepals 4–5 (often absent); stamens 4–5; pistil 1. **Fruit:** Drupe purplish red to crimson, 5–7 mm across, on stalks about 3 cm long; seeds 4–5.

Notes: The genus name comes from the Greek words *nema*, "thread," and *pous*, "foot," a reference to the slender flower and fruit stalks. The species name *mucronatus* means "sharp-pointed," a reference to the leaf tip.

STAFF-TREE FAMILY · Celastraceae

Family description on p. 59.

Oriental Bittersweet
Celastrus orbiculatus

Habitat: Open woods; introduced from China and Japan. **General:** Woody vine to 20 m tall/long; bark greyish brown, smooth when young, becoming scaly with age; lenticels present. **Leaves:** Alternate, deciduous, oval to round, 2–13 cm long, 1.5–8 cm wide, glossy and green above, pale below, tip blunt; margins finely toothed; stalks 1–3 cm long; turning yellow in autumn. **Flower Cluster:** Cyme 3–7-flowered, in leaf axils; appearing after leaves. **Flowers:** Yellowish green, 2–7 mm across; petals 5; sepals 5; stamens 5; pistil 1. **Fruit:** Capsule globe-shaped, 0.6–1 cm across, green turning yellow with maturity; seeds 3–6, 4–5 mm long, in a fleshy, bright red aril.

Notes: *Celastrus* comes from the Greek word *kelastros*, the name of an evergreen tree. The species name *orbiculatus* means "round," a reference to the leaf shape.

BUCKTHORN FAMILY · Rhamnaceae

Family description on p. 66.

New Jersey Tea · *Ceanothus americanus*

Habitat: Dry, open clearings in woods, along riverbanks and on hillsides. **General:** Shrub 30–100 cm tall; branches grey to reddish brown, somewhat hairy. **Leaves:** Alternate, deciduous, oval to oblong or lance-shaped, 2–9 cm long, 1–5 cm wide, prominently 3-ribbed, dark green above, greyish green below; margins finely toothed; stalks less than 1 cm long; stipules small, densely hairy, soon deciduous. **Flower Cluster:** Panicle thimble-shaped, 1–4 cm long; stalks 15–18 cm long; 1–2 small leaves below cluster. **Flowers:** White, 2–6 mm across; petals 5; sepals 5; stamens 5; pistil 1; stalks 0.6–1 cm long. **Fruit:** Capsule 3-lobed, 3–6 mm across, brown.

Notes: *Ceanothus* is the Greek name for a spiny shrub. The species name means "of America." • The common name refers to the leaves being used as a tea substitute during the American Revolution.

Alder-leaved Buckthorn
Rhamnus alnifolia

Habitat: Moist, shady woods and swamps. **General:** Shrub 50–200 cm tall; young branches hairy; older twigs hairless. **Leaves:** Alternate, deciduous, oval, 2–10 cm long, 1–5 cm wide, prominently veined; margins wavy or toothed; short-stalked. **Flower Cluster:** Umbel 1–3-flowered, in leaf axils.

Flowers: Yellowish green, 2–4 mm across; unisexual or perfect; petals 5; sepals 5; stamens 5; pistil 1. **Fruit:** Drupe berry-like, black, 6–9 mm across, poisonous; seeds 3.

Notes: *Rhamnus* comes from the Greek word *rhamnos*, the name given to a species of spiny shrub. The species name *alnifolia* means "alder-leaved."

GRAPE FAMILY
Vitaceae

The grape family is composed of deciduous climbing vines with alternate leaves. The leaves are usually palmately veined or lobed, or pinnately compound. Tendrils are present and occur opposite the flower cluster. Flowers are borne in cymes or panicles. The perfect flowers have 4–5 petals, 4–5 sepals, 4–5 stamens and 1 pistil. The fruit is a berry.

The best-known species from this family is *Vitis vinifera*, the wine grape. Species of the genera *Ampelopsis* and *Parthenocissus* are often grown as ornamentals.

Virginia Creeper
Parthenocissus quinquefolia

Habitat: Escaped from cultivation; native to the eastern U.S. **General:** Woody vine to 15 m tall/long; bark greyish brown, rough owing to aerial roots and tendrils. **Leaves:** Alternate, deciduous, stalks 15–20 cm long, palmately compound with 5 leaflets; leaflets elliptic, 6–12 cm long, 2.5–5 cm wide, shiny and green above, pale below, margins coarsely toothed, stalks 0.5–1.5 cm long; tendrils opposite leaves, much-branched, the tips with adhesive discs; turning purplish red in autumn. **Flower Cluster:** Panicle with distinct main axis, 10–16 cm across, branches ending in umbel-like clusters. **Flowers:** Greenish white, 4–6 mm across; petals 5; sepals 5; stamens 5, short; pistil 1. **Fruit:** Berry bluish black with white, powdery film, 5–7 mm across, reportedly poisonous; seeds 1–4.

Notes: *Parthenocissus* comes from the Greek words *parthenos*, "virgin," and *kissos*, "ivy." The species name *quinquefolia* means "5-leaved."

Thicket Creeper
Parthenocissus vitacea

Habitat: Forest edges, thickets and fence lines. **General:** Woody vine 3–10 m tall/long; young stems green, older stems ridged. **Leaves:** Alternate, deciduous, stalks 5–20 cm long, palmately compound with 5 leaflets; leaflets elliptic to oval or oblong, 5–12 cm long, 1–7 cm wide, dark green above, pale below, margins coarsely toothed; tendrils opposite leaves, sparingly branched, adhesive discs absent. **Flower Cluster:** Panicle umbel-like, wider than long, dichotomously branched. **Flowers:** Yellowish green, 4–6 mm wide; petals 5; sepals 5; stamens 5; pistil 1. **Fruit:** Berry dark blue to black, 8–10 mm across; fruiting stalk turning bright red at maturity; seeds 1–3.

Notes: The species name *vitacea* means "like *Vitis* (grape)," a reference to the plant's growth habit.

Riverbank Grape
Vitis riparia

Habitat: Riverbanks, thickets, fence lines and roadsides. **General:** Woody vine to 10 m tall/long; young stems reddish green, becoming smooth and finely ridged; older bark grey, shaggy, peeling. **Leaves:** Alternate, deciduous, heart-shaped to oval, 10–15 cm long, 7–15 cm wide, 3–5-lobed, dark green above, pale below, base U-shaped; margins coarsely toothed; stalks 2–15 cm long; tendrils sparingly branched, coiling. **Flower Cluster:** Panicle many-flowered, 2.5–12.5 cm long, opposite leaves. **Flowers:** Greenish white to cream-coloured, fragrant, 2–7 mm across; petals 5, attached at tips and falling off as flower opens; sepals 5, scale-like; stamens 5; pistil 1. **Fruit:** Berry dark blue to black with powdery white film, 6–12 mm across; edible but sour; seeds 2–4.

Notes: *Vitis* is the Latin name for grapevine. The species name *riparia* means "along streambanks."

TAMARISK FAMILY
Tamaricaceae

The tamarisk or salt-cedar family is a small family of deciduous trees and shrubs found in moist, alkaline soils. The alternate, scale-like leaves are reported to secrete salts, thereby increasing the salinity of the soil. Flowers are borne in racemes or panicles. The perfect flowers have 4–5 petals, 4–5 sepals, 4–10 stamens and 1 pistil. The fruit is a capsule.

Several species have been introduced into North America as ornamental shrubs.

Tamarisk · *Tamarix ramosissima*

Habitat: Roadsides and landscaped areas; introduced from Ukraine, Iraq, China and Korea. **General:** Shrub to 8 m tall; bark reddish brown to grey; young branches reddish to orangey yellow. **Leaves:** Alternate, deciduous, scale-like, 1.5–3.5 mm long; often covered in a salty bloom. **Flower Cluster:** Panicle 2–8 cm long, 3–5 mm wide; branches with about 20 flowers per 2.5 cm of stem. **Flowers:** Pink, 2–3 mm across; petals 5; sepals 5; stamens 5; pistil 1. **Fruit:** Capsule oval to lance-shaped, 3–5 mm long; seeds 8–20.

Notes: *Tamarix* is the Latin name for salt-cedar, another common name for this plant. The species name *ramosissima* means "much-branched."
• The leaves contain salt-secreting glands that increase soil salinity, thereby reducing competition from adjacent plants.

OLEASTER FAMILY · Elaeagnaceae

Family description on p. 69.

Autumn Olive
Elaeagnus umbellata

Habitat: Abandoned fields and forest edges; introduced from China. **General:** Shrub 1–3 m tall; branches erect, spreading; young stems silvery, scaly. **Leaves:** Alternate, deciduous, oblong to oval, 2.2–8 cm long, 1–2.5 cm wide, underside densely covered in white scales; veins 5–8 per side; margins smooth to wavy; stalks 3–10 mm long. **Flower Cluster:** Umbel-like, 1–7-flowered; stalks 3–8 mm long, elongating to 12 mm in fruit. **Flowers:** Silvery white, tube to funnel-shaped, 5–7 mm long; petals absent; sepals 4, lobes triangular; stamens 4; pistil 1. **Fruit:** Drupe red, globe-shaped, 6–9 mm across.

Notes: The species name means "umbel," a reference to the flower cluster.

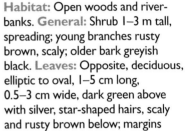

Canada Buffalo-berry
Shepherdia canadensis

Habitat: Open woods and riverbanks. **General:** Shrub 1–3 m tall, spreading; young branches rusty brown, scaly; older bark greyish black. **Leaves:** Opposite, deciduous, elliptic to oval, 1–5 cm long, 0.5–3 cm wide, dark green above with silver, star-shaped hairs, scaly and rusty brown below; margins smooth; stalks 0.7–1.2 cm long. **Flower Cluster:** Cyme 2–8-flowered, in leaf axils; appearing before leaves. **Flowers:** Yellowish brown, 3–5 mm across; male or female; petals absent; sepals 4; stamens 8; pistil 1. **Fruit:** Berry red, 4–6 mm across; edible but ill-tasting.

Notes: The genus name commemorates English botanist John Shepherd (1764–1836). The species name means "of Canada." • Soapberry is another common name for this plant because the juice of the berries contains saponin, a substance with a soapy flavour and texture.

DOGWOOD FAMILY · Cornaceae
Family description on p. 70.

Silky Dogwood · *Cornus amomum*

Habitat: Low-lying ground near marshes and streams. **General:** Shrub to 3 m tall; branches grey, finely haired, becoming reddish to purplish brown with age. **Leaves:** Opposite, deciduous, elliptic to oval or lance-shaped, 5–10 cm long, 1–5 cm wide, dark green above, pale greyish green below, somewhat drooping; veins 4–6 per side; margins smooth; stalks less than 2 cm long. **Flower Cluster:** Cyme flat-topped to slightly rounded, 3–8 cm across, many-flowered. **Flowers:** Creamy white, 5–7 mm across; petals 4; sepals 4; stamens 4; pistil 1. **Fruit:** Drupe berry-like, 0.6–1 cm across, blue to bluish white.

Notes: This species is the last dogwood to bloom in our region.

Grey Dogwood
Cornus racemosa

Habitat: Moist soil along streams and roads. **General:** Shrub to 2.5 m tall; branches smooth, green to light brown. **Leaves:** Opposite, deciduous, numerous, elliptic to oval, 5–10 cm long, 2–4 cm wide, dark green above, greyish green below; veins 3–4 per side; margins smooth; stalks 6–12 mm long. **Flower Cluster:** Cyme longer than broad, 2–9 cm long, 1–5 cm wide, often panicle- or raceme-like. **Flowers:** Creamy white, ill-scented, 2–4 mm across; petals 4; sepals 4; stamens 4; pistil 1. **Fruit:** Drupe berry-like, white, 5–6 mm across; stalk often bright red at maturity.

Notes: The species name *racemosa*, "raceme," refers to the type of flower cluster.

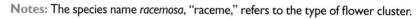

Round-leaved Dogwood
Cornus rugosa

Habitat: Open woods, often on sandy or rocky soil. **General:** Shrub, sometimes tree-like, to 3 m tall; branches warty, yellowish green to pink, dotted or streaked with reddish or purplish brown; older stems purplish. **Leaves:** Opposite, deciduous, oval to rounded, 7–15 cm long, 5–12 cm wide, pale to dark green above, densely grey-hairy below, tip pointed; veins 7–8 per side; margins smooth; stalks 1.2–1.8 cm long. **Flower Cluster:** Cyme flat-topped, 3–8 cm wide, densely flowered. **Flowers:** White to creamy white, 2–4 mm across; petals 4; sepals 4; stamens 4; pistil 1. **Fruit:** Drupe berry-like, greenish white to pale blue, 5–7 mm across.

Notes: The species name *rugosa* mean "rough," a reference to the surface of the branches.

Red-osier Dogwood
Cornus sericea ssp. *sericea*

Habitat: Riverbanks, moist woods and meadows. **General:** Shrub 1–3 m tall; bark bright red, occasionally purplish or green; lower branches rooting when in contact with soil. **Leaves:** Opposite, deciduous, oval to lance-shaped, 3–8 cm long, 2.5–9 cm wide, dark green above, light green below; margins smooth; stalks 0.6–2.5 cm long. **Flower Cluster:** Cyme flat-topped, 3–6 cm across, many-flowered. **Flowers:** White, 3–6 mm across; petals 4; sepals 4; stamens 4; pistil 1. **Fruit:** Drupe berry-like, white or blue-tinged, 4–6 mm across, juicy; inedible.

Notes: The species name *sericea* means "silky." • The bright red stems are often woven into baskets or used in winter floral displays.

HEATHER FAMILY · Ericaceae
Family description on p. 307.

Bearberry · *Arctostaphylos uva-ursi*

Habitat: Dry woods and sandhills. **General:** Evergreen shrub, stems 0.5–4 m long, trailing, forming large mats; branches numerous, seldom more than 15 cm tall; bark reddish brown to greyish black, often peeling. **Leaves:** Alternate, evergreen, thick, leathery, oblong to oval or spatula-shaped, 1.5–3 cm long, 0.6–1.2 cm wide, net-veined below; margins smooth, slightly rolled; stalks short. **Flower Cluster:** Raceme 3–10-flowered. **Flowers:** Pinkish white, urn-shaped, nodding, 4–6 mm long; petals 4–5; sepals 4–5; stamens 8 or 10; pistil 1. **Fruit:** Drupe berry-like, bright red, 6–10 mm across, often remaining on stem until following summer; dry and tasteless.

Notes: The genus name *Arctostaphylos* comes from the Greek words *arktos*, "bear," and *staphule*, "bunch of grapes," a direct translation into "bearberry." The species name means "grape (*uva*) of the bear (*ursi*)."

Leatherleaf
Chamaedaphne calyculata

Habitat: Bogs and lakeshores. **General:** Evergreen shrub 20–100 cm tall, much-branched; older bark greyish, scurfy. **Leaves:** Alternate, evergreen, leathery, oblong to elliptic, 1–4.5 cm long, 0.3–1.5 cm wide, underside scurfy brown; margins smooth to wavy-toothed; stalks short. **Flower Cluster:** Raceme leafy, 1-sided, many-flowered. **Flowers:** White, urn-shaped, 5–7 mm long; petals 5; sepals 5; stamens 10; pistil 1. **Fruit:** Capsule round, 3–5 mm wide, remaining attached to stem for several seasons.

Notes: The genus name *Chamaedaphne* comes from the Greek words *chamai*, "on the ground," and *daphne*, "laurel." The species name *calyculata* refers to the 2 bracts below the flower that resemble an outer calyx. • The evergreen leaves have a leathery texture, hence the common name.

Wintergreen, Teaberry
Gaultheria procumbens

Habitat: Sandy, mossy woods. **General:** Evergreen shrub to 15 cm tall; upper branches leafy; rhizome woody, creeping. **Leaves:** Alternate, evergreen, leathery, oval to elliptic or oblong, 1–5 cm long, 0.7–3 cm wide, dark green and shiny above, pale below; margins toothed but obscured by rolled edges; stalks 2–5 mm long, reddish. **Flower Cluster:** Solitary, in leaf axils; stalks 5–10 mm long, curved. **Flowers:** White, barrel-shaped, 5–10 mm long; petals 5; sepals 5; stamens 10; pistil 1. **Fruit:** Capsule berry-like, surrounded by fleshy, red calyx, 9–11 mm across; edible, wintergreen-flavoured.

Notes: The genus name commemorates Canadian physician and botanist Jean-François Gauthier (1708–58). The species name *procumbens* means "lying on the ground," a reference to the prostrate growth habit.

Pale Laurel · *Kalmia polifolia*

Habitat: Peat bogs and swamps. **General:** Evergreen shrub 30–60 cm tall; twigs with sharp angles; young branches pale brown, older bark dark brown to black. **Leaves:** Opposite, evergreen, oval to elliptic or oblong, 1–4 cm long, 0.6–1.2 cm wide, upper surface dark green, underside finely white-hairy; margins rolled; stalkless. **Flower Cluster:** Umbel terminal, 6–9-flowered. **Flowers:** Pink to purplish, saucer-shaped, 1–1.7 cm across; petals 5; sepals 5; stamens 10; pistil 1. **Fruit:** Capsule egg-shaped, 4–7 mm long.

Notes: The genus name honours Swedish explorer and botanist Pehr Kalm (1716–79), who made the first botanical collections in Canada. The species name *polifolia* means "whitish grey leaves." • All parts of this plant are poisonous.

Labrador Tea
Ledum groenlandicum

Habitat: Bogs and moist, coniferous forests. **General:** Evergreen shrub 40–120 cm tall, straggling; twigs covered with rusty brown, velvety hairs; older bark reddish brown to purplish, hairless. **Leaves:** Alternate, evergreen, oval to elliptic, 1–5 cm long, 0.5–2 cm wide, underside covered with woolly, rust-coloured hairs; margins rolled; stalks short. **Flower Cluster:** Umbel terminal, 5–20-flowered; flower stalks 1–2 cm long.

Flowers: White, 9–12 mm across; petals 5; sepals 5; stamens 5–7; pistil 1. **Fruit:** Capsule oval, 5–7 mm long; style persistent.

Notes: *Ledum* comes from the Greek word *ledon*, the common name for an evergreen species in the genus *Cistus*. The species name means "of Greenland." • The leaves of this plant can be used to make tea, hence the common name.

Blueberry
Vaccinium myrtilloides

Habitat: Dry, sandy woods. **General:** Shrub 10–40 cm tall; young branches greenish brown, covered with white hairs; old bark reddish brown to black, slightly peeling. **Leaves:** Alternate, deciduous, elliptic to lance-shaped, 1–4 cm long, 1–2.5 cm wide, both surfaces with short, soft hairs. **Flower Cluster:** Umbel 2–10-flowered, on short branches; appearing with leaves. **Flowers:** Greenish to pinkish white, cylindric to bell-shaped, 4–6 mm long; petals 5; sepals 5; stamens 8–10; pistil 1. **Fruit:** Berry blue, 4–7 mm across; edible.

Notes: The species name *myrtilloides* means "like a little myrtle."

OLIVE FAMILY · Oleaceae
Family description on p. 72.

Lilac · *Syringa vulgaris*

Habitat: Edges of woods and along roadsides; introduced ornamental from Europe. **General:** Shrub to 6 m tall, spreading by roots, forming thickets. **Leaves:** Opposite, deciduous, somewhat leathery, oval, 5–10 cm long, 2–6 cm wide, bright green; margins smooth; stalks 1.5–2.5 cm long. **Flower Cluster:** Panicle 10–20 cm long, densely flowered. **Flowers:** Lilac to white, fragrant, 8–12 mm long, trumpet-shaped; petals 4–5; sepals 4; stamens 2; pistil 1. **Fruit:** Capsule brown, 1–1.5 cm long, 4–6 mm across; seeds 2, winged.

Notes: *Syringa* comes from the Greek word *syrinx*, meaning "pipe," a reference to the hollow stems. The species name *vulgaris* means "common."

NIGHTSHADE FAMILY · Solanaceae

Family description on p. 343.

Climbing Nightshade, Bittersweet Nightshade · *Solanum dulcamara*

Habitat: Waste places, fence lines and roadsides; introduced from Europe and Asia. **General:** Semi-woody climbing vine 1–4 m tall/long, ill-scented; bark light grey, shredding with age; stems branched; rhizomes creeping. **Leaves:** Alternate, deciduous, oval to heart-shaped, 4–10 cm long, 1–5 cm wide, dark green (often purplish); upper leaves with 1–2 small lobes or leaflets; stalks 1–2 cm long; leaf shape variable, often on the same plant. **Flower Cluster:** Raceme 2–8 cm long, 3–20-flowered, on stalks 6–8 mm long on internodes or opposite leaves. **Flowers:** Purple, wheel-shaped, 8–12 mm across; petals 5; sepals 5; stamens 5, yellow; pistil 1. **Fruit:** Berry elliptic, 8–12 mm long, bright red at maturity, 40–60-seeded; often remaining on stem throughout winter.

Notes: The species name comes from the Latin words *dulcis*, "sweet," and *amarus*, "bitter," which combined form "bittersweet." • The leaves and berries contain toxic alkaloids that are reportedly poisonous to livestock and humans.

BEDSTRAW OR MADDER FAMILY · Rubiaceae

Family description on p. 360.

Buttonbush · *Cephalanthus occidentalis*

Habitat: Streamsides, marshes, ponds and ditches. **General:** Shrub to 3 m tall; branches green turning brown; old stems greyish brown to purplish grey; lenticels present. **Leaves:** Opposite or in whorls of 3–4, deciduous, elliptic to oval or lance-shaped, 5–18 cm long, 1–7.5 cm wide, bright green and shiny above, pale below; margins smooth to wavy; stalks 1–2.5 cm long; stipules triangular, 2–3 mm long. **Flower Cluster:** Head globe-shaped, 2–4 cm across, 100–200-flowered. **Flowers:** Creamy white, tubular, 5–8 mm long; petals 4; sepals 4; stamens 4; pistil 1; style exserted. **Fruit:** Nutlet cone-shaped, 4–6 mm long; in head-like clusters.

Notes: The genus name comes from the Greek words *kephale*, "head," and *anthos*, "flower," a reference to the head-like flower clusters. The species name means "western."

HONEYSUCKLE FAMILY
Caprifoliaceae

The honeysuckle family has over 400 species worldwide. Members of this family are herbs, shrubs and woody vines with simple, opposite leaves. Stipules are absent. Only one genus, *Sambucus* (elderberry), has compound leaves. The flowers, composed of 5 united petals, 5 sepals, 4–5 stamens and 1 style, are wheel-shaped to tubular. The fruit may be a berry, drupe, capsule or achene.

Species such as honeysuckles (*Lonicera* spp.) and bush cranberries (*Viburnum* spp.) are grown for their ornamental value.

Bush Honeysuckle
Diervilla lonicera

Habitat: Dry, sandy to rocky woods. **General:** Shrub to 1 m tall; young branches reddish green; older stems brown to grey. **Leaves:** Opposite, deciduous, oval to oblong or lance-shaped, 5–13 cm long, 1.5–6 cm wide, dark green above, pale below; margins sharply toothed; stalks 0.3–1.2 cm long. **Flower Cluster:** Raceme 2–7-flowered; in leaf axils. **Flowers:** Pale yellow turning orangey red with age, tubular, 1.2–2 cm long; petals 5; sepals 5, 2–7 mm long; stamens 5; pistil 1. **Fruit:** Capsule brown, 1–1.5 cm long.

Notes: The genus name commemorates 17th-century French surgeon Dièreville, who collected several plant species while exploring eastern Canada. The species name *lonicera* refers to the shrub's resemblance to members of the genus *Lonicera*.

Fly Honeysuckle
Lonicera canadensis

Habitat: Damp, shady woods near swamps and bogs. **General:** Shrub to 1.5 m tall, straggling; branches green to purplish, smooth; older stems brown to grey with shredding bark. **Leaves:** Opposite, deciduous, oval to oblong or elliptic, 2.5–12 cm long, 2.5–5 cm wide, base somewhat heart-shaped; margins smooth; stalks with fine hairs. **Flower Cluster:** Raceme 2-flowered, on stalks 2–3 cm long from leaf axils. **Flowers:** Pale yellow, tubular to funnel-shaped, 1.2–2.2 cm long; petals 5; sepals 5; stamens 5; pistil 1. **Fruit:** Berries red, in pairs; seeds 3–4.

Notes: The genus name commemorates German botanist Adam Lonitzer (1528–86). The species name means "of Canada."

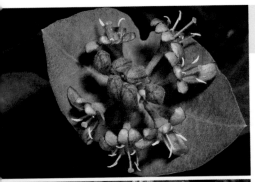

Twining Honeysuckle
Lonicera dioica

Habitat: Open forests and fence lines. **General:** Woody vine, stems 1–3 m tall/long, twining, hollow; bark shredding. **Leaves:** Opposite, deciduous, oval to oblong, 5–8 cm long, 1–6 cm across, smooth above, hairy below; lower leaves short-stalked, upper leaves stalkless; upper pair of leaves may be fused, forming a cup around the stem. **Flower Cluster:** Spike 3–9-flowered. **Flowers:** Yellow, turning reddish orange with age, tubular, 2-lipped, 1.5–2.5 cm long; petals 5; sepals 5; stamens 5; pistil 1. **Fruit:** Berry red, 5–8 mm across; inedible.

Notes: The species name *dioica* means "dioecious," even though the plants have perfect (bisexual) flowers.

Tatarian Honeysuckle
Lonicera tatarica

Habitat: Moist, wooded areas; introduced from Asia as a garden shrub.
General: Shrub 1–2 m tall; bark brownish green, shredding with age.
Leaves: Opposite, deciduous, oblong to oval, 2–6 cm long, 1–4 cm wide; margins smooth; stalks 2–6 mm long
Flower Cluster: Raceme 2-flowered, in leaf axils. **Flowers:** White or pink, tubular, 1.5–2 cm long; petals 5; sepals 5; stamens 5; pistil 1. **Fruit:** Berry red, orange or yellow, 5–7 mm across, in pairs; inedible, possibly poisonous.

Notes: The species name *tatarica* refers to the shrub's native region, Tataria, an area of northern and central Asia that includes Siberia, Turkestan, Mongolia and Manchuria.

Canada Elderberry
Sambucus nigra ssp. *canadensis*

Habitat: Low-lying ground, swamps and forest edges.
General: Shrub to 3 m tall; branches yellowish green; older stems with greyish brown bark.
Leaves: Opposite, deciduous, stalks 2.5–5 cm long, compound with 5–11 (usually 7) leaflets; leaflets elliptic, 5–15 cm long, 2.5–5.5 cm wide, margins sharply toothed, stalks short; ill-scented when bruised. **Flower Cluster:** Panicle of cymes

5–18 cm across, rays 5, flat to dome-shaped, long-stalked; appearing after leaves.
Flowers: White, 3–5 mm across; petals 5; sepals 5; stamens 5; pistil 1. **Fruit:** Drupe berry-like, purplish black, 4–6 mm across; edible and juicy.

Notes: *Sambucus* is the Latin name for elderberry. The species name means "of Canada," the native range of this shrub.

Red Elder, Elderberry
Sambucus racemosa var. *racemosa*

Habitat: Moist woods. **General:** Shrub 1–3 m tall; bark greyish brown. **Leaves:** Opposite, deciduous, stalks 2.5–5 cm long, pinnately compound with 5–7 leaflets; leaflets elliptic to lance-shaped, 5–17 cm long, 2–6 cm wide; margins sharply toothed. **Flower Cluster:** Panicle or raceme 4–10 cm long, 3–5 cm wide, crowded; appearing with leaves. **Flowers:** Yellowish to greenish white, 3–6 mm across; petals 5; sepals 5, small; stamens 5; pistil 1. **Fruit:** Drupe berry-like, red to black, 5–6 mm wide; inedible, possibly poisonous.

Notes: The species name means "raceme," a reference to the type of flower cluster.

Snowberry
Symphoricarpos albus

Habitat: Moist, wooded areas. **General:** Shrub 30–100 cm tall, broom-like; branches numerous, delicate. **Leaves:** Opposite, deciduous, oval, 1–4 cm long, 1–3 cm across, underside with soft hairs; margins smooth; stalks less than 1 cm long. **Flower Cluster:** Raceme 2–3-flowered, in upper leaf axils. **Flowers:** Pinkish white, bell-shaped, 4–7 mm long; petals 4–5; sepals 4–5; stamens 4–5; pistil 1. **Fruit:** Berry white, waxy, 6–12 mm across; edible but a strong laxative.

Notes: *Symphoricarpos* comes from the Greek words *symphorein*, "to bear together," and *karpos*, "fruit," a reference to the clustered berries. The species name means "white," referring to the colour of the fruit.

Buckbrush
Symphoricarpos occidentalis

Habitat: Dry slopes and open forests; introduced from western Canada as an ornamental shrub. **General:** Shrub to 1 m tall; stems copper-coloured; rhizomes creeping, often forming large colonies. **Leaves:** Opposite, deciduous, oval to oblong, greyish green, 2–6 cm long, 1.5–5.5 cm wide, thick, somewhat leathery; margins wavy-toothed to lobed; stalks 0.8–1.2 cm long. **Flower Cluster:** Spike densely flowered, terminal and axillary. **Flowers:** Pinkish white, funnel-shaped, 5–9 mm long; petals 4–5; sepals 4–5; stamens 4–5, long-stalked; pistil 1. **Fruit:** Berry whitish green to purplish black, 0.8–1 cm across; inedible.

Notes: The species name *occidentalis* means "western," a reference to this plant's native range.

Wayfaring Tree
Viburnum lantana

Habitat: Waste areas and roadsides; introduced from Europe and Asia as an ornamental shrub. **General:** Shrub to 4 m tall; young stems covered in soft, grey hairs, older stems hairless. **Leaves:** Opposite, deciduous, oblong to oval, 5–10 cm long, 2–5 cm wide, all surfaces covered in soft hairs; margins finely toothed; stalks 1–3 cm long. **Flower Cluster:** Cyme 5–15 cm across, 7-rayed, short-stalked. **Flowers:** Creamy white, 3–5 mm across; petals 5; sepals 5; stamens 5; pistil 1. **Fruit:** Drupe berry-like, 0.8–1 cm long, green turning red then black at maturity.

Notes: The species name *lantana* is the old name for *Viburnum*.

Hobblebush
Viburnum lantanoides

Habitat: Moist woods and ravines. **General:** Shrub to 2 m tall; stems sprawling to prostrate, rooting at nodes; young branches with cinnamon-coloured hairs. **Leaves:** Opposite, deciduous, oval to round, 10–20 cm long, 7–18 cm wide; dark green above, paler below with light brown hairs, tip pointed, base heart-shaped; margins finely toothed; stalks 1–6 cm long. **Flower**

Cluster: Cyme 8–13 cm across; rays usually 5; flowers of 2 types, large and sterile or small and fertile. **Flowers:** Sterile flowers pinkish white, 2–3.5 cm across; fertile flowers white, 3–5 mm across; petals 5; sepals 5; stamens 5; pistil 1. **Fruit:** Drupe berry-like, 0.8–1 cm across, crimson to purplish black.

Notes: The species name means "like *Lantana*," referring to the leaves, which resemble those of the tropical shrub *Lantana camara*. • The common name refers to being hobbled by the difficulty of walking through a patch of this shrub.

Wild Raisin
Viburnum nudum var. *cassinoides*

Habitat: Moist, sandy soil. **General:** Shrub to 5 m tall; young branches scurfy, becoming purplish and ridged with age. **Leaves:** Opposite, deciduous, oval to oblong or diamond-shaped, 4.5–9 cm long, 2.5–5 cm wide, dull dark green above, pale and brown-hairy below; margins smooth to wavy; stalks grooved, 0.5–2 cm long. **Flower Cluster:** Cyme flat-topped, 5–10 cm across; on stalks 5–25 cm long. **Flowers:** Creamy white, 3–7 mm across, ill-scented; petals 5; sepals 5; stamens 5; pistil 1. **Fruit:** Drupe berry-like, 0.6–1 cm across, whitish yellow turning pink then bluish black.

Notes: The species name *nudum* means "naked" or "nude." The varietal name *cassinoides* means "resembling *Ilex cassine* or *Cassinia* spp." • The common name refers to the resemblance of the dry fruit to raisins.

Highbush Cranberry
Viburnum opulus var. *americanum*

Habitat: Moist forests. **General:** Shrub or small tree 1–4 m tall; bark grey. **Leaves:** Opposite, deciduous, dark green, 6–12 cm long, 5–10 cm wide, prominently 3-lobed; margins coarsely toothed; stalks 1–4 cm long; stipules thickened at tip; turning brilliant reddish purple in autumn. **Flower Cluster:** Cyme flat-topped, 5–15 cm across, containing both sterile and fertile flowers. **Flowers:** White; sterile flowers (attract insect pollinators) 1–2 cm across, on margins of flower cluster; fertile flowers (fruit producing) 6–8 mm across; petals 5; sepals 5; stamens 5; pistil 1. **Fruit:** Drupe berry-like, reddish orange, 0.8–1 cm across; edible but tart.

Notes: The species name *opulus* is the Latin name for a kind of maple, a reference to the leaf shape.

Guelder-rose
Viburnum opulus var. *opulus*

Habitat: Moist, open areas; introduced from Europe as an ornamental shrub. **General:** Shrub or small tree 1–5 m tall; bark grey. **Leaves:** Opposite, deciduous, dark green, 6–12 cm long, 5–10 cm wide, prominently 3-lobed, base with club-shaped glands; margins coarsely toothed; stalks 1–4 cm long; stipules bristle-like; turning brilliant reddish purple in autumn. **Flower Cluster:** Cyme flat-topped, 5–15 cm across, containing both sterile and fertile flowers. **Flowers:** White; sterile flowers (attract insect pollinators) 1–2 cm across, on margins of flower cluster; fertile flowers (fruit producing) 6–8 mm across; petals 5; sepals 5; stamens 5; pistil 1. **Fruit:** Drupe berry-like, reddish orange, 0.8–1 cm across; edible but tart.

Notes: This variety is suspected to hybridize with our native highbush cranberry (*V. o.* var. *americanum*).

Wildflowers

Pale touch-me-not (*Impatiens pallida*)

Key to the Wildflowers

Leaf Arrangement	Leaf Complexity		Flower Colour		Group
Absent	–		Various		Group 1, p. 130
Alternate	Simple	Dissected	Various		Group 2, p. 130
		Entire	Blue		Group 3, p. 131
			Green		Group 4, p. 132
			Pink, Purple, Red or Orange		Group 5, p. 132
			White	Petals 4 or fewer	Group 6, p. 134
				Petals 5	Group 7, p. 134
				Petals 6 or more	Group 8, p. 135
			Yellow	Petals 5 or fewer	Group 9, p. 136
				Petals 6 or more	Group 10, p. 137
		Lobed	Various		Group 11, p. 137
	Compound	Palmate	Various		Group 12, p. 139
		Bi- or Trifoliate	Various		Group 13, p. 139
		Pinnate	Pink, Red, Purple or Green		Group 14, p. 139
			Yellow		Group 15, p. 140
			White		Group 16, p. 140
Basal	Simple	Dissected or Lobed	Various		Group 17, p. 141
		Entire	Blue, Purple, Green or Pink		Group 18, p. 141
			Yellow, Red or Orange		Group 19, p. 142
			White		Group 20, p. 143
	Compound		Various		Group 21, p. 144
Opposite	Simple	Entire	Blue, Purple or Green		Group 22, p. 144
			Red, Orange or Yellow		Group 23, p. 145
			Pink		Group 24, p. 146
			White		Group 25, p. 147
		Lobed	Various		Group 26, p. 148
	Compound		Various		Group 27, p. 148
Whorled	–		Various		Group 28, p. 148

WILDFLOWERS

In this guide, the wildflower group includes herbaceous plants with non-woody stems and broad leaves and/or showy flowers. Occasionally these plants have a woody crown, but the stems die back to the ground each fall.

Key to the Wildflowers

Species in this section are identified by leaf arrangement (i.e., absent, alternate, compound, basal, opposite or whorled), flower colour and number of petals.

GROUP 1: LEAVES ABSENT

Flower colour various

Corallorhiza maculata p. 180

Corallorhiza striata p. 181

Corallorhiza trifida p. 181

Arceuthobium pusillum p. 193

Cuscuta gronovii p. 324

Opuntia fragilis p. 290

Flower colour various

Monotropa hypopithys p. 309

Monotropa uniflora p. 310

Conopholis americana p. 350

Epifagus virginiana p. 351

GROUP 2: LEAVES ALTERNATE, SIMPLE, DISSECTED

Flower colour various

Corydalis aurea p. 226

Corydalis sempervirens p. 226

Fumaria officinalis p. 228

Chelidonium majus p. 225

Achillea millefolium p. 371

Ambrosia artemisiifolia p. 372

Flower colour various

Anthemis cotula p. 373

Artemisia campestris p. 375

Artemisia vulgaris p. 376

Centaurea diffusa p. 379

Centaurea stoebe ssp. micranthos p. 380

Matricaria discoidea p. 394

Flower colour various

Tripleurospermum perforata
p. 407

Flowers blue

| *Commelina communis* p. 161 | *Tradescantia virginiana* p. 162 | *Linum perenne* p. 267 | *Viola adunca* p. 286 | *Echium vulgare* p. 328 | *Lappula squarrosa* p. 329 |

Flowers blue

     (and second row images)

| *Mertensia paniculata* p. 330 | *Mertensia virginica* p. 330 | *Myosotis sylvatica* p. 331 | *Chaenorrhinum minus* p. 353 | *Campanula rotundifolia* p. 367 | *Campanulastrum americanum* p. 367 |

Flowers blue

| *Lobelia inflata* p. 368 | *Lobelia kalmii* p. 369 | *Lobelia siphilitica* p. 369 | *Cichorium intybus* p. 380 | *Echinops sphaerocephalus* p. 384 | *Eurybia macrophylla* p. 385 |

Flowers blue

| *Symphyotrichum cordifolium* p. 403 | *Symphyotrichum laeve* p. 404 | *Symphyotrichum lanceolatum* p. 404 |

GROUP 4: LEAVES ALTERNATE, SIMPLE, ENTIRE

Flowers green

Scheuchzeria
palustris
p. 153

Sparganium
eurycarpum
p. 150

Coeloglossum
viride
p. 180

Platanthera
hyperborea
p. 187

Laportea
canadensis
p. 190

Parietaria
pensylvanica
p. 191

Flowers green

Amaranthus
blitoides
p. 204

Amaranthus
retroflexus
p. 205

Atriplex
hortensis
p. 200

Axyris
amaranthoides
p. 201

Chenopodium
album
p. 201

Chenopodium
glaucum var.
salinum, p. 202

Flowers green

Chenopodium
simplex
p. 202

Kochia
scoparia
p. 203

Salsola
kali
p. 203

Penthorum
sedoides
p. 247

Acalypha
rhomboidea
p. 274

Ambrosia
trifida
p. 372

Flowers green

Artemisia
ludoviciana
p. 376

Xanthium
strumarium
p. 409

GROUP 5: LEAVES ALTERNATE, SIMPLE, ENTIRE

Flowers pink, purple, red or orange

Lilium
philadelphicum
p. 168

Cypripedium
arietinum
p. 182

Cypripedium
reginae
p. 183

Epipactus
helleborine
p. 184

Fallopia
convolvulus
p. 196

Rumex
crispus
p. 199

132

Flowers pink, purple, red or orange

*Rumex
obtusifolius*
p. 199

*Polygonum
amphibium* var.
emersum, p. 198

*Hesperis
matronalis*
p. 239

*Lepidium
densiflorum*
p. 240

*Chamerion
angustifolium*
p. 294

*Ipomoea
hederacea*
p. 325

Flowers pink, purple, red or orange

*Viola
rostrata*
p. 288

*Viola
tricolor*
p. 289

*Polygala
paucifolia*
p. 273

*Vaccinium
oxycoccus*
p. 312

*Cynoglossum
officinale*
p. 328

*Symphytum
asperum*
p. 331

Flowers pink, purple, red or orange

*Castilleja
coccinea*
p. 350

*Pedicularis
canadensis*
p. 351

*Campanula
rapunculoides*
p. 366

*Lobelia
cardinalis*
p. 368

*Arctium
minus*
p. 374

*Arctium
tomentosum*
p. 374

Flowers pink, purple, red or orange

*Carduus
acanthoides*
p. 378

*Centaurea
jacea*
p. 379

*Centaurea stoebe
ssp. micranthos*
p. 380

*Erigeron
philadelphicus*
p. 384

*Prenanthes
racemosa*
p. 395

*Symphyotrichum
novae-angliae*
p. 405

*Flowers pink, purple,
red or orange*

*Vernonia
gigantea*
p. 409

GROUP 6: LEAVES ALTERNATE, SIMPLE, ENTIRE

Flowers white, petals 4 or fewer

Saururus cernuus p. 189	*Maianthemum canadense* p. 169	*Alliaria petiolata* p. 231	*Berteroa incana* p. 232	*Cakile edentula* p. 233	*Capsella bursa-pastoris* p. 234

Flowers white, petals 4 or fewer

Cardamine douglasii p. 236	*Cardamine pensylvanica* p. 236	*Hesperis matronalis* p. 239	*Thlapsi arvense* p. 243	*Epilobium ciliatum* p. 295

GROUP 7: LEAVES ALTERNATE, SIMPLE, ENTIRE

Flowers white, petals 5

Fallopia cilinodis p. 196	*Fallopia japonica* p. 197	*Persicaria lapathifolia* p. 197	*Polygonum aviculare* p. 198	*Phytolacca americana* p. 206	*Comandra umbellata* p. 193

Flowers white, petals 5

Geocaulon lividum p. 194	*Viola arvensis* p. 286	*Viola canadensis* p. 287	*Malva rotundifolia* p. 282	*Polygala senega* p. 273	*Calystegia sepium* p. 323

Flowers white, petals 5

Convolvulus arvensis p. 324	*Epigaea repens* p. 308	*Lithospermum officinale* p. 329	*Chaenorrhinum minus* p. 353	*Datura inoxia* p. 344	*Datura stramonium* p. 344

*Flowers white,
petals 5*

| *Solanum carolinense* p. 345 | *Solanum nigrum* p. 346 | *Digitalis lanata* p. 354 |

*Flowers white,
petals 6 or more*

| *Convallaria majalis* p. 166 | *Maianthemum racemosum* p. 169 | *Maianthemum stellatum* p. 170 | *Maianthemum trifolium* p. 170 | *Platanthera dilatata* p. 186 | *Spiranthes cernua* p. 188 |

*Flowers white,
petals 6 or more*

| *Spiranthes romanzoffiana* p. 188 | *Streptopus amplexifolius* p. 172 | *Anaphalis margaritacea* p. 373 | *Arnoglossum plantagineum* p. 375 | *Centaurea diffusa* p. 379 | *Conyza canadensis* p. 382 |

*Flowers white,
petals 6 or more*

| *Doellingeria umbellata* p. 383 | *Erigeron philadelphicus* p. 384 | *Leucanthemum vulgare* p. 393 | *Prenanthes alba* p. 395 | *Prenanthes racemosa* p. 395 | *Solidago bicolor* p. 398 |

*Flowers white,
petals 6 or more*

*Symphyotrichum
ericoides*
p. 403

Flowers yellow,
petals 5 or fewer

| *Euphorbia* *cyparissias* p. 275 | *Euphorbia* *esula* p. 276 | *Euphorbia* *helioscopia* p. 277 | *Cypripedium* *parviflorum* p. 183 | *Caltha* *palustris* p. 220 | *Portulaca* *oleraceae* p. 208 |

Flowers yellow,
petals 5 or fewer

| *Brassica rapa* var. *rapa* p. 233 | *Camelina* *microcarpa* p. 234 | *Conringia* *orientalis* p. 237 | *Erysimum* *cheiranthoides* p. 239 | *Neslia* *paniculata* p. 240 | *Raphanus* *raphanistrum* p. 241 |

Flowers yellow,
petals 5 or fewer

| *Rorippa* *palustris* p. 241 | *Sinapis* *alba* p. 242 | *Oenothera* *biennis* p. 296 | *Chrysosplenium* *alternifolium* ssp. *iowense*, p. 249 | *Sedum* *acre* p. 248 | *Abutilon* *theophrasti* p. 281 |

Flowers yellow,
petals 5 or fewer

| *Helianthemum* *canadense* p. 285 | *Impatiens* *capensis* p. 279 | *Impatiens* *pallida* p. 280 | *Viola* *pubescens* p. 288 | *Physalis* *heterophylla* p. 345 | *Solanum* *rostratum* p. 346 |

Flowers yellow,
petals 5 or fewer

| *Aureolaria* *flava* p. 349 | *Linaria* *vulgaris* p. 355 | *Verbascum* *blattaria* p. 347 | *Verbascum* *thapsus* p. 348 |

*Flowers yellow,
petals 6 or more*

*Heteranthera
dubia*
p. 163

*Polygonatum
pubescens*
p. 172

*Uvularia
grandiflora*
p. 175

*Caltha
palustris*
p. 220

*Crepis
tectorum*
p. 383

*Euthamia
graminifolia*
p. 386

*Flowers yellow,
petals 6 or more*

*Grindelia
squarrosa*
p. 387

*Helenium
autumnale*
p. 388

*Helianthus
annuus*
p. 388

*Helianthus
tuberosus*
p. 389

*Inula
helenium*
p. 391

*Lactuca
serriola*
p. 392

*Flowers yellow,
petals 6 or more*

*Rudbeckia
hirta*
p. 396

*Solidago
caesia*
p. 399

*Solidago
canadensis*
p. 399

*Solidago
flexicaulis*
p. 400

*Solidago
juncea*
p. 400

*Solidago
nemoralis*
p. 401

*Flowers yellow,
petals 6 or more*

*Tragopogon
dubius*
p. 406

*Tragopogon
pratensis*
p. 407

*Tussilago
farfara*
p. 408

*Verbesina
alternifolia*
p. 408

Flower colour various

*Echinocystis
lobata*
p. 365

*Chelidonium
majus*
p. 225

*Erucastrum
gallicum*
p. 238

*Raphanus
raphanistrum*
p. 241

*Hibiscus
trionum*
p. 281

*Hydrophyllum
virginianum*
p. 327

Flower colour various

Sicyos
angulatus
p. 365

Sinapis
alba
p. 242

Barbarea
vulgaris
p. 232

Descurainia
sophia
p. 237

Diplotaxis
muralis
p. 238

Ranunculus
abortivus
p. 221

Flower colour various

Ranunculus
acris
p. 221

Ranunculus
sceleratus
p. 222

Rorippa
palustris
p. 241

Sinapis
arvensis
p. 242

Sisymbrium
altissimum
p. 243

Carduus
acanthoides
p. 378

Flower colour various

Carduus
nutans
p. 378

Cirsium
arvense
p. 381

Cirsium
muticum
p. 381

Cirsium
vulgare
p. 382

Echinops
sphaerocephalus
p. 384

Onopordum
acanthium
p. 394

Flower colour various

Jacobaea
vulgaris
p. 392

Ratibida
pinnata
p. 396

Senecio
viscosus
p. 397

Senecio
vulgaris
p. 397

Sonchus arvensis
ssp. arvensis
p. 401

Sonchus
asper
p. 402

Flower colour various

Sonchus
oleraceus
p. 402

Tanacetum
vulgare
p. 405

Flower colour various

*Polanisia
dodecandra*
p. 230

*Tarenaya
hassleriana*
p. 230

*Lupinus
perennis*
p. 263

Flower colour various

*Potentilla
norvegica*
p. 256

*Desmodium
canadense*
p. 259

*Desmodium
nudiflorum*
p. 260

*Lathyrus
latifolius*
p. 261

*Lathyrus
tuberosus*
p. 262

*Medicago
lupulina*
p. 263

Flower colour various

*Medicago
sativa*
p. 264

*Melilotus
alba*
p. 264

*Trifolium
pratense*
p. 265

*Trifolium
repens*
p. 266

*Oxalis
stricta*
p. 269

*Heracleum
mantegazzianum*
p. 303

*Flower colour
various*

*Heracleum
sphondylium* ssp.
montanum, p. 303

Flowers pink, red, purple or green

*Caulophyllum
thalictroides*
p. 223

*Aquilegia
brevistyla*
p. 219

*Aquilegia
canadensis*
p. 219

*Comarum
palustre*
p. 252

*Geum
rivale*
p. 254

*Geum
triflorum*
p. 255

GROUP 14: LEAVES ALTERNATE, COMPOUND, PINNATE

Flowers pink, red, purple or green

Amphicarpa bracteata p. 258   *Apios americana* p. 259   *Lathyrus tuberosus* p. 262   *Vicia cracca* p. 266   *Sanicula marilandica* p. 305

GROUP 15: LEAVES ALTERNATE, COMPOUND, PINNATE

Flowers yellow

Agrimonia striata p. 251   *Geum aleppicum* p. 253   *Potentilla paradoxa* p. 256   *Potentilla recta* p. 257   *Lotus corniculatus* p. 262   *Pastinaca sativa* p. 304

Flowers yellow

Taenidia integerrima p. 306   *Zizia aurea* p. 306

GROUP 16: LEAVES ALTERNATE, COMPOUND, PINNATE

Flowers white

Actaea pachypoda p. 215   *Actaea rubra* p. 216   *Thalictrum dasycarpum* p. 223   *Geum canadense* p. 254   *Potentilla arguta* p. 255   *Rubus pubescens* p. 257

Flowers white

Galega officinalis p. 260   *Lathyrus ochroleucus* p. 261   *Securigera varia* p. 265   *Aralia racemosa* p. 298   *Angelica atropurpurea* p. 299   *Anthriscus sylvestris* p. 299

Flowers white

Carum
carvi
p. 300

Cicuta
bulbifera
p. 300

Cicuta
maculata
p. 301

Conium
maculatum
p. 301

Cryptotaenia
canadensis
p. 302

Daucus
carota
p. 302

Flowers white

Osmorhiza
claytoni
p. 304

Sium
suave
p. 305

GROUP 16: LEAVES ALTERNATE, COMPOUND, PINNATE

Flower colour various

Dicentra
canadensis
p. 227

Dicentra
cucullaria
p. 227

Jeffersonia
diphylla
p. 224

Anemone
canadensis
p. 217

Anemone
cylindrica
p. 218

Anemone
quinquefolia
p. 218

Flower colour various

Anemone
acutiloba
p. 216

Anemone
americana
p. 217

Tiarella
cordifolia
p. 251

Hypochaeris
radicata
p. 391

Leontodon
autumnalis
p. 393

Taraxacum
officinale
p. 406

GROUP 17: LEAVES BASAL, SIMPLE, DISSECTED OR LOBED

Flowers blue, purple, green or pink

Pontederia
cordata
p. 163

Iris
lacustris
p. 176

Iris
versicolor
p. 177

Sisyrinchium
montanum
p. 178

Viola
sororia
p. 289

Pinguicula
vulgaris
p. 352

GROUP 18: LEAVES BASAL, SIMPLE, ENTIRE

141

GROUP 18: LEAVES BASAL, SIMPLE, ENTIRE

Flowers blue, purple, green or pink

Acorus americanus p. 158	*Triglochin maritima* p. 152	*Typha angustifolia* p. 151	*Typha latifolia* p. 151	*Plantago lanceolata* p. 356	*Plantago major* p. 356

Flowers blue, purple, green or pink

Butomus umbellatus p. 157	*Bellis perennis* p. 377	*Allium schoenoprasum* p. 164	*Calopogon tuberosus* p. 179	*Calypso bulbosa* p. 179	*Cypripedium acaule* p. 182

Flowers blue, purple, green or pink

Pogonia ophioglossoides p. 187	*Claytonia virginica* p. 208	*Primula mistassinica* p. 316	*Pyrola asarifolia* p. 311	*Symplocarpus foetidus* p. 160

GROUP 19: LEAVES BASAL, SIMPLE, ENTIRE

Flowers yellow, red or orange

Xyris montana p. 154	*Clintonia borealis* p. 165	*Erythronium americanum* p. 167	*Hemerocallis fulva* p. 168	*Iris pseudacorus* p. 177	*Piperia unalascensis* p. 186

Flowers yellow, red or orange

Caltha palustris p. 220	*Ranunculus ficaria* p. 222	*Diplotaxis muralis* p. 238	*Hieracium caespitosum* p. 390	*Tussilago farfara* p. 408	*Taraxacum officinale* p. 406

Flowers yellow, red or orange

Sarracenia
purpurea
p. 244

Hieracium
aurantiacum
p. 390

Asarum
canadense
p. 195

Flowers white

Alisma
plantago-aquatica
p. 155

Allium
tricoccum
p. 165

Erythronium
albidum
p. 166

Galanthus
nivalis
p. 167

Ornithogalum
umbellatum
p. 171

Triantha
glutinosa
p. 173

Flowers white

Zigadenus
elegans
p. 175

Goodyera
oblongifolia
p. 184

Malaxis
monophyllos
p. 185

Calla
palustris
p. 160

Sagittaria
cuneata
p. 156

Sagittaria
latifolia
p. 156

Flowers white

Drosera
linearis
p. 245

Drosera
rotundifolia
p. 246

Podophyllum
peltatum
p. 224

Sanguinaria
canadensis
p. 228

Mitella
diphylla
p. 249

Mitella
nuda
p. 250

Flowers white

Saxifraga
virginiensis
p. 250

Parnassia
palustris
p. 278

Orthilia
secunda
p. 310

Pyrola
chlorantha
p. 311

Viola
mackloskeyi
var. pallens, p. 287

Aralia
nudicaulis
p. 297

GROUP 20: LEAF... SIMPLE, ENTIRE

Flowers white

*Bellis
perennis*
p. 377

GROUP 21: LEAVES BASAL, COMPOUND

Flower colour various

| *Aquilegia brevistyla* p. 219 | *Aquilegia canadensis* p. 219 | *Coptis trifolia* p. 220 | *Cardamine diphylla* p. 235 | *Argentina anserina* p. 252 | *Fragaria vesca* p. 253 |

Flower colour various

| *Waldsteinia fragarioides* p. 258 | *Aralia nudicaulis* p. 297 |

GROUP 22: LEAVES OPPOSITE, SIMPLE, ENTIRE

Flowers blue, purple or green

| *Gentiana andrewsii* p. 317 | *Gentianopsis crinita* p. 318 | *Halenia deflexa* p. 318 | *Phlox divaricata* p. 326 | *Vinca minor* p. 322 | *Vincetoxicum rossicum* p. 322 |

Flowers blue, purple or green

| *Euonymus obovatus* p. 277 | *Agastache foeniculum* p. 336 | *Glechoma hederacea* p. 337 | *Scutellaria galericulata* p. 342 | *Chaenorrhinum minus* p. 353 | *Mimulus ringens* p. 332 |

144

GROUP 22: LEAVES OPPOSITE, SIMPLE, ENTIRE

Flowers blue, purple or green

| *Penstemon hirsutus* p. 355 | *Veronica americana* p. 357 | *Veronica arvensis* p. 358 | *Veronica officinalis* p. 358 | *Veronica persica* p. 359 | *Phryma leptostachya* p. 333 |

Flowers blue, purple or green

| *Verbena hastata* p. 334 | *Verbena stricta* p. 334 | *Pilea pumila* p. 191 | *Urtica dioica* p. 192 | *Chamaesyce serpyllifolia* p. 275 | *Euphorbia dentata* p. 276 |

Flowers blue, purple or green

| *Plantago psyllium* p. 357 | *Xanthium strumarium* p. 409 |

GROUP 23: LEAVES OPPOSITE, SIMPLE, ENTIRE

Flowers red, orange or yellow

| *Listera cordata* p. 185 | *Arisaema triphyllum* p. 159 | *Asclepias tuberosa* p. 321 | *Anagallis arvensis* p. 313 | *Heteranthera dubia* p. 163 | *Portulaca oleracea* p. 208 |

Flowers red, orange or yellow

| *Hypericum majus* p. 283 | *Hypericum perforatum* p. 283 | *Lysimachia ciliata* p. 313 | *Lysimachia nummularia* p. 314 | *Lysimachia terrestris* p. 314 | *Lysimachia thyrsiflora* p. 315 |

SIMPLE, ENTIRE

Flowers red, orange or yellow

| *Lysimachia vulgaris* p. 315 | *Aureolaria flava* p. 349 | *Bidens cernua* p. 377 | *Helianthus divaricatus* p. 389 | *Silphium perfoliatum* p. 398 |

GROUP 24: LEAVES OPPOSITE, SIMPLE, ENTIRE

Flowers pink

| *Saponaria officinalis* p. 210 | *Spergularia salina* p. 213 | *Vaccaria hispanica* p. 214 | *Geranium bicknellii* p. 271 | *Impatiens glandulifera* p. 279 | *Epilobium ciliatum* p. 295 |

Flowers pink

| *Epilobium hirsutum* p. 296 | *Rhexia virginica* p. 293 | *Triadenum fraseri* p. 284 | *Apocynum androsaemifolium* p. 319 | *Asclepias incarnata* p. 320 | *Asclepias syriaca* p. 321 |

Flowers pink

| *Vincetoxicum rossicum* p. 322 | *Decodon verticillatus* p. 291 | *Lythrum salicaria* p. 292 | *Clinopodium vulgare* p. 336 | *Galeopsis tetrahit* p. 337 | *Lamium amplexicaule* p. 338 |

Flowers pink

| *Leonurus cardiaca* p. 338 | *Mentha arvensis* p. 339 | *Monarda fistulosa* p. 340 | *Phystostegia virginiana* p. 341 | *Prunella vulgaris* p. 341 | *Stachys palustris* p. 343 |

Flowers pink

Agalinis tenuifolia p. 349   *Mimulus ringens* p. 332   *Linnaea borealis* p. 363   *Dipsacus fullonum* ssp. *sylvestris* p. 364   *Eutrochium maculatum* p. 386

Flowers white

Circaea alpina p. 294   *Circaea lutetiana* ssp. *canadensis* p. 295   *Cerastium arvense* p. 209   *Gypsophila paniculata* p. 210   *Scleranthus annuus* p. 211   *Silene latifolia* ssp. *alba* p. 211

Flowers white

Silene noctiflora p. 212   *Silene vulgaris* p. 212   *Stellaria longifolia* p. 213   *Stellaria media* p. 214   *Apocynum cannabinum* p. 320   *Euonymus obovatus* p. 277

Flowers white

Chimaphila umbellata p. 308   *Moneses uniflora* p. 309   *Gentiana andrewsii* (white form) p. 317   *Lycopus asper* p. 339   *Nepeta cataria* p. 340   *Pycanthemum virginianum* p. 342

Flowers white

Chelone glabra p. 354   *Mitchella repens* p. 362   *Phryma leptostachya* p. 333   *Verbena urticifolia* p. 335   *Galinsoga quadriradiata* p. 387   *Ageratina altissima* var. *altissima*, p. 371

GROUP 25: LEAVES OPPOSITE, SIMPLE, ENTIRE

Flowers white

*Eupatorium
perfoliatum*
p. 385

GROUP 26: LEAVES OPPOSITE, SIMPLE, LOBED

Flower colour various

| *Stylophorum diphyllum* p. 229 | *Cardamine diphylla* p. 235 | *Ambrosia trifida* p. 372 |

GROUP 27: LEAVES OPPOSITE, COMPOUND

Flower colour various

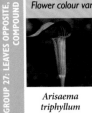

| *Arisaema triphyllum* p. 159 | *Cardamine diphylla* p. 235 | *Erodium cicutarium* p. 270 | *Geranium maculatum* p. 271 | *Geranium robertianum* p. 272 | *Tribulus terrestris* p. 268 |

GROUP 28: LEAVES WHORLED

Flower colour various

| *Lilium philadelphicum* p. 168 | *Medeola virginiana* p. 171 | *Trillium erectum* p. 173 | *Trillium grandiflorum* p. 174 | *Trillium undulatum* p. 174 | *Mollugo verticillata* p. 207 |

Flower colour various

| *Cardamine concatenata* p. 235 | *Asclepias tuberosa* p. 321 | *Impatiens glandulifera* p. 279 | *Decodon verticillatus* p. 291 | *Lythrum salicaria* p. 292 | *Chimaphila umbellata* p. 308 |

Flower colour various

| *Moneses uniflora* p. 309 | *Cornus canadensis* p. 307 | *Trientalis borealis* p. 316 | *Veronicastrum virginicum* p. 359 | *Plantago psyllium* p. 357 | *Galium aparine* p. 360 |

Flower colour various

| *Galium boreale* p. 361 | *Galium trifidum* p. 361 | *Galium triflorum* p. 362 | *Eutrochium maculatum* p. 386 | *Silphium perfoliatum* p. 398 |

CATTAIL FAMILY
Typhaceae

Cattails are erect, semi-aquatic herbs with creeping roots and flat, upright leaves. The terminal flower clusters are differentiated into male and female, with the uppermost cluster composed of male flowers. Two large spathes, one surrounding each flower cluster, fall off soon after opening. Petals are absent. The sepals of both types of flowers are reduced to small scales or bristles. Male flowers have 1–7 stamens, and female flowers have a single pistil. The wind-dispersed fruit is an achene-like follicle.

The bur-reed family (Sparganiaceae) is now included in the cattail family.

Giant Bur-reed
Sparganium eurycarpum

Habitat: Muddy shores or shallow water of ponds and lakes. **General:** Semi-aquatic; stems 0.5–2.5 m tall, zigzagged; rhizomes creeping. **Leaves:** Alternate, grass-like, stiff, 50–80 cm long, 0.6–2 cm wide, slightly keeled, sheathing at the base; underwater leaves collapsing when removed from water. **Flower Cluster:** Spike of globe-shaped heads; male heads 2–40, 1–2 cm across, near top of stem; female heads 1–6, 2–2.5 cm across, lower on stem. **Flowers:** Yellowish green, inconspicuous, numerous; sepals 3–6; male flowers with 5 stamens; female flowers with 1 pistil and 2 stigmas. **Fruit:** Achene beaked, 5–8 mm long; fruiting head 2–2.5 cm across, resembling a knight's mace.

Notes: *Sparganium* is believed to have been derived from the Greek word *sparganon*, meaning "a band," a reference to the strap-like leaves. The species name *eurycarpum* means "broad-fruited."

Narrow-leaved Cattail

Typha angustifolia

Habitat: Shallow water and wet soil along lakes, ponds and streams. General: Semi-aquatic; flowering stems 1.5–3 m tall, 5–12 mm thick near the base. Leaves: Basal (technically opposite), grass-like, 4–12 mm wide, upper leaves exceeding the flower cluster, sheathing at the base. Flower Cluster: Spike with male flowers above and female flowers below, 10–20 cm long, 0.8–1.5 cm wide; male and female sections 2–8 cm apart. Flowers: Green to brown, inconspicuous, numerous; male flowers 4–6 mm wide, sepals scaly and thread-like, stamens 1–7; female flowers 2 mm wide, pistil 1. Fruit: Follicle achene-like, 5–8 mm long, covered in white hairs.

Notes: *Typha* is the Greek name for cattail. The species name *angustifolia* means "narrow-leaved."

Common Cattail

Typha latifolia

Habitat: Marshes, ditches, sloughs and lakeshores. General: Semi-aquatic; stems 1–3 m tall, 1–2 cm thick; rhizomes white, thick, creeping. Leaves: Basal (technically opposite), sword-like, linear, flat, 0.3–3 m long, 1–2.5 cm wide, sheathing at the base; spongy when crushed. Flower Cluster: Spike with male flowers above, female flowers below, less than 1 cm apart; male portion greenish brown, 8–15 cm long, 2 cm wide, at top of stem; female portion brown, 7–20 cm long, 4–5 cm wide, on lower part of spike. Flowers: Yellowish or brownish green, inconspicuous, numerous; male flowers 5–12 mm wide, stamens 1–7; female flowers 2–3 mm wide, pistil 1. Fruit: Follicle achene-like; surface with scattered fluffy, white hairs.

Notes: The species name *latifolia* means "broad-leaved."

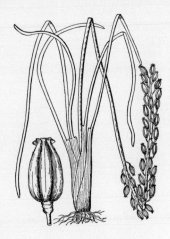

ARROW-GRASS FAMILY
Juncaginaceae

Members of this family are herbaceous plants of moist, alkaline areas. The dark green leaves are basal and somewhat fleshy. The erect flowering stem is leafless with numerous small, greenish purple flowers. The flowers have 3 petals and 3 sepals, all similar in appearance, 6 short-stalked stamens and 1 pistil with 3 or 6 feathery stigmas. The petals and sepals often fall off when the flower opens. The feathery stigmas are a common trait of wind-pollinated plants.

Species of this family contain triglochinin, which is poisonous to humans and livestock.

Seaside Arrow-grass · *Triglochin maritima*

Habitat: Moist, alkaline areas and marshes. **General:** Grass-like; flowering stems 10–100 cm tall, leafless; poisonous. **Leaves:** Basal, grass-like, 4–10, dark green, fleshy, 10–50 cm long, 2–6 mm wide; sheaths of previous year's growth present. **Flower Cluster:** Spike 10–50 cm long, bracts absent. **Flowers:** Greenish purple, 2–5 mm wide; petals 3; sepals 3; stamens 6; pistil 1; stigmas 6, feathery, collecting pollen from the wind. **Fruit:** Follicle 5–7 mm long, in clusters of 3–6, separating at maturity.

Notes: The genus name *Triglochin* comes from the Greek words *treis*, "three," and *glochin*, "point," a reference to the 3-pointed fruiting cluster. The species name means "maritime" or "seaside." • When ingested, the leaves of seaside arrow-grass release cyanide gas, which results in respiratory failure.

SCHEUCHZERIA FAMILY · Scheuchzeriaceae

The scheuchzeria family is represented by a single species worldwide. An inhabitant of cold peat bogs in the Northern Hemisphere, it is a low plant with a zigzagged stem. The linear leaves are primarily basal with a few alternate stem leaves. The flowers, borne in terminal clusters, are composed of 3 petals, 3 sepals, 6 stamens and 3 or 6 pistils. The fruit is a cluster of follicles.

Scheuchzeria, Rannoch-rush

Scheuchzeria palustris

Habitat: Cold peat bogs and marshes. **General:** Grass-like; stems 20–40 cm tall, zigzagged. **Leaves:** Basal, 10–40 cm long, somewhat tubular, sheaths membranous, 2–10 cm long; stem leaves alternate, 2–13 cm long. **Flower Cluster:** Raceme 3–10-flowered; bracts leaf-like; flower stalks 2–4 mm long. **Flowers:** Greenish white, 2–4 mm long; petals 3; sepals 3; stamens 6; pistils 3. **Fruit:** Follicle 4–8 mm long; seeds 1–2, black, 4–5 mm long.

Notes: The genus name commemorates Johann Jakob Scheuchzer (1672—1733), a Swiss botanist. The species name *palustris* means "of the marsh," a reference to the plant's habitat.

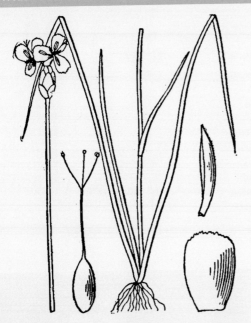

YELLOW-EYED GRASS FAMILY
Xyridaceae

The yellow-eyed grass family is a small family of tropical to subtropical, wetland plants. The alternate, parallel-veined leaves are 2-ranked and appear basal. Flowers are borne in short spikes on leafless stems. The flowers are composed of 3 petals, 3 sepals, 6 stamens (3 fertile and 3 sterile) and 1 pistil. The fruit, a capsule, contains numerous seeds.

Yellow-eyed Grass

Xyris montana

Habitat: Peat bogs, muskeg, fens and shorelines. **General:** Flowering stems 5–30 cm tall. **Leaves:** Alternate but appearing basal, tufted, 2-ranked, 4–15 cm long, 1–3 mm wide, sheaths reddish; margins smooth. **Flower Cluster:** Spikes 4–8 mm long; bracts 3–4, 1–4.5 mm long. **Flowers:** Yellow, 7–11 mm across; petals 3, clawed; sepals 3, boat-shaped; stamens 6 (staminodes 3, clawed and bearded); pistil 1. **Fruit:** Capsule 3-valved; seeds numerous.

Notes: The genus name comes from the Greek word *xyron*, "razor," a reference to the 2-edged leaves. The species name *montana* means "of the mountains."

WATER-PLANTAIN FAMILY · Alismataceae

The water-plantain family is a small group of aquatic or marsh plants with basal, parallel-veined leaves. The long-stalked leaves may be erect or floating. The flowers are composed of 3 white petals, 3 green sepals, 6 to many stamens and numerous styles. The petals fall off soon after the flower opens. The fruit is an achene.

Broad-leaved Water-plantain

Alisma plantago-aquatica

Habitat: Marshes and ditches. **General:** Semi-aquatic; flowering stems to 1.5 m tall; roots producing tubers. **Leaves:** Basal, narrowly oval to elliptic or lance-shaped, 5–30 cm long, 3–8 cm wide, long-stalked, distinctly parallel-veined, base rounded to heart-shaped. **Flower Cluster:** Panicle open, to 1 m long; branches appearing in 3–10 whorls. **Flowers:** White, 7–12 mm across; petals 3, white; sepals 3, green; stamens 6–9, yellow; pistils several. **Fruit:** Achene oblong to oval, flat, 2–3 mm long; in heads 4–6.5 mm across.

Notes: The genus name *Alisma* is Greek for "water plant." The species name *plantago-aquatica* means "water-plantain," a reference to the resemblance of the leaves to those of common plantain (*Plantago major*).

Northern Arrowhead
Sagittaria cuneata

Habitat: Shallow water along shores of lakes and ponds. **General:** Semi-aquatic; flowering stems 20–40 cm tall; juice milky. **Leaves:** Basal, arrowhead-shaped, 10–15 cm long, 1–4 cm wide, deep green above, pale below. **Flower Cluster:** Raceme 14–21 cm long, 2–10 cm wide, stem triangular; flowers in 2–10 whorls, the whorls 3-flowered. **Flowers:** White, 6–12 mm across; male or female; petals 3, white; sepals 3, green; stamens 10–18; pistils numerous; male flowers on upper part of stem. **Fruit:** Achene 1–2 mm long, in dense, head-like clusters 1–1.5 cm across.

Notes: The species name *arifolia* means "arum-leaf."

Broadleaf Arrowhead
Sagittaria latifolia

Habitat: Moist lakeshores and shallow waters. **General:** Semi-aquatic; flowering stems 30–120 cm tall; juice milky; rhizome short, bearing tubers. **Leaves:** Basal, arrow-head-shaped, 5–40 cm long, 2–25 cm wide; stalks 6–50 cm long, angular in cross-section. **Flower Cluster:** Raceme 4–29 cm long, stalk 10–59 cm long; flowers in 3–9 whorls; whorls 2–15-flowered. **Flowers:** White, 2–4 cm across; male, female or perfect; petals 3, white; sepals 3, green; stamens 25–40; pistils numerous; male flowers on upper part of stem; bracts 8–10 mm long. **Fruit:** Achene 2.5–3.5 mm long, in clusters 1–3 cm across.

Notes: The genus name *Sagittaria* comes from the Latin word *sagitta*, meaning "arrow," a reference to the leaf shape. The species name *latifolia* means "broad-leaved."

FLOWERING-RUSH FAMILY · Butomaceae

The flowering-rush family is represented by a single species worldwide. This rhizome-bearing, semi-aquatic perennial has basal leaves with parallel veins. Flowers are borne in bracted umbels. The perfect flowers have 3 petals, 3 sepals, 9 stamens and 6 pistils. The fruit is a follicle.

Flowering-rush

Butomus umbellatus

Habitat: Edges of streams, lakes and rivers; introduced from Europe into Québec in 1897 as an ornamental garden plant. **General:** Semi-aquatic; flowering stems to 1.5 m tall, triangular; rhizome stout. **Leaves:** Basal, erect or floating, 1–2.7 m long, 0.5–1 cm wide, sheathing. **Flower Cluster:** Umbel 20–30-flowered; bracts 3, purplish green, triangular; flower stalks 5–10 cm long. **Flowers:** Pink, 2–2.5 cm wide; petals 9–12 mm long; sepals 5–7.5 mm long; stamens 9, in 2 whorls; pistils 6. **Fruit:** Follicle 8–12 mm long, leathery.

Notes: The genus name comes from the Greek words *bous*, "cow," and *temno*, "to cut," a reference to the leaves, which are sharp enough to cut the mouths of cattle. The species name *umbellatus* refers to the type of flower cluster.

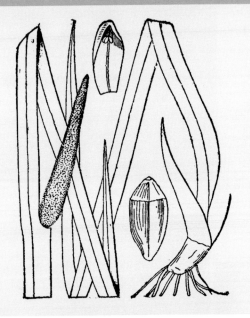

SWEET-FLAG FAMILY · Acoraceae

The sweet-flag family is a small family of wetland perennials with branched rhizomes. The sword-like leaves are parallel-veined and set edgewise to the stem. Flowers are borne in a spadix subtended by a bract (spathe) that resembles the leaves. The perfect flowers have 6 tepals, 6 stamens and 1 pistil. The fruit is a leathery berry.

Sweet-flag
Acorus americanus

Habitat: Swamps and marshes.
General: Grass-like; flowering stems 50–100 cm tall; rhizome thick, aromatic. **Leaves:** Basal, sword-like, bright green, 60–200 cm long, 0.8–2.5 cm wide, prominently 2–6-veined; margins often crisped; aromatic when crushed or bruised. **Flower Cluster:** Spadix (fleshy spike) 4–10 cm long, 2 cm thick; flowering stem to 2 m tall, resembling the leaves and producing an erect, green spathe 20–60 cm long. **Flowers:** Yellowish brown, 2–5 mm across; tepals 6, scale-like; stamens 6; pistil 1. **Fruit:** Berry 2–5 mm across, brown, hard outside, jelly-like inside.

Notes: The species name means "of America."

ARUM FAMILY · Araceae

Members of the arum family are terrestrial and aquatic
plants, and are easily identified by their characteristic
flower cluster. Flowers are produced in a fleshy spike
(spadix), and a petal-like bract (spathe) is found below
each spadix. The flowers may be perfect or unisexual.
The perfect flowers are composed of 0–6 petals and
sepals, 6 stamens and 1 pistil. Unisexual male flowers
have 1 stamen, whereas female flowers have 1 pistil.
The fruit is a fleshy berry or utricle.

The 3 subfamilies in Ontario are distinguished below:

Aroideae and Orontoideae	Lemnoideae
Plants terrestrial	Plants aquatic, free-floating
Stem and leaves present	Stem and leaves absent (plants thallus-like)
Fruit a berry	Fruit a utricle

Ornamental members of the arum family include philodendrons (*Philodendron* spp.),
dieffenbachia (*Dieffenbachia* spp.), calla-lilies (*Calla* spp.) and anthurium (*Anthurium*
spp.). Taro (*Colocasia esculenta*) is grown in the tropics for its edible tubers. A few
species, including water-lettuce (*Pistia stratiotes*), are noxious tropical weeds.

Jack-in-the-pulpit

Arisaema triphyllum

Habitat: Moist to wet, shady woods.
General: Stem erect, 20–100 cm tall; rhizome fleshy, globe-shaped, tuber-bearing.
Leaves: Opposite, 1–3 (usually 2), stalks
30–60 cm tall at maturity, compound with
3 leaflets; leaflets dark green, elliptic to oval
or diamond-shaped, 8–15 cm long, 3–7 cm
wide; margins wavy. **Flower Cluster:**
Spadix (the "jack") cylindric to club-shaped,
5–7.5 cm long, yellow; spathe (the "pulpit")
green-and-brown-striped, 10–25 cm
long, 4.5–9 cm wide, surrounding and arching over the spadix; stalk 3–20 cm long.
Flowers: Yellow; male flowers at the top,
female flowers lower on the stalk; petals
and sepals absent; stamens 2–5; pistil 1.
Fruit: Berry orange to bright red, shiny, 8–12 mm across; seeds 1–6.

Notes: The genus name comes from the Greek words *aris,* "arum," and *haema,*
"blood," a reference to the red leaves of some species. The species name *triphyllum*
means "3-leaved." • All parts of this plant are poisonous.

Water Arum
Calla palustris

Habitat: Bogs and lakeshores.
General: Semi-aquatic, 10–30 cm tall; leaves and flowering stem attached to an underwater stem; rooting at nodes; free-floating or rooted in mud or muskeg. **Leaves:** Basal, oval to round or heart-shaped, 5–30 cm long, 4–14 cm wide; stalks long. **Flower Cluster:** Spadix 1.5–2.5 cm long; spathe white, leaf-like, 2.5–8 cm long.

Flowers: White, 2–5 mm across; upper flowers male, lower flowers female or perfect; petals and sepals absent; stamens 6–12; pistil 1. **Fruit:** Berry red, pear-shaped, 6–12 mm wide, in a compact cluster 1.5–3.5 cm long.

Notes: *Calla* comes from the Greek word for "beautiful." The species name *palustris* means "of the marsh," a reference to this plant's habitat.

Eastern Skunk-cabbage
Symplocarpus foetidus

Habitat: Swamps, wet woods and low-lying areas.
General: Plant with strong, skunk-like odour; rhizome to 30 cm long; roots fleshy. **Leaves:** Basal, oval to heart-shaped, 10–60 cm long, 7–40 cm wide, pale to dark green; stalks 5–57 cm long; appearing after flowers. **Flower Cluster:** Spadix globe-shaped, 1–3 cm long, 1.5–3 cm wide, yellow; spathe hood-like, fleshy, yellowish green to purplish or reddish brown, spotted to mottled or striped, 8–15 cm long, tip pointed and incurved; appearing March–April, often when snow is still on the ground. **Flowers:** Yellowish brown, 4–7 mm across; tepals 4, yellowish to reddish purple; stamens 4; pistil 1. **Fruit:** Berry dark purplish green to reddish brown, 6–10 mm across, embedded in the fleshy spadix; in clusters 8–12 cm wide.

Notes: The genus name comes from the Greek words *symplokos*, "connected," and *karpos*, "fruit," a reference to the compound fruiting cluster. The species name *foetidus* refers to the plant's unpleasant, skunk-like odour.

SPIDERWORT FAMILY
Commelinaceae

The spiderwort family is a small family of herbs with showy flowers and parallel-veined leaves. The alternate, stalkless leaves are somewhat fleshy and form a sheath around the stem, giving the node a swollen appearance. The flowers, which last only a few hours, are borne in cymes or umbels and are subtended by spathe-like bracts. The flowers have 3 petals, 3 sepals, 3 or 6 stamens and 1 pistil. The stalk of the stamen is often covered in coloured hairs. The fruit is a capsule with 3 compartments.

Well-known ornamental species of this family include spiderworts (*Tradescantia* spp.), dayflowers (*Commelina* spp.) and wandering-Jew (*Zebrina pendula*).

Common Dayflower, Asiatic Dayflower
Commelina communis

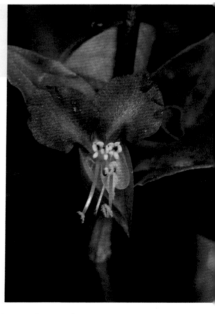

Habitat: Gardens, roadsides and waste areas; introduced ornamental from Asia.
General: Stems erect to creeping, 20–75 cm long, thick; nodes swollen, smooth to hairy, often rooting when in contact with soil; roots fibrous; susceptible to frost.
Leaves: Alternate, parallel-veined, oval to lance-shaped, 3–12 cm long, 1–4 cm wide, often hairy, underside pale green; stalkless and clasping the stem; sheath 1–2 cm long, hairy at the leaf base. **Flower Cluster:** Cyme enclosed by a spathe, on a stalk 0.8–3.5 cm long and opposite the leaf; spathe 1.5–3 cm long, with prominent, dark green veins. **Flowers:** Blue, lasting a single day; petals 3 (2 blue, 1 small and white); sepals 3, unequal, 2 partly fused; stamens 6 (3 fertile, 3 sterile); pistil 1; spathe leaf-like. **Fruit:** Capsule 4.5–8 mm long; seeds 4, brown, 2.5–4.5 mm long.

Notes: The genus name commemorates Dutch botanists Johan (1629–92) and Caspar (1667–1731) Commelin. The species name *communis* means "common."
• The common name "dayflower" refers to the flowers lasting only a single day.

Virginia Spiderwort
Tradescantia virginiana

Habitat: Open woods, fields and roadsides. **General:** Stems erect, 5–40 cm tall, rarely rooting at nodes; roots fleshy. **Leaves:** Alternate, grass-like, linear to lance-shaped, 13–37 cm long, 4–25 mm wide, usually 2–4 per stem; stalkless and sheathing at the base. **Flower Cluster:** Cyme terminal, umbel-like; bracts leaf-like; flower stalks 1.5–3.5 cm long. **Flowers:** Blue to purple (occasionally white or rose pink), showy, 1.2–2 cm across; petals 3, 1.2–1.8 cm long; sepals 3, leaf-like, 1–1.5 cm long; stamens 6, filaments bearded; pistil 1. **Fruit:** Capsule 4–7 mm long; seeds 2–3 mm long.

Notes: The genus name honours John Tradescant the Younger, a 17th-century English gardener. The species name means "of Virginia."

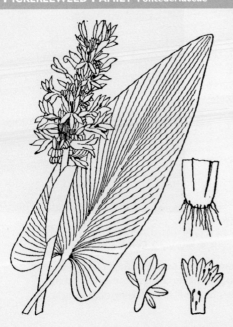

PICKERELWEED FAMILY · Pontederiaceae

The pickerelweed family is a small family of perennial, aquatic and marshland plants with creeping rhizomes. The leaves are alternate or basal and of various shapes. The showy flowers, borne singly in leaf axils or in terminal spikes, often have a bract-like spathe. Flowers have 3 petals, 3 sepals, 3 or 6 stamens and 1 pistil. The ovary is superior. The fruit is a capsule.

A well-known member of this family is water-hyacinth (*Eichornia crassipes*), an aquatic ornamental often used in residential ponds. It is a noxious weed in tropical regions of the world.

Mud Plantain
Heteranthera dubia

Habitat: Shallow water and muddy shorelines. **General:** Semi-aquatic; flowering stems less than 5 mm long. **Leaves:** Alternate but appearing opposite, crowded near the base, linear, 2–4 cm long, 1–3 mm wide; underwater leaves (when aquatic) 3–15 cm long, 1–5 mm wide. **Flower Cluster:** Solitary; spathes 1.2–4.5 cm long. **Flowers:** Yellow, trumpet-shaped, 1.5–11 cm long; petals 3; sepals 3; stamens 3 (2 short, 1 long); pistil 1; opening at dawn and wilting by dusk. **Fruit:** Capsule elongated; seeds 7–30.

Notes: The genus name comes from the Greek words *heteros*, "different," and *antheros*, "anther," a reference to the 2 types of stamens.

Pickerelweed
Pontederia cordata

Habitat: Muddy shorelines and shallow water. **General:** Semi-aquatic; flowering stems 20–120 cm tall, with 1 leaf; rhizome creeping. **Leaves:** Basal, heart to lance-shaped, thick, 6–25 cm long, 1–15 cm wide; stalks long; bases sheathing. **Flower Cluster:** Spike 2–15 cm long, densely flowered, hairy; spathe green, leaf-like, 3–17 cm long. **Flowers:** Violet to blue, showy, funnel-shaped, 7–10 mm long; petals 3, upper lip with 2 yellow dots on middle lobe; sepals 3; stamens 6 (3 long-stalked, 3 shorter); pistil 1. **Fruit:** Utricle bladder-like, 7–10 mm long; single-seeded.

Notes: The genus name commemorates Giuilo Pontedera (1688–1757), an Italian botany professor. The species name *cordata* means "heart-leaved," a reference to the leaf shape.

LILY FAMILY · Liliaceae

The lily family includes over 4000 species worldwide. Although some taxonomists recognize a single, large family, others divide the group into 20–75 smaller families. Most species have underground, overwintering parts such as bulbs, rhizomes or corms. The parallel-veined leaves may be alternate, basal, opposite or whorled. The leaf bases often form a sheath around the stem. Flowers are borne singly or in racemes, panicles or umbels. These flowers usually have 3 petals, 3 sepals, 6 stamens and 1 pistil. The petals and sepals are often the same shape and colour, making it difficult to distinguish between the two. In the genus *Trillium,* the petals and sepals are different colours. One genus, *Maianthemum,* may have 2 petals, 2 sepals, 4 stamens and 1 pistil. The fruit is a capsule with 3 compartments or a fleshy berry.

Many genera and species are grown for their ornamental value, and include lilies (*Lilium* spp.), tulips (*Tulipa* spp.), daffodils (*Narcissus* spp.), hyacinths (*Hyacinthus* spp.), hostas (*Hosta* spp.) and lily of-the-valley (*Convallaria majalis*). A few species such as asparagus (*Asparagus officinalis*) and onions, chives, leeks and garlic (*Allium* spp.) are grown for their food value.

Wild Chives
Allium schoenoprasum

Habitat: Wet meadows and lakeshores; native but widely grown as an ornamental. General: Grasslike; flowering stems 20–60 cm tall; bulb cylindric, 1.5–2 cm long, outer coats greyish brown; distinctive onion odour when crushed. Leaves: Basal, usually 2, 10–60 cm long, 2–7 mm wide, oval in cross-section, hollow. Flower Cluster: Umbel globe-shaped, 2.5–5 cm across, 30–50-flowered; flower stalks 2–6 mm long; spathes 2, 3–7-veined. Flowers: Pink to purple, bell-shaped, 8–14 mm long; petals 3; sepals 3; stamens 6, anthers purple; pistil 1; petals and sepals with dark midribs. Fruit: Capsule 3-lobed; seeds black.

Notes: Allium is the Greek name for garlic (*A. sativum*). The species name *schoenoprasum* comes from the Greek words *schoinos*, "rush," and *prasum*, "leek," a reference to the rush-like leaves.

Wild Leek
Allium tricoccum

Habitat: Rich, deciduous woods.
General: Flowering stem
10–60 cm tall; roots fibrous from
an onion-like bulb 2–6 cm long;
outer coats of bulbs brown to grey;
plant with distinctive onion odour.
Leaves: Basal, 2–3, elliptic to
lance-shaped, 20–30 cm
long, 1.5–9 cm wide; margins
smooth; absent at flowering time.
Flower Cluster: Umbel

30–50-flowered; flowering stem leafless; bracts 2, leaf-
like, below cluster; flower stalks 1–2 cm long. **Flowers:**
White to yellowish, bell-shaped, 4–7 mm long; petals 3;
sepals 3; stamens 6; pistil 1. **Fruit:** Capsule 3-lobed,
4–5 mm long; seeds 3; fruiting stem remaining erect
through winter.

Notes: The species name *tricoccum* means "3-sided."

Blue Bead-lily,
Yellow Clintonia,
Corn-lily
Clintonia borealis

Habitat: Cool, moist
woods. **General:** Flowering
stems 15–40 cm tall; rhizome
short. **Leaves:** Basal, 2–4,
oblong to elliptic, dark green,
glossy, 15–30 cm long,
4–10 cm wide; sheathing at
the base. **Flower Cluster:**
Umbel 3–10-flowered; bracts

1–3, narrow, leaf-like. **Flowers:** Greenish yellow, bell-shaped, nodding, 1.5–1.8 cm
long; petals 3; sepals 3; stamens 6; pistil 1. **Fruit:** Berry shiny, dark blue, 7–9 mm
across; fruiting stalks erect; poisonous.

Notes: The genus name commemorates DeWitt Clinton (1769–1828), governor of
New York State in the early 1800s. The species name *borealis* means "northern,"
a reference to the plant's geographic range.

European Lily of-the-valley
Convallaria majalis

Habitat: Disturbed and waste areas and forest edges; introduced ornamental from Europe.
General: Flowering stem 10–30 cm tall; rhizome creeping, often forming large colonies. **Leaves:** Alternate, 2–3 per stem, elliptic, 9–26 cm long, 2.5–13 cm wide; stalks 4–30 cm long; a few scale-like leaves may occur near the stem base.
Flower Cluster: Raceme 1-sided; bracts papery, 4–20 mm long; flower stalks 7–11 mm long.
Flowers: White, bell-shaped, nodding, fragrant, 6–9 mm long; petals 3; sepals 3; stamens 6; pistil 1.
Fruit: Berry globe-shaped, 9–11 mm across; seeds 1–5; poisonous.

Notes: The genus name comes from the ancient name for this plant, *Lilium convallium*, which was commonly called "lily-of-the-valleys." The species name *majalis* means "blooming in May."

White Trout-lily
Erythronium albidum

Habitat: Moist woods and valley bottoms. **General:** Flowering stem 7–20 cm tall; bulb 1.5–3 cm long; stolons producing large colonies of non-flowering plants with 1 basal leaf. **Leaves:** Basal, 1–2 per stem, elliptic to oval or lance-shaped, 8–22 cm long, green, irregularly mottled; margins smooth. **Flower Cluster:** Solitary, terminal, produced on plants with 2 basal leaves. **Flowers:** White tinged with pink, blue or lavender, yellow-spotted at the base, nodding, 2.2–4 cm long; petals 3, reflexed; sepals 3, reflexed; stamens 6, anthers yellow; pistil 1. **Fruit:** Capsule erect, 1–2.2 cm long.

Notes: The species name *albidum* means "white," a reference to the flower colour.

Yellow Trout-lily
Erythronium americanum

Habitat: Moist, wooded areas.
General: Flowering stems 10–18 cm
tall; bulb 1.5–2.8 cm long; stolons producing non-flowering plants with 1 basal
leaf. **Leaves:** Basal, 1–2 per stem, elliptic to lance-shaped, 5–23 cm long,
1–5 cm wide, green, mottled with
brown and purple; margins smooth.
Flower Cluster: Solitary, terminal,
produced on plants with 2 basal leaves.
Flowers: Yellow, nodding, 2–2.5 cm
wide, often purple-spotted near the base; petals 3;
sepals 3; stamens 6, anthers yellow or reddish brown;
pistil 1. **Fruit:** Capsule erect, 1.2–1.5 cm long.

Notes: The genus name comes from the Greek
word *erythros*, "red," a reference to the flower colour
of several European species. The species name means
"of America." • The speckled appearance of the
leaves, thought to resemble the coloration of a trout,
gave rise to the common name "trout-lily."

Snowdrop
Galanthus nivalis

Habitat: Moist woods; introduced ornamental from Europe and southwestern
Asia. **General:** Flowering stems 7–20 cm
tall; bulb 1.5–2.5 cm long. **Leaves:** Basal,
2–3, linear, 5–15 cm long, 3–7 mm wide;
sheathing at the base. **Flower Cluster:**
Solitary; spathe 2–3.5 cm long; flower
stalk 1.2–4 cm long. **Flowers:** Greenish
white, nodding, fragrant; petals 3, white
with green blotch at tip; sepals 3, white;
stamens 6; pistil 1. **Fruit:** Capsule globe-
shaped; fruiting stem prostrate at maturity.

Notes: The genus name comes from the
Greek words *gala*, "milk," and *anthos*, "flower," a reference to the colour of the
blossoms. The species name *nivalis* means "of the snow," alluding to the blooming
period in March and April.

Orange Day-lily, Tawny Day-lily · *Hemerocallis fulva*

Habitat: Moist woods and forest edges; introduced ornamental from Europe and Asia in the 1600s. **General:** Flowering stems to 1.5 m tall; rhizomes thick, creeping. **Leaves:** Basal, numerous, grass-like, 60–90 cm long, 1–2 cm wide, channelled or V-shaped. **Flower Cluster:** Raceme 10–20-flowered; flower stalks 3–6 mm long. **Flowers:** Orange, 7–9 cm across, lasting a single day; petals 3, often reflexed; sepals 3; stamens 6; pistil 1. **Fruit:** Capsule cylindric, 2–2.5 cm long, 1.2–1.5 cm wide; fruit and seeds rarely produced.

Notes: The genus name comes from the Greek words *hemera*, "day," and *kallos*, "beauty," a reference to the flowers lasting only one day. The species name *fulva* means "tawny," referring to the flower colour.

Wood Lily
Lilium philadelphicum

Habitat: Dry, open woods and meadows. **General:** Stems 20–90 cm tall, leafy; bulb scaly, 2–3 cm across. **Leaves:** Alternate or in whorls of 3–8, stalkless, lance-shaped, 5–10 cm long, 1–2.5 cm wide. **Flower Cluster:** Solitary (occasionally 2–5). **Flowers:** Reddish orange, bell-shaped, 5.5–8.5 cm across; petals and sepals 3 each, similar in appearance, prominently ribbed, spotted with purplish brown near the base; stamens 6; pistil 1. **Fruit:** Capsule 1–2 cm long.

Notes: The species name mean "of Philadelphia," a reference to the location where the plant was first collected.

Wild Lily of-the-valley
Maianthemum canadense

Habitat: Dry to moist forests. **General:** Stems 10–25 cm tall, zigzagged, leafy; rhizome slender, creeping. **Leaves:** Alternate, 1–3 (commonly 2), oval to oblong or lance-shaped, 3–10 cm long, 3–4.5 cm wide, base heart-shaped; stalks 1–7 mm long; non-flowering plants with 1 basal leaf. **Flower Cluster:** Raceme terminal, 2–5 cm long, 12–25-flowered; flower stalks 3–7 mm long. **Flowers:** White, 4–6 mm wide; petals 2; sepals 2; stamens 4; pistil 1. **Fruit:** Berry pale to deep red, speckled, 3–6 mm across; edible but should not be eaten in large quantities.

Notes: *Maianthemum* comes from the Greek words *maios*, "May," and anthemon, "flower."

False Solomon's-seal, False Spikenard · *Maianthemum racemosum*

Habitat: Moist, wooded areas. **General:** Stems 30–125 cm tall, slightly zigzagged; rhizome thick, fleshy, creeping. **Leaves:** Alternate, 5–12, elliptic to oval or broadly lance-shaped, 5–17 cm long, 2–8 cm wide; short-stalked to stalkless, clasping the stem. **Flower Cluster:** Panicle oval to pyramidal, 3–15 cm long, 70–250-flowered. **Flowers:** White, numerous, 2–6 mm across; petals 3; sepals 3; stamens 6; pistil 1. **Fruit:** Berry red, dotted with purple, 3–6 mm across; edible but should not be eaten in large quantities.

Notes: The species name *racemosum* means "having racemes," a reference to the panicle being composed of racemes.

Star-flowered Solomon's-seal
Maianthemum stellatum

Habitat: Moist, open areas.
General: Stems arched, zigzagged, 20–50 cm tall; rhizome fleshy, creeping.
Leaves: Alternate, 6–12, oblong to lance-shaped, 3–15 cm long, 1–5 cm wide, pale green; stalkless and clasping. **Flower Cluster:** Raceme terminal, 1.5–5 cm long, 5–15-flowered; flower stalks 0.6–1.2 cm long. **Flowers:** White, star-like, 0.6–1 mm across; petals 3; sepals 3; stamens 6; pistil 1. **Fruit:** Berry 0.6–1 cm across, green with purple or black stripes when young, maturing to red or black.

Notes: The species name means "star," a reference to the shape of the flowers.

Three-leaved Solomon's-seal
Maianthemum trifolium

Habitat: Moist forests. **General:** Stems 5–20 cm tall, slender; rhizome branched, rooting only at nodes. **Leaves:** Alternate, 2–4 (commonly 3), oval, 4–12 cm long, 2.5–4 cm wide; margins smooth; stalkless. **Flower Cluster:** Raceme 3–15-flowered; stem zigzagged; flower stalks 1–3 mm long. **Flowers:** White, 4–6 mm across; petals 3; sepals 3; stamens 6; pistil 1. **Fruit:** Berry 4–6 mm across, green with red spots when young, maturing to dark red.

Notes: The species name *trifolium*, "3-leaved," refers to the number of stem leaves.

Indian Cucumber-root
Medeola virginiana

Habitat: Moist woods and slopes. **General:** Stems 20–90 cm tall, unbranched; rhizome 3–8 cm long, 1 cm wide, white, tuber-like. **Leaves:** Whorled; lowest whorl with 5–9 leaves, 6–16 cm long, 1.5–5 cm wide; upper whorl usually 3 (occasionally 5) leaves, 2.5–5 cm long, 1.5–4 cm wide; margins smooth; stalkless. **Flower Cluster:** Umbel 2–9-flowered. **Flowers:** Yellowish green, 1–1.5 cm long; petals 3, reflexed; sepals 3, reflexed; stamens 6, anthers reddish; pistil 1. **Fruit:** Berry 5–14 mm across, dark purple to black; seeds 2–4 mm long, shiny, brown.

Notes: The edible rhizomes of this species have the taste and scent of cucumber, hence the common name.

Star-of-Bethlehem, Sleepydick
Ornithogalum umbellatum

Habitat: Abandoned lots and farms; introduced ornamental from Europe and Asia. **General:** Flowering stem 20–30 cm tall; bulb ovoid, 1–2 cm long, 1–2.5 cm wide, outer coats white to pale brown, papery, poisonous. **Leaves:** Basal, 4–9, linear to lance-shaped, 20–30 cm long, 3–5 mm wide. **Flower Cluster:** Corymb flat-topped, 8–20-flowered; bracts white, papery, 1–4 cm long. **Flowers:** White; sepals and petals 3 each, green-striped, 1.5–2.2 cm long, 7–8 mm wide; stamens 6, the outer often shorter than the inner; pistil 1; opening on sunny days at noon and closing at sunset. **Fruit:** Capsule oblong to ovoid, 6-angled; seeds numerous.

Notes: The species name *umbellatum* refers to the umbel-like cluster of flowers.

Hairy Solomon's-seal
Polygonatum pubescens

Habitat: Moist, rich forests.
General: Stems erect, 50–110 cm tall;
rhizome horizontal, gnarled, 1–1.8 cm
thick. **Leaves:** Alternate, elliptic to
oval, 4–15 cm long, 1–7.5 cm wide,
underside prominently 3–9-nerved
and hairy; margins smooth; stalkless.
Flower Cluster: Raceme axillary,
drooping, 1–3-flowered. **Flowers:**
Greenish yellow, tubular, 0.7–1.5 cm
long; petals 3; sepals 3; stamens 6; pistil 1. **Fruit:** Berry dark blue to black,
6–9 mm across; seeds several; fruiting
stalk to 1.3 cm long.

Notes: The genus name comes from the Greek words *poly*, "many," and *gony*,
"knee," referring to the jointed rhizome. The species name *pubescens* means "hairy,"
a reference to the hairs along the veins on the underside of the leaves.

Twisted-stalk · *Streptopus amplexifolius*

Habitat: Moist, wooded areas. **General:**
Stems stout, branched, 30–100 cm tall; rhizome
short, thick. **Leaves:** Alternate, oval to lance-
shaped, 4–12 cm long, 1–6 cm wide, base heart-
shaped and clasping the stem; stalkless. **Flower
Cluster:** Solitary or in pairs, on twisted and
bent stalks in leaf axils. **Flowers:** Whitish
green, bell-shaped, 0.8–1.2 cm long; petals 3;
sepals 3; stamens 6; pistil 1. **Fruit:** Berry red,
1–1.5 cm long, tasteless.

Notes: The genus name comes from the Greek
word *streptos*, meaning "bent" or "twisted."

The species name
amplexifolius means
"clasping leaves."

False Asphodel · *Triantha glutinosa*

Habitat: Calcareous marshes and shorelines.
General: Stems 10–50 cm tall, sticky near the top;
rootstock short. **Leaves:** Basal, linear, 5–20 cm long,
3–8 mm wide, somewhat 2-ranked; may have
1 smaller leaf near midstem. **Flower Cluster:**
Raceme terminal, 2–5 cm long, 3–30-flowered, the
flowers in groups of 3; stalks sticky. **Flowers:** White,
3–5 mm long; petals 3;
sepals 3; stamens 6; pistil 1.
Fruit: Capsule yellow or
red, 5–6 mm long.

Notes: The genus name
comes from the Greek
words *tri*, "three," and
anthos, "flower," alluding
to the flowers appearing
in groups of 3.

Wake-robin, Red Trillium

Trillium erectum

Habitat: Moist, wooded areas. **General:**
Stems 15–60 cm tall; rhizome short, thick,
unpleasant smelling. **Leaves:** Whorl of
3 atop stem, broadly diamond-shaped,
5–20 cm long, 5–20 cm wide, dark green,
net-veined; margins smooth. **Flower
Cluster:** Solitary, arising from whorl of
leaves. **Flowers:** Reddish brown to maroon,
6–6.5 cm wide, ill-scented; petals 3; sepals 3,
green; stamens 6, anthers purple; pistil 1.
Fruit: Berry dark maroon, globe-shaped
to pyramidal, 1–1.6 cm long; juicy.

Notes: The genus name comes from the
Latin word *trilix*, "triple," a reference to the
3 sepals and petals. The species name *erectum* refers to the erect flower. • The
flowers are pollinated by carrion flies, which are attracted by the foul odour.

White Trillium
Trillium grandiflorum

Habitat: Rich woods. **General:** Stems 15–30 cm tall; rhizome short. **Leaves:** Whorl of 3 atop stem, round to diamond-shaped, 8–20 cm long, 8–15 cm wide; stalkless. **Flower Cluster:** Solitary, arising from whorl of leaves; flower stalks 5–8 cm long. **Flowers:** White, fading to pink with age, erect, waxy, 5–10 cm across; petals 3; sepals 3, green; stamens 6, 0.9–2.5 cm long, anthers yellow; pistil 1. **Fruit:** Berry pale green to red, 6-lobed, 12–16 mm long, mealy textured.

Notes: The species name *grandiflorum* means "large-flowered." • White trillium has been the provincial flower of Ontario since 1937.

Painted Trillium
Trillium undulatum

Habitat: Deep humus in shaded, coniferous forests. **General:** Stems 10–40 cm tall; rhizome short. **Leaves:** Whorl of 3 atop stem, oval, 12–18 cm long, 8–20 cm wide, glossy; stalks 0.4–1.7 cm long. **Flower Cluster:** Solitary, arising from whorl of leaves; flower stalk 2–5 cm long. **Flowers:** White with V-shaped, pinkish red throat; petals 3, 2–4 cm long, margins wavy; sepals 3; stamens 6; pistil 1. **Fruit:** Berry scarlet, 3-lobed, 1–2 cm long, fleshy.

Notes: The species name *undulatum* means "wavy," a reference to the petal margins.

Large-flowered Bellwort
Uvularia grandiflora

Habitat: Rich woods. General: Stems forked, 20–75 cm tall, hairless; rhizome about 1 cm long, bearing a cluster of fleshy roots. Leaves: Alternate, elliptic to oval or oblong, 6–13.5 cm long, 2–6.5 cm wide, clasping; 1–2 sheath-like leaves below the stem fork; non-flowering branch with 4–8 leaves, flowering stem with several.
Flower Cluster: Raceme drooping, 1–4-flowered; stalk 1–2.5 cm long, with 1 clasping bract. Flowers: Yellow, bell-shaped, 2.5–5 cm long; petals 3; sepals 3; stamens 6, 1–2.5 cm long; pistil 1. Fruit: Capsule oblong to pyramidal, 3-lobed, 1–1.5 cm long.

Notes: The genus name comes from the Latin word *uvula*, the hanging structure at the back of the human throat, and refers to the hanging flowers. The species name *grandiflora* means "large-flowered."

White Camas
Zigadenus elegans

Habitat: Dry to moist meadows and open woods. General: Flowering stem 20–60 cm tall; bulb oval. Leaves: Primarily basal, grass-like, pale green, V-shaped in cross-section, 10–20 cm long, 1–6 mm wide; flowering stem may have a few small, scale-like leaves. Flower Cluster: Raceme terminal, 5–20 cm long. Flowers: Greenish white, showy, 1.4–2 cm wide; petals 3; sepals 3; stamens 6; pistil 1; dark gland at base of each petal and sepal. Fruit: Capsule egg-shaped, 1.5–2 cm long.

Notes: This plant is slightly poisonous to humans and livestock, and large doses can be lethal.

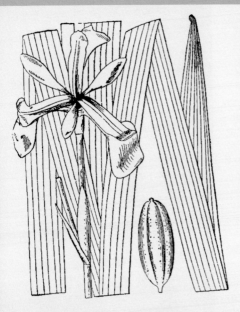

IRIS FAMILY · Iridaceae

Irises are herbaceous plants with rhizomatous or bulb-like roots. The basal and alternate leaves are set edgewise to the stem, a condition technically referred to as equitant. These sheathing leaves have parallel veins. Flowers occur in cymes, umbels or spikes, or are solitary. The irregular flowers have 3 petals, 3 sepals, 3 stamens and 1 pistil with 3 carpels. In the genus *Iris*, the styles are often petal-like. The fruit is a capsule.

Well-known horticultural members of this family include irises (*Iris* spp.), gladiolas (*Gladiolus* spp.), freesias (*Freesia* spp.) and crocuses (*Crocus* spp.).

Dwarf Lake Iris

Iris lacustris

Habitat: Moist, sandy areas along lakeshores. **General:** Stems 1–4 cm tall; rhizome cord-like, to 2 m long, scaly. **Leaves:** Basal, grass-like, 8–12 per plant, linear, 4–6 cm long, 6–8 mm wide, enlarging to 16 cm long and 1 cm wide at maturity. **Flower Cluster:** Solitary; spathes green, to 4.5 cm long. **Flowers:** Blue with yellow tube, funnel-shaped, 1–2 cm long; petals 3; sepals 3; stamens 3; pistil 1. **Fruit:** Capsule 1.2–8 mm long; seeds dark brown.

Notes: The species name *lacustris* means "growing by the lake."

Yellow Iris, Yellow Flag
Iris pseudacorus

Habitat: Swamps, riverbanks and lake-shores; introduced from Europe in about 1900 for erosion control. **General:** Clump-forming; stems green, solid, 40–150 cm tall; rhizome pink, branched, 1–4 cm across, roots fleshy, 10–30 cm long; fibrous remains of old leaves at the base. **Leaves:** Basal, about 10, set edgewise to stem, green, sword-like, 40–100 cm long, 1–3 cm wide, midrib prominent and raised. **Flower Cluster:** Cyme 4–12-flowered; bracts green with brown margins; outer bract keeled, 4.9–5.1 cm long, 0.7–1 cm wide; inner bract 6–9 cm long. **Flowers:** Yellow, showy, 8–10 cm across; petals 3, yellow, 2–3 cm long; sepals 3, yellow or creamy white streaked with purple or brown, 5–7.5 cm long, 3–4 cm wide; stamens 3; pistil 1; styles 3, 3–4 cm long, branches yellow. **Fruit:** Capsule 3-angled, 3.5–8 cm long; seeds D-shaped, pale to dark brown, 6–7 mm across.

Notes: The species name *pseudacorus* means false acorus, a reference to the plant's resemblance to sweet-flag (*Acorus americanus*), p. 158, when not in flower.

Blue Flag · *Iris versicolor*

Habitat: Wet shorelines, marshes and swamps. **General:** Grass-like; stems 20–80 cm tall; rhizome pinkish white, 1–2.5 cm across, poisonous. **Leaves:** Basal and alternate on the stem, sword-shaped, 20–80 cm long, 1–3 cm wide, base sheathing; stem leaves 1–2, linear to lance-shaped; old leaves remaining at the base. **Flower Cluster:** Raceme 2–4-flowered; spathes 3–6 cm long, the outer shorter than the inner; flower stalks 2–8 cm long. **Flowers:** Violet to blue, 6–8 cm wide; petals 3, erect, smaller than sepals; sepals 3, petal-like, striped yellowish green at the base; stamens 3; pistil 1, style 3–3.5 cm long. **Fruit:** Capsule 1.5–6 cm long, remaining closed at maturity; seeds dark brown, D-shaped, 5–8 mm long.

Notes: The genus is named for Iris, the Greek goddess of rainbows. The species name *versicolor* means "various colours," a reference to the flower colours. • The common name "flag" is from the Old English *flagge*, meaning "rush" or "reed," which the leaves resemble.

Blue-eyed Grass
Sisyrinchium montanum

Habitat: Moist, open areas.
General: Grass-like, inconspicuous when not in flower; stems 10–30 cm tall, winged, 2–4 mm wide; roots fibrous. **Leaves:** Basal, grass-like, 2–6 cm long, 1–4 mm wide, set edgewise to the stem. **Flower Cluster:** Umbel 2–8-flowered; bract spathe-like, pale green or purplish. **Flowers:** Blue to violet with yellow throat, 5–9 mm wide; petals 3; sepals 3; stamens 3; pistil 1. **Fruit:** Capsule globe-shaped, tan to dark brown, 4–7 mm across.

Notes: The genus name *Sisyrinchium* is the Greek name for a bulbous plant described by Theophrastus (371–286 BCE), a student of Aristotle. The species name *montanum* means "of the mountains."

ORCHID FAMILY
Orchidaceae

The orchid family is the largest family of flowering plants in the world. The leaves are parallel-veined and may be alternate, basal, opposite or whorled. They vary in shape, from grass-like to broad, and in a few genera are reduced to bladeless sheaths surrounding the stem. The flowers are highly specialized and are often only pollinated by a single insect species. The irregular flowers are composed of 3 petals, 3 sepals, 1–2 stamens and 1 pistil. Orchid petals are of 2 types: 2 lateral petals and a labellum or lip, usually located at the bottom of the flower. The labellum may be pouch- or sac-like, and a spur must be present as an extension of the labellum. Sepals are green or brightly coloured and may be fused. The stamens produce pollen sticky clusters called pollinia. Unlike most other plants, the stamens are fused to the stigma and the style of the ovary, forming a structure called the column. The underside of the column is the called the stigmatic surface. An insect must deposit pollen on this surface to ensure pollination. The fruit is a capsule with numerous tiny seeds.

Orchids have a symbiotic association with mycorrhizal fungi in the soil. The thickened roots of orchids are covered in fungal tissue that provides nutrients and water to the plant. Some species require the presence of a specific fungus in order to survive, and if the fungus is absent in the soil at the time of germination, the seedling dies.

Grass-pink
Calopogon tuberosus

Habitat: Moist, coniferous swamps.
General: Flowering stems 10–70 cm tall, leafless; corm bulb-like, 0.8–3.1 cm long.
Leaves: Basal, 1, grass-like, 2–50 mm long, 2–35 mm wide; may have 1–2 small, scale-like leaves at the base. **Flower Cluster:** Raceme 2–15-flowered. **Flowers:** Pink, showy, 2–4 cm wide; petals 3, the lip 1.5–2 cm long, fan-shaped, covered in white hairs with pink and yellow tips; sepals 3; stamen 1; pistil 1. **Fruit:** Capsule ovoid to elliptic, 1.3–3 cm long.

Notes: Unlike most orchid species, the lip of grass-pink is at the top of the flower.
• The genus name comes from the Greek words *kallos*, "beautiful," and *pogon*, "beard," a reference to the showy, bearded lip. The species name refers to the tuberous root.

Venus'-slipper, Calypso Orchid
Calypso bulbosa

Habitat: Coniferous forests.
General: Flowering stems 5–15 cm tall; roots few, fleshy.
Leaves: Basal, 1, oval, 2–4 cm long. **Flower Cluster:** Solitary, at top of flowering stem.
Flowers: Whitish pink to purple, 1.2–2.5 cm long; lateral petals 2, pale purple; lip whitish pink, sac-like, with purple markings and yellow hairs within; sepals 3, pale purple. **Fruit:** Capsule about 1.2 cm long.

Notes: This species is very rare in Ontario.

Bracted Green Orchid
Coeloglossum viride

Habitat: Moist, wooded areas.
General: Stems 20–50 cm tall, leafy; rhizome fleshy. **Leaves:** Alternate, grass-like, 5–12 cm long, 1–5 cm wide, reduced in size upward, sheathing. **Flower Cluster:** Raceme 5–20 cm long; bracts leaf-like, the lower to 5 cm long. **Flowers:** Green, often tinged reddish brown or purple, 4–9 mm across; petals 3, the lip 6–10 mm long and 3-toothed at the tip; spur 2–3 mm long; sepals 3; stamen 1; pistil 1. **Fruit:** Capsule 7–14 mm long.

Notes: *Coeloglossum* means "hollow tongue," a reference to the flower spur. The species name *viride* means "green." • The common name refers to the large, green bracts below each flower.

Spotted Coralroot
Corallorhiza maculata

Habitat: Moist, deciduous and coniferous forests.
General: Stems brownish purple, 20–50 cm tall; rhizome coral-like. **Leaves:** Reduced to bladeless sheaths around the stem base. **Flower Cluster:** Raceme 10–65 cm long, 6–41-flowered. **Flowers:** Purplish white, spotted; lateral petals 2, purplish white; lip white, purple-spotted, 0.8–1.2 cm long; spur short; sepals 3, purplish white; stamen 1; pistil 1. **Fruit:** Capsule 0.9–2.4 cm long.

Notes: *Corallorhiza* means "coral root," a reference to the underground rhizome. The species name *maculata* means "spotted."

Striped Coralroot
Corallorhiza striata

Habitat: Moist, coniferous and deciduous forests. **General:** Stems purplish to yellowish brown, 15–40 cm tall; rhizome coral-like. **Leaves:** Reduced to bladeless sheaths around the stem base. **Flower Cluster:** Raceme 5–67 cm long, 2–35-flowered. **Flowers:** Purplish yellow, 1–1.5 cm long; 3 sepals and 2 lateral petals 8–16 mm long, yellowish to purplish with 3 conspicuous purple stripes; lip 3–17 mm long, white to pink with 5 purple stripes; stamen 1; pistil 1. **Fruit:** Capsule elliptic, 1.1–3 cm long.

Notes: The species name *striata* refers to the striped corolla.

Pale Coralroot · *Corallorhiza trifida*

Habitat: Moist, deciduous forests. **General:** Flowering stem greenish yellow, 10–30 cm tall; rhizome coral-like. **Leaves:** Basal, reduced to scale-like sheaths. **Flower Cluster:** Raceme 2–7 cm long, 3–18-flowered. **Flowers:** Pale yellowish green, 4–6 mm across; lateral petals 2; lip white, spotted red or purple; spur inconspicuous, fused to ovary; sepals 3; stamen 1; pistil 1. **Fruit:** Capsule elliptic, 4.5–15 mm long.

Notes: The genus name *Corallorrhiza* refers to the coral-like rhizome of this plant. The species name *trifida* means "3-parted."
• This species is the most common orchid in North America.

Moccasin Flower, Stemless Lady's-slipper · *Cypripedium acaule*

Habitat: Moist to dry forests. **General:** Flowering stem 20–40 cm tall; rhizome short. **Leaves:** Basal, 2, nearly opposite, elliptic, 10–20 cm long, 2–8 cm wide, dark green above, pale below, prominently ribbed. **Flower Cluster:** Solitary (occasionally 2–3). **Flowers:** Pink, showy, fragrant; 3 sepals and 2 lateral petals yellowish green to greenish brown, 3–5 cm long; lip sac-like, 3–6 cm long, pink with red veins; stamen 1; pistil 1. **Fruit:** Capsule 4–5 cm long.

Notes: The genus name *Cypripedium* comes from the Greek words *kypris*, "Venus," and *pedilon*, "slipper," a reference to the shape of the flowers.

Ram's-head Lady's-slipper

Cypripedium arietinum

Habitat: Dry to moist, open, coniferous forests. **General:** Stems 10–35 cm tall. **Leaves:** Alternate, 3–4 per stem, elliptic to lance-shaped, 5–11 cm long, 1.3–3.5 cm wide. **Flower Cluster:** Solitary. **Flowers:** Reddish brown and white, 1–1.6 cm long; lateral petals 2, reddish brown, 1.1–2.4 cm long; lip sac-like, white with red nerves and green tip, 1–1.6 cm long; sepals 3, reddish brown, 1.2–2.5 cm long; stamen 1; pistil 1. **Fruit:** Capsule 0.5–1.2 cm long.

Notes: The species name *arietinum* means "like a ram's head."

Yellow Lady's-slipper
Cypripedium parviflorum

Habitat: Moist woods and meadows.
General: Stems 10–40 cm tall, leafy; rhizome stout, roots coarse. **Leaves:** Alternate, 3–4 per stem, oval to lance-shaped, 5–15 cm long, 2.5–9 cm wide, prominently veined and plaited; stalks sheathing at the base. **Flower Cluster:** Solitary (occasionally 2), terminal. **Flowers:** Yellow, showy, 2.2–3.4 cm long; 3 sepals and 2 lateral petals twisted, 3–6 cm long, greenish brown, purple-striped; lip bright yellow with red or purple spots, slipper- or pouch-like, 2–5 cm long. **Fruit:** Capsule 1–2.5 cm long.

Notes: The species name *parviflorum* means "small-flowered."

Showy Lady's-slipper
Cypripedium reginae

Habitat: Moist, coniferous forests.
General: Stems 21–90 cm tall; rhizome with numerous fibrous roots.
Leaves: Alternate, 3–9 per stem, elliptic to lance-shaped, 10–27 cm long, 5–16 cm wide; clasping at the base. **Flower Cluster:** Solitary or raceme of 2–4 flowers.
Flowers: Pale pink, 2.5–5.3 cm long; lateral petals 2, white, 2.5–4.7 cm long; lip sac-like, pale pink to magenta, 2.5–5.3 cm long; sepals 3, white, 2.4–4.5 cm long; stamen 1; pistil 1. **Fruit:** Capsule 1–4.5 cm long.

Notes: The species name *reginae* means "queen," a reference to the tall stems and large flowers.

Helleborine · *Epipactus helleborine*

Habitat: Open woods, roadsides and waste areas; introduced ornamental from Europe. **General:** Stems 25–80 cm tall; roots thickened. **Leaves:** Alternate, 3–10 per stem, elliptic to lance-shaped, 2–18 cm long, 1.5–8.5 cm wide, reduced in size upward; margins smooth; stalk-less and clasping. **Flower Cluster:** Raceme often 1-sided, 15–50-flowered, 10–30 cm long; bracts 2–6 cm long, 1–2 cm wide. **Flowers:** Greenish purple; 3 sepals and 2 lateral petals 10–14 mm long, 5–6 mm wide, purple-veined; lip greenish purple, 8–11 mm long, 4–6 mm wide, bowl-shaped. **Fruit:** Capsule oblong to oval, 9–14 mm long.

Notes: The genus and species names refer to the plant's resemblance to hellebores (*Helleborus* spp.), showy members of the buttercup family (Ranunculaceae).

Rattlesnake-plantain
Goodyera oblongifolia

Habitat: Coniferous woods. **General:** Flowering stems 7–38 cm tall, leafless; rhizome short. **Leaves:** Basal, evergreen, broadly oval, 2.5–10 cm long, 1.3–3.5 cm wide, dark green with white veins. **Flower Cluster:** Spike 5–48-flowered; bracts sheathing. **Flowers:** White tinged with green, 5–8 mm long; 3 sepals and 2 lateral petals forming a hood 5–10 mm long; lip boat-shaped, 5–8 mm long; stamen 1; pistil 1. **Fruit:** Capsule erect.

Notes: The genus name commemorates John Goodyer (1592–1664), a British botanist. • The common name refers to the resemblance of the leaves to the skin of a snake.

Heart-leaved Twayblade

Listera cordata

Habitat: Moist woods and bogs. **General:** Stems 5–33 cm tall, green to reddish purple. **Leaves:** Opposite, 2, heart- to kidney-shaped, 1–4 cm long, attached below midstem. **Flower Cluster:** Raceme 2–10 cm long, 5–25-flowered; bracts 1–1.5 mm long. **Flowers:** Reddish purple to yellowish green; lateral petals 2, 2–3 mm long; lip oblong with 2 small lobes at the base, purplish, 4–5 mm long; sepals 3, 2–3 mm long; stamen 1; pistil 1. **Fruit:** Capsule globe-shaped, 4–6 mm across.

Notes: The genus name honours English naturalist and physician Martin Lister (1639–1712). The species name *cordata* means "heart-shaped."

Adder's-mouth

Malaxis monophyllos

Habitat: Moist woods and bogs. **General:** Stems 3–25 cm tall; corm 1–10 mm across. **Leaves:** Basal, 1–2, yellowish green, 1.5–10 cm long, 0.5–4 cm wide, clasping at or below midstem. **Flower Cluster:** Raceme 2–10 cm long; flower stalks 1–2 mm long; bracts 1–4 mm long. **Flowers:** Greenish white, about 2 mm long; lateral petals 2, 1.5–2.5 mm long; lip 3-lobed, tapered, 2–3 mm long; sepals 3, 1–2.5 mm long; stamen 1; pistil 1. **Fruit:** Capsule erect.

Notes: The species name *monophyllos* means "one-leaved," a reference to the single leaf. • The common name comes from the shape of the lip, which resembles an adder's tongue.

Alaska Bog Orchid
Piperia unalascensis

Habitat: Moist woods and meadows.
General: Leafless flowering stem
20–50 cm tall; roots fleshy, tuber-
bearing. **Leaves:** Basal, 2–4, lance-
shaped, 6–12 cm long, 1–3 cm wide;
yellow and withered at flowering time.
Flower Cluster: Raceme 10–30 cm
long. **Flowers:** Yellowish green with
purple marks, 3–6 mm across, odour
unpleasant; 3 sepals and 2 lateral petals oval to lance-shaped; lip 3–5 mm
long, oval to lance-shaped, widest at
the base; spur club-shaped, 1–3 mm
long. **Fruit:** Capsule erect.

Notes: This species is rare in
Ontario.

Tall White Bog Orchid
Platanthera dilatata

Habitat: Bogs and wet woods.
General: Stems 11–130 cm tall, leafy,
hollow; roots fleshy. **Leaves:** Alternate, 3–12 per stem, linear to lance-
shaped, 3.5–32 cm long, 0.3–7 cm
wide. **Flower Cluster:** Raceme
10–30 cm long; bracts 1.5–4 cm long.
Flowers: White to yellowish white,
spicy-scented; lateral petals 2, white,
sickle-shaped; lip 5–8 mm long, widest
at the base; spur 3–8 mm long; sepals 3,
white; stamen 1; pistil 1. **Fruit:**
Capsule erect.

Notes: The species name *dilatata*
means "expanded" or "widened,"
a reference to the widened base of
the lip.

Northern Green Orchid
Platanthera hyperborea

Habitat: Bogs, wet meadows and moist woods. **General:** Stems 7–70 cm tall, leafy; roots fleshy. **Leaves:** Alternate, lance-shaped, 2–14 cm long, 0.4–4 cm wide, reduced in size upward and gradually becoming floral bracts. **Flower Cluster:** Spike or raceme, crowded, 2–10 cm long. **Flowers:** Yellowish green, 6–9 mm wide; upper sepal oval, 3 sepals and 2 lateral petals lance-shaped; lip lance-shaped, 4–7 mm long; spur cylindric to club-shaped, 4–6 mm long. **Fruit:** Capsule erect.

Notes: The genus name *Platanthera* means "wide or broad anther." The species name *hyperborea* means "northern."

Rose Pogonia
Pogonia ophioglossoides

Habitat: Open, boggy meadows. **General:** Stems 5–70 cm tall; rhizome short. **Leaves:** Single, clasping leaf near midstem, lance-shaped, 1.4–12 cm long, 0.4–3.2 cm wide. **Flower Cluster:** Solitary (occasionally 2); bract 0.7–3.7 cm long. **Flowers:** Pink, 3 cm long, 1.5 cm wide; 2 lateral petals whitish pink, 1.3–2.5 cm long; lip 1.2–2.5 cm long, pink with red veins and 3 rows of yellow hairs; sepals 3, whitish pink, 1.4–2.3 cm long; stamen 1; pistil 1. **Fruit:** Capsule erect, 1.4–3 cm long.

Notes: The genus name comes from the Greek word *pogon*, "beard," a reference to the hairy lip. The species name *ophioglossoides* means "like *Ophioglossum*," a fern-like plant with a single leaf near the middle of the stem.

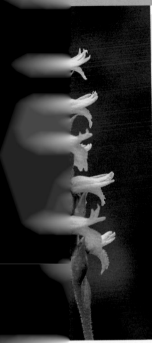

Nodding Ladies'-tresses
Spiranthes cernua

Habitat: Damp, open woods and moist meadows.
General: Stem 10–50 cm tall; roots fleshy, thickened.
Leaves: Basal and alternate, oblong to lance-shaped,
5–30 cm long, 0.5–2 cm wide; upper leaves reduced
to small scales. **Flower Cluster:** Raceme densely
flowered, 4–15 cm long, 3–4 flowers per spiral.
Flowers: Creamy white, nodding, fragrant, spirally
arranged in 3–4 rows; 3 sepals and 2 lateral petals triangular, 8–14 mm long; upper sepal and lateral petals
fused to form a hood; lip oval to oblong, 7–15 mm
long, crinkled near the tip; stamen 1; pistil 1. **Fruit:**
Capsule erect, many-seeded.

Notes: The genus name comes from the Greek
words *speira*, "coil," and *anthos*, "flower," a reference
to the twisted flower cluster. The species name *cernua*
means "nodding," referring to the flowers.

Ladies'-tresses
Spiranthes romanzoffiana

Habitat: Wet meadows. **General:** Stems
8–55 cm tall, leafy; roots fleshy. **Leaves:**
Alternate, linear to elliptic or lance-shaped,
5–15 cm long, less than 3 cm wide, reduced in
size upward. **Flower Cluster:** Raceme spike-
like, crowded, 3–8 cm long, 3 flowers per spiral. **Flowers:** White, fragrant, 6–10 mm long,
spirally arranged; 3 sepals and 2 lateral petals
partly fused to form a hood; lip fiddle-shaped,
5–12 mm long, bent downward; spur absent;
stamen 1; pistil 1. **Fruit:** Capsule erect, many-
seeded.

Notes: The species name commemorates
Count Nikolai Romanzoff (1754–1826),
a Russian statesman and a supporter of
scientific exploration. • The spirally arranged
flowers resemble a braid of hair, giving the
plant its common name "ladies'-tresses."

LIZARD'S-TAIL FAMILY

Saururaceae

Members of this family are perennial wetland herbs with creeping rhizomes. The alternate leaves are simple, with the stipules fused to the petiole. Flowers are produced in compact, terminal spikes. Each flower consists of a bract, 6 stamens and 1 pistil. Sepals and petals are absent. The fruit is a capsule or schizocarp.

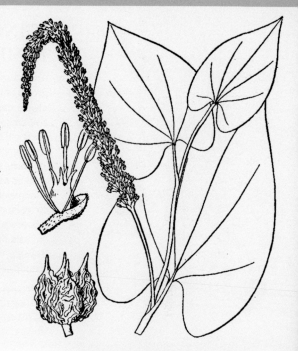

Lizard's-tail

Saururus cernuus

Habitat: Wet soil and shallow water.
General: Stems 15–120 cm tall; rhizomes creeping. **Leaves:** Alternate, lance-shaped to nearly triangular, 2–17 cm long, 1–10 cm wide, base heart-shaped; stalks 1–10 cm long. **Flower Cluster:** Racemes erect to nodding, 5–35 cm long, opposite upper leaves, 175–350-flowered; bracts green, boat-shaped, 1.5–3 mm long. **Flowers:** Creamy white, fragrant; sepals and petals absent; stamens 6; pistil 1. **Fruit:** Schizo-carp 1.5–3 mm long, surface wrinkled; seeds brown, less than 1.3 mm long.

Notes: The genus name comes from the Greek words *sauros*, "lizard," and *oura*, "tail."

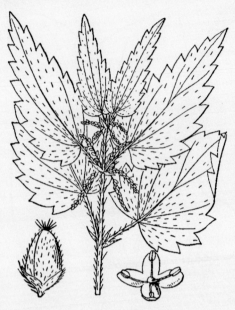

STINGING NETTLE FAMILY · Urticaceae

The stinging nettle family is a small group of herbs and shrubs. Many species in this family are distinguished by their stinging hairs, which have bulbous bases filled with a skin irritant. The simple leaves are alternate or opposite. Stipules are usually present. The flowers, borne in panicles, racemes or cymes, may be unisexual or perfect. Male flowers are composed of 3–5 sepals and 3–5 stamens. Female flowers are composed of 3–5 sepals and 1 style. Petals are absent. The fruit is an achene or drupe.

The best-known species in this family is stinging nettle (*Urtica dioica*), which has stinging hairs that can cause severe skin irritation.

Wood Nettle, Canada Nettle

Laportea canadensis

Habitat: Rich, moist woods. **General:** Stems 4-sided, 30–150 cm tall, covered in stinging hairs; rhizomes with tuberous roots. **Leaves:** Alternate, oval, 6–30 cm long, 3–18 cm wide; margins coarsely toothed; stalks long. **Flower Cluster:** Panicle terminal and axillary; male or female **Flowers:** Green; male flowers 1–2 mm across, sepals 5, stamens 5; female flowers less than 1 mm wide, sepals 2–4, pistil 1, style feathery. **Fruit:** Achene crescent or D-shaped, 2–4 mm long, 2 sepals remaining attached to fruit.

Notes: The genus name commemorates François Louis de Laporte (1810–80), a French entomologist. The species name means "of Canada."

American Pellitory
Parietaria pensylvanica

Habitat: Dry, shaded places and waste areas. **General:** Stems 4–60 cm tall, simple to branched, stinging hairs absent. **Leaves:** Alternate, elliptic to oblong or lance-shaped, 2–9 cm long, 0.4–3 cm wide; margins smooth; stipules absent. **Flower Cluster:** Cyme axillary; male, female and perfect flowers in the same cyme; bracts linear to lance-shaped, 1.8–5 mm long. **Flowers:** Green, 1–3 mm wide; sepals 4; stamens 4; pistil 1. **Fruit:** Achene oval, light brown, 1–1.5 mm long, enclosed by the 4 sepals.

Notes: *Parietaria* comes from the Latin word *paries*, "wall," a reference to the habitat of many species. The species name means "of Pennsylvania."

Clearweed
Pilea pumila

Habitat: Moist to wet woods and streambanks. **General:** Stems 7–70 cm tall, erect, unbranched, hairless, slightly succulent. **Leaves:** Opposite, elliptic to oval, 2–13 cm long, 1–9 cm wide; margins with 3–17 teeth per side; stalks 0.4–4.5 cm long; stipules soon withering, 2–3 mm long. **Flower Cluster:** Cyme axillary; male, female and perfect flowers in the same cyme; bracts triangular to linear. **Flowers:** Green; male flowers with 4 sepals, 4 stamens and 1 sterile pistil; female flowers with 3 sepals (1 often hood-like and hairy), 3 sterile stamens and 1 pistil. **Fruit:** Achene teardrop-shaped, 1.3–1.7 mm long, light-coloured or streaked with purple.

Notes: *Pilea* comes from the Latin word *pileus*, "felt cap," a reference to the hairy sepals surrounding the fruit. The species name *pumila* means "dwarf."

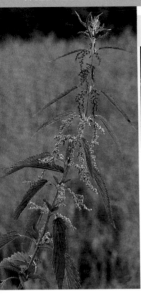

Common Stinging Nettle · *Urtica dioica*

Habitat: Moist, open areas and woods. **General:** Stem 0.5–3 m tall, 4-angled; stem and leaves covered with short, stinging hairs; rhizome creeping; plants male or female. **Leaves:** Opposite, oval to lance-shaped, 5–20 cm long, 2–13 cm wide, prominently 3–7-veined; margins toothed; stalks 1–1.5 cm long; stipules linear to lance-shaped, 0.5–1.2 cm long. **Flower Cluster:** Panicle of racemes, each raceme 1–7 cm long, originating in leaf axils. **Flowers:** Yellowish to purplish green; male flowers with 4 sepals and 4 stamens; female flowers with 4 sepals (2 small outer and 2 large inner) and 1 pistil. **Fruit:** Achene lens to oval-shaped, 1–1.5 mm long, enclosed by the 2 inner sepals.

Notes: The genus name *Urtica* comes from the Latin word *uro*, meaning "to burn," a reference to the stinging hairs. The species name *dioica* means "2 houses," referring to plants being male or female. • Each sharp-pointed hair has a bulb-shaped base filled with an irritating fluid. The hairs break when they penetrate the skin, releasing the fluid beneath the skin's surface, causing irritation and inflammation.

SANDALWOOD FAMILY
Santalaceae

Members of the sandalwood family range from herbs to trees. Some species are partly parasitic on the roots of other plants, attaching themselves with sucker-like organs to obtain water and nutrients from a host plant, whereas others are completely parasitic on the stems of a host plant. The leaves are simple and alternate or opposite. Flowers are borne singly or in cymes or racemes. The unisexual or bisexual flowers have 3–6 sepals, 3–6 stamens and 1 pistil with 3–5 carpels. Petals are absent. The fruit is a nut or drupe.

The mistletoe family (formerly Loranthaceae) is now included in the sandalwoods under the subfamily Visceae.

A well-known species of this family is sandalwood (*Santalum album*), whose aromatic wood is sought after for perfumes, herbal medicines and incense. Mistletoes are known to cause damage in coniferous woodlots and plantations.

Eastern Dwarf Mistletoe
Arceuthobium pusillum

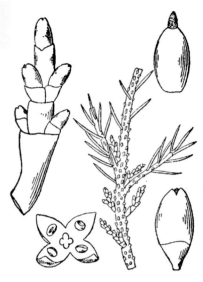

Habitat: Parasitic on the branches of trees, primarily black spruce (*Picea mariana*), p. 21. General: Stems green, 5–20 mm tall, fragile, attached by root-like structures called haustoria; plants male or female. Leaves: Reduced to scales. Flower Cluster: Solitary in leaf axils. Flowers: Green, 2–5 mm across; male flowers with 3 sepals and 3–4 stamens; female flowers with 2 sepals and 1 pistil. Fruit: Capsule 1–3 mm long, exploding at maturity; seed 1, covered with sticky pulp, dispersed up to 5 m from mother plant.

Notes: The genus name comes from the Greek words *arkeuthos*, "juniper," and *bios*, "life." The species name *pusillum* means "little."

Bastard Toadflax
Comandra umbellata

Habitat: Dry hillsides and gravel slopes. General: Semi-parasitic; stems green, 15–40 cm tall, usually branched; rhizome creeping; roots attaching themselves to other plants by small, sucker-like organs. Leaves: Alternate, numerous, linear to oblong or lance-shaped, 1–3.8 cm long, 0.5–1.2 cm wide; stalkless. Flower Cluster: Panicle of numerous, 3–5-flowered, terminal cymes. Flowers: Greenish white to pinkish, 4–6 mm long; sepals 5, green outside, white inside; stamens 5, anthers with tufts of hairs; pistil 1. Fruit: Drupe berry-like, globe-shaped, 4–8 mm long; sepals persistent, forming a crown.

Notes: The genus name comes from the Greek words *kome*, "hair," and *aner*, "man," a reference to the tufts of hairs on the anthers. The species name *umbellata* means "umbel-like."

Northern Comandra
Geocaulon lividum

Habitat: Mossy areas in moist, coniferous woods.
General: Semi-parasitic on roots of heathers (Ericaceae), pines (Pinaceae) and willows (Salicaceae); stems 10–25 cm tall, unbranched; rhizome creeping. **Leaves:** Alternate, oval, 1–3 cm long, bright green or variegated yellow, short-stalked. **Flower Cluster:** Cyme 1–4-flowered, in leaf axils. **Flowers:** Greenish white, 2–5 mm across; sepals 5; stamens 5; pistil 1. **Fruit:** Drupe berry-like, red or orange, 2–4 mm across; edible but tasteless.

Notes: The genus name *Geocaulon* means "earth stem," alluding to the creeping rhizome. The species name *lividum* means "bluish," a reference to the colour of the leaves.

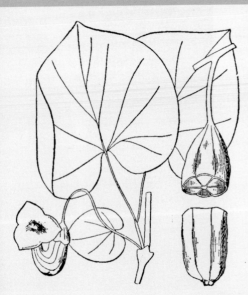

BIRTHWORT FAMILY · Aristolochiaceae

The birthwort family is a small family of herbs and vines. The basal or alternate leaves may be evergreen or deciduous. Flowers are produced in terminal or axillary racemes, or are solitary. The perfect flowers have 3 petal-like sepals, 6 or 12 stamens and a single pistil. Petals are absent. The fruit is a capsule with 6 compartments.

The well-known ornamental Dutchman's-pipe (*Aristolochia macrophylla*) is a woody vine to 20 m tall/long and is grown for its unusual flowers.

Wild Ginger

Asarum canadense

Habitat: Rich woods. **General:** Leaves and flowers arising from a slender, horizontal rhizome up to 50 cm long. **Leaves:** Basal (technically alternate), 2, heart-shaped, 4–20 cm long, 4–21.5 cm wide, covered in soft, downy hairs; stalk 6–20 cm long. **Flower Cluster:** Solitary, erect, arising from the rhizome tip; flower stalk 1.5–3 cm long. **Flowers:** Reddish to greenish brown, foul-smelling and attracting beetles for pollination, 3–4 cm wide, 2–4 cm long; sepals 3, petal-like, pointed; stamens 12; pistil 1. **Fruit:** Capsule fleshy, purplish brown; seeds large, wrinkled.

Notes: *Asarum* is the Latin and Greek name for wild ginger. The species name means "of Canada." • The rhizome has a ginger-like smell. Early settlers cooked it with sugar as a substitute for ginger.

BUCKWHEAT FAMILY

Polygonaceae

The buckwheat family is a small family of herbs and shrubs with alternate, opposite, basal or whorled leaves. The 2 stipules at the base of each simple leaf are fused to form a sheath. This collar-like sheath, called an ocrea, is a distinctive feature of this family. Flowers are produced in terminal or axillary cymes that resemble racemes, spikes, umbels or heads. The flowers are composed of 2–6 petal-like sepals appearing in 2 whorls, 3–9 stamens and 1 pistil. Petals are absent. The fruit is an achene.

Well-known members of this family include rhubarb (*Rheum rhabarbarum*), garden sorrel (*Rumex acetosa*) and buckwheat (*Fagopyrum esculentum*). Several species are grown for their commercial and ornamental value, whereas others are noxious weeds.

Fringed Black Bindweed
Fallopia cilinodis

Habitat: Dry, open woods.
General: Stems trailing, 1–5 m long, branched, often red. Leaves: Alternate, heart- to arrowhead-shaped, 2–6 cm long, 2–5 cm wide; margins wavy; stalks 1–6 cm long, often reddish; ocrea light brown, 3–4 mm long, fringed at the base. Flower Cluster: Panicle 4–15 cm long, on stalk 1–12 cm long and bristly haired; flowers 4–7 per fascicle, stalks 3–4 mm long. Flowers: Greenish white to white, 1.5–2 mm long; sepals 5, the outer 3 keeled; stamens 6–8; pistil 1. Fruit: Achene brownish black to black, 3–4 mm long, shiny, smooth.

Notes: The species name *cilinodis* refers to the fringed base of the ocrea.

Wild Buckwheat, Black Bindweed
Fallopia convolvulus

Habitat: Cultivated crops, roadsides and waste areas; introduced from Europe as a contaminant in seed in 1873. General: Stem 20–200 cm long, twining or trailing, branched, hairless. Leaves: Alternate, arrow-head-shaped, basal lobes backward-pointing, 2–15 cm long, 1–10 cm wide, tip pointed; margins wavy; stalks 0.5–5 cm long; ocrea papery, 2–4 mm long. Flower Cluster: Raceme spike-like, axillary, 2–15 cm long, in groups of 3–6; flower stalks 1–3 mm long. Flowers: Greenish pink, 3–5 mm long; sepals 4–6; stamens 8; pistil 1. Fruit: Achene black, 3-sided, 3–6 mm long.

Notes: The species name *convolvulus* refers to wild buckwheat's climbing growth habit, which is similar to that of morning-glories (Convolvulaceae), pp. 322–24.

Japanese Knotweed
Fallopia japonica

Habitat: Waste areas, roadsides and disturbed areas; introduced from Japan or China as an ornamental plant in the late 1800s. **General:** Stems hollow and jointed (bamboo-like), 1–3 m tall, often mottled red or purple, arched near the top; rhizomes to 6 m long, creeping. **Leaves:** Alternate, broadly ovate, 5–15 cm long, 5–12 cm wide, base abruptly cut to wedge-shaped; margins smooth; stalks 1–3 cm long; ocrea 4–6 mm long, reddish to pale green, turning brown with age. **Flower Cluster:** Raceme or panicle 5–15 cm long, 3–8 flowers per fascicle, in axils of upper leaves; flower stalks 1–25 mm long. **Flowers:** Greenish white, 2–4 mm long; male flowers with 5 sepals and 5 stamens; female flowers with 5 sepals and 1 pistil. **Fruit:** Achene triangular, 2–4 mm long, shiny, black to brown; outer sepals enlarged and forming a wing.

Notes: The genus name commemorates Italian botanist Gabriello Fallopia (1523–62). The species and common names refer to Japan, this plant's native range.

Pale Smartweed
Persicaria lapathifolia

Habitat: Waste areas, gardens and roadsides; introduced from western North America. **General:** Stems 20–150 cm tall, erect to spreading, branched, often rooting at lower nodes. **Leaves:** Alternate, lance-shaped, 5–20 cm long, 0.7–5 cm wide, dark blotch or "thumbprint" near midleaf, underside of upper leaves with sticky, yellow dots or hairs; stalks 1–16 mm long; ocrea papery, 4–35 mm long, margins truncate, hairy. **Flower Cluster:** Raceme spike-like, terminal or axillary, 1–10 cm long, 0.5–1.2 cm wide, 4–14 flowers per fascicle; stem 0.2–2.5 cm long. **Flowers:** Greenish white to pink, 1–2.5 mm across; sepals 5; stamens 3–9; pistil 1. **Fruit:** Achene brown, heart-shaped, 2–3 mm long.

Notes: The species name *lapathifolia* is the Latin word for "dock" or "sorrel," a reference to the leaves, which resemble those of dock and sorrel.

Marsh Smartweed
Polygonum amphibium var. *emersum*

Habitat: Shorelines and marshy wetlands. **General:** Stem 30–100 cm long, erect to trailing; roots white, fleshy. **Leaves:** Alternate, lance-shaped, 5–20 cm long, 1–5 cm wide; aerial leaves short-stalked, hairy; floating leaves (when present) long-stalked, hairless; ocrea with numerous stiff, white hairs. **Flower Cluster:** Raceme spike-like, 4–18 cm long, 0.7–1.5 cm thick. **Flowers:** Pink, 4–5 mm long; sepals 4–6, petal-like; stamens 3–9; pistil 1. **Fruit:** Achene brownish black, lens-shaped, 2.5–3 mm long.

Notes: The species name *amphibium* means "amphibious, suited to growing on land or water."

Yard Knotweed, Doorweed
Polygonum aviculare

Habitat: Yards, roadsides and waste areas; introduced from Europe. **General:** Stem 5–200 cm long, trailing, profusely branched, forming large mats. **Leaves:** Alternate, bluish green, elliptic to oval or lance-shaped, 6–50 mm long, 1–22 mm wide; margins smooth; stalks 0.5–9 mm long; ocrea silvery, translucent with jagged edges, 3–15 mm long. **Flower Cluster:** Cyme 1–8-flowered, in leaf axils; flower stalks 1.5–5 mm long. **Flowers:** Green with pinkish white margins, 1–3 mm across; sepals 5; stamens 3–9; pistil 1. **Fruit:** Achene 3-sided, 2–4.2 mm long, dull brown.

Notes: The genus name comes from the Greek words *poly*, "many," and *gony*, "knee," a reference to the jointed stem. The species name *aviculare* means "pertaining to birds," referring to the seeds, which are relished by birds.

Curled Dock, Yellow Dock

Rumex crispus

Habitat: Moist meadows, waste areas and roadsides; introduced from Europe in the mid-1700s. **General:** Stems 30–150 cm tall, reddish green, enlarged at nodes; taproot fleshy with yellow centre. **Leaves:** Alternate, lance-shaped, 10–30 cm long, 2–6 cm wide; margins wavy or crinkled; stalks 2–5 cm long; ocrea papery, 0.5–5 cm long, turning brown and papery with age. **Flower Cluster:** Panicle 10–60 cm long; flowers in whorls of 10–25. **Flowers:** Greenish red, 3–4 mm across; sepals 6 (3 outer, 3 inner); stamens 6; pistil 1. **Fruit:** Achene 2–3 mm long, reddish brown, 3-sided, enclosed by the enlarged inner sepals.

Notes: The species name *crispus*, "crisped," refers to the leaf margins. The common name "curled dock" also alludes to this.

Blunt-leaved Dock, Bitter Dock

Rumex obtusifolius

Habitat: Waste areas, roadsides and fields; introduced from Europe and Asia. **General:** Stems erect, 60–150 cm tall, hairless; taproot spindle-shaped. **Leaves:** Alternate, oblong to oval, 20–40 cm long, 10–15 cm wide; margins smooth to wavy; stalks long; ocrea deciduous with age. **Flower Cluster:** Panicle terminal; flowers in whorls of 10–25; flower stalks thread-like, 2.5–10 mm long. **Flowers:** Greenish red, 2–4 mm across; sepals 6 (3 outer, 3 inner); stamens 6; pistil 1. **Fruit:** Achene reddish brown, 2–3 mm long.

Notes: The species name *obtusifolius* means "blunt-leaved."

GOOSEFOOT FAMILY
Chenopodiaceae

Members of the goosefoot family are herbs and shrubs, often found on dry, alkaline sites. The alternate leaves are simple. Stipules are absent. The unisexual or perfect flowers are borne in cymes. These regular flowers have no petals, 2–5 united sepals, 2–5 stamens opposite the sepals and 1 pistil. The fruit is an achene, utricle or nutlet and is often surrounded by the calyx.

Economically important members of this family include spinach (*Spinacia oleracea*), beets (*Beta vulgaris*) and quinoa (*Chenopodium quinoa*). Many species are considered noxious weeds and compete with agricultural crops for light, moisture and nutrients.

Garden Orache
Atriplex hortensis

Habitat: Roadsides and waste areas; introduced garden vegetable from Europe and southwestern Asia. **General:** Stems 60–250 cm tall, yellowish green to reddish purple, hairless. **Leaves:** Opposite or alternate, lance-shaped to triangular, 5–25 cm long, 0.8–15 cm wide, green above, whitish green below, base heart- to arrowhead-shaped; margins smooth; stalks 3–40 mm long. **Flower Cluster:** Panicle of spikes 10–30 cm long. **Flowers:** Reddish green, inconspicuous; male flowers with 5 sepals and 5 stamens; female flowers of 2 types: no sepals and 2 bracts (to 12 mm long at maturity) or 3–5 sepals and no bracts, pistil 1. **Fruit:** Achenes of 2 types: brown and 3–5 mm across or black and 1–2 mm across.

Notes: *Atriplex* is the Greek name for this plant. The species name *hortensis* means "of gardens."

Russian Pigweed, Upright Axyris

Axyris amaranthoides

Habitat: Waste areas, roadsides and edge of fields; introduced from Siberia. **General:** Stems 40–120 cm tall, pale green, turning yellow at maturity **Leaves:** Alternate, oval to lance-shaped, 2–10 cm long, 1–3 cm wide, upper surface hairless, lower surface with numerous star-shaped hairs; margins smooth to wavy; stalks short. **Flower Cluster:** Spike or cyme, compact, densely flowered; often male or female. **Flowers:** Yellowish green, inconspicuous; male flowers at branch ends, sepals 3–5 and papery, stamens 2–5; female flowers in leaf axils or mixed with male flowers, sepals 3–5 and papery, pistil 1. **Fruit:** Achenes of 2 types: dark brown, oval and 2–3 mm long with a lobed, papery wing, or grey, round, less than 2 mm across and wingless.

Notes: The genus name *Axyris* comes from the Greek words *a*, "without," and *xuron*, "razor," a reference to the blunt-tipped sepals. The species name *amaranthoides* refers to this plant's resemblance to members of the amaranth family (Amaranthaceae), pp. 204–05.

Lamb's-quarters, Pigweed

Chenopodium album

Habitat: Waste areas, roadsides and disturbed sites; introduced from Europe and Asia. **General:** Highly variable; stems 0.4–2.5 m tall, bluish green, grooved, often blotched or striped with red or purple, branched. **Leaves:** Alternate, triangular to diamond- or lance-shaped, 2–12 cm long, 1–8 cm wide, often with reddish purple blotches, green above, mealy white below; margins wavy; stalks 1–2.5 cm long. **Flower Cluster:** Panicle 2–19 cm long, of dense, spike-like clusters 3–4 mm across. **Flowers:** Bluish green, 2–3 mm wide; sepals 5, partly fused; stamens 5; pistil 1. **Fruit:** Achene black, shiny, 1–1.5 mm long, surrounded by a papery sac.

Notes: The genus name comes from the Greek words *cheno*, "goose," and *podium*, "foot," referring to the shape of the leaves. The species name *album* means "whitish," referring to the whitish bloom on the lower surface of the leaves.

Oak-leaved Goosefoot
Chenopodium glaucum var. *salinum*

Habitat: Moist, alkaline soils; introduced from Europe and Asia. **General:** Stems 5–40 cm long, spreading to erect, reddish green, somewhat fleshy. **Leaves:** Alternate, oblong to lance-shaped, resembling oak leaves, 1–2.5 cm long, 0.3–1.5 cm wide, often spotted red or purple, green above, powdery white below; margins wavy to toothed; stalks 0.8–1.1 cm long. **Flower Cluster:** Spike 5–10 cm long, composed of head-like clusters 1–3 mm across, in leaf axils **Flowers:** Green; sepals 3–4, partly fused; stamen 1; pistil 1. **Fruit:** Achene dark brown, 0.5–1.1 mm across, surrounded by a papery sac.

Notes: The species name *salinum* means "growing in salty places," a reference to this plant's habitat.

Maple-leaved Goosefoot
Chenopodium simplex

Habitat: Waste areas and roadsides. **General:** Stems 30–150 cm tall, bright green, hairless, grooved. **Leaves:** Alternate, oval to triangular, resembling maple leaves, 3.5–20 cm long, 2–9 cm wide, bright green above, paler below; stalks 1–5 mm long; margins toothed to wavy, 1–5 large teeth per side. **Flower Cluster:** Panicle 6–15 cm long, of numerous compact spikes 1–2 mm across. **Flowers:** Green, 0.5–2 mm wide; sepals 5, partly fused; stamens 5; pistil 1. **Fruit:** Achene black, lens-shaped, 1–2 mm across, surrounded by a papery sac.

Notes: The species name *simplex* means "simple" or "unbranched," a reference to the stems.

Kochia, Mexican Fireweed, Summer Cypress
Kochia scoparia

Habitat: Roadsides and waste areas; introduced ornamental from Europe and Asia. General: Plant bushy, pyramid-shaped; stems 0.3–2 m tall; turning reddish purple in autumn; becoming a tumbleweed at maturity. Leaves: Alternate, numerous, linear to lance-shaped, 1–7.5 cm long, 2–15 mm wide, 3–5-veined; margins smooth; leaf undersides and margins with long hairs; stalks short, hairy. Flower Cluster: Solitary or in spike-like clusters of 1–6 flowers, in leaf axils. Flowers: Green, 2–3 mm across, male or female; sepals 5, partly fused, covered in long hairs; stamens 3–5; pistil 1. Fruit: Achene reddish brown to black, 1–2 mm across, often enclosed by the sepals.

Notes: The genus name honours German botanist W.D.J. Koch (1771–1849). The species name *scoparia* means "broom-like."

Russian Thistle, Common Saltwort
Salsola kali

Habitat: Waste areas, railway grades and roadsides; introduced from Russia as a seed contaminant in 1894. General: Plant pyramid-shaped; stems 10–120 cm tall, spiny; branches numerous, often red-striped; turning red in autumn; becoming a tumbleweed at maturity. Leaves: Alternate; lower leaves thread-like, 2–6 cm long, 1–2 mm wide; upper leaves awl-shaped, spine-tipped, becoming stiff and bristly with age. Flower Cluster: Solitary, in leaf axils. Flowers: Green, inconspicuous, 1–2 mm wide; male flowers with 5 sepals and 5 stamens; female flowers with 5 sepals and 1 pistil; bracts 2, spine-tipped, 3–7 mm long. Fruit: Achene dull brown to grey, 1–3 mm long, top-shaped, coiled.

Notes: The genus name comes from the Latin word *salsus*, "salty"; when the plant is burned, the resulting ash is rich in sodium. • The prickly nature and native range of this plant gave rise to the common name "Russian thistle."

AMARANTH FAMILY
Amaranthaceae

The amaranth family consists of herbs and subshrubs. All members have stalked, alternate or opposite leaves. The plants may be male, female or both. The small, inconspicuous flowers are usually green-ish purple and occur in terminal or axillary clusters. The 3 bracts found below each flower are a distinguishing feature of this family. The flowers have no petals, 2–5 sepals, 2–5 stamens opposite the sepals and 1 pistil with 2–3 stigmas. The fruit is a single-seeded capsule.

Well-known ornamental species include cockscomb (*Celosia cristata*) and love-lies-bleeding (*Amaranthus caudatus*).

Prostrate Amaranth
Amaranthus blitoides

Habitat: Waste areas, roadsides and disturbed sites; introduced from southern Europe and northern Africa. **General:** Stems trailing to erect, 15–90 cm long, green to reddish purple, branches numerous, often form-ing mats. **Leaves:** Alternate, oval to spoon-shaped, 0.5–5 cm long, 0.3–4 cm wide, dark green, shiny, underside with prominent white veins, tip slightly indented; stalks 0.5–2.5 cm long. **Flower**

Cluster: Raceme axillary, 2–8-flowered; bracts lance-shaped, 1.5–2 mm long. **Flowers:** Green, inconspicu-ous, 1.5–3 mm across; male flowers with 3 sepals and 3 stamens; female flowers with 3 sepals and 1 pistil; bracts 1.5–5 mm long, soft, sharp-pointed. **Fruit:** Capsule lens-shaped, 1.5–2.5 mm long; seed 1, dark reddish brown to black.

Notes: *Amaranthus* comes from the Greek word *amarantos*, "unfading," a reference to the lasting quality of some of the flowers in this genus.

Redroot Pigweed
Amaranthus retroflexus

Habitat: Waste areas, road-sides and disturbed sites; introduced from the southern U.S. **General:** Stems 30–100 cm tall, pale green or reddish, branched, rough textured; tap-root pinkish red, short, fleshy; plants male or female. **Leaves:** Alternate, oval to diamond-shaped, 2–15 cm long, 1–7 cm wide; margins smooth to wavy; stalks 1–15 cm long. **Flower Cluster:** Panicle 5–20 cm long, of numerous spikes; bracts 1–3 per flower, 2.5–6 mm long, spine-tipped. **Flowers:** Green, 2–4 mm wide; male flowers with 5 greenish purple, spine-tipped sepals and 4–5 stamens; female flowers with 5 sepals and 1 pistil. **Fruit:** Capsule 1.5–2.5 mm long; seed 1, shiny, black.

Notes: The species name *retroflexus* means "reflexed" or "bent," a reference to the flower cluster, which is often nodding.

POKEWEED FAMILY · Phytolaccaceae

The pokeweed family consists of tropical and subtropical trees, shrubs and herbs. The alternate leaves are simple and have smooth margins. Flowers are borne in terminal and axillary racemes. The flowers have 4–5 free or fused sepals, 4 to many stamens and 1–12 free or fused pistils. Petals are absent. The fruit is an aggregate, berry or schizocarp.

The best-known species from this family is pokeweed (*Phytolacca americana*), whose young leaves and shoots can be eaten, though the water must be changed several times during the cooking process.

Pokeweed, Pigeon-berry
Phytolacca americana

Habitat: Moist, open areas and edge of thickets. **General:** Stems reddish green, 0.5–3 m tall, hollow, branched, hairless; taproot white, fleshy, to 15 cm across; poisonous. **Leaves:** Alternate, oblong to egg- or lance-shaped, 10–50 cm long, 3–18 cm wide, veins prominent and pinkish; margins smooth; stalks 1–6 cm long; ill-scented when crushed. **Flower Cluster:** Raceme 6–30 cm long, 5–100-flowered, stalk red, 2–15 cm long; flower stalks 3–13 mm long. **Flowers:** Greenish white to pinkish, 5–7 mm across; sepals 5; stamens 10; pistils 6–12, partly fused. **Fruit:** Berry dark purple, 6–11 mm across, juice crimson; clusters often drooping.

Notes: The genus name comes from the Greek words *phyton,* "plant," and *lacac,* "crimson lake," referring to the red juice of the berries. The species name means "of America."

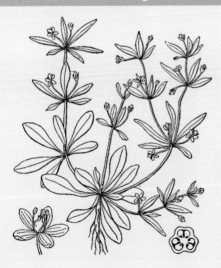

CARPETWEED FAMILY · Molluginaceae

The carpetweeds are a small family of herbaceous plants that are often included in the ice-plant family (Aizoaceae). Carpetweeds have alternate, opposite or whorled leaves. The simple and entire leaves do not have stipules. Flowers are borne in terminal or axillary cymes or umbels. The perfect flowers have 5 sepals, 2 to many stamens and 1 pistil with 1–5 styles. Petals are absent. The fruit is a capsule.

Indian Chickweed

Mollugo verticillata

Habitat: Dry, sandy roadsides; introduced from tropical North America. **General:** Stems trailing, 3–45 cm long, yellowish green, hairless, profusely branched, forming mats. **Leaves:** Whorls of 3–8, linear to elliptic or spatula-shaped, 1–2.5 cm long, dull green above, pale below; stalks 0.5–4 mm long. **Flower Cluster:** Umbel 2–6-flowered, in leaf axils; flower stalks 3–20 mm long. **Flowers:** Greenish white, 3–5 mm across; sepals 5; stamens 3; pistil 1. **Fruit:** Capsule oval, 2–4 mm long; seeds 15–35, orangey brown.

Notes: The genus name *Mollugo* refers to this plant's resemblance to *Galium mollugo*, false baby's-breath. The species name *verticillata* means "whorled," describing the leaf arrangement.

PURSLANE FAMILY · Portulacaceae

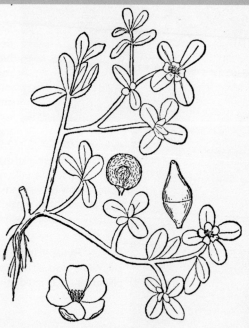

The purslane family is a small family of succulent plants with alternate or opposite leaves. Flowers are borne in racemes. The perfect flowers have 3–6 petals (occasionally absent), 2 sepals, 2–5 stamens and 1 pistil with 2–8 styles. The fruit is a capsule whose top falls off and releases numerous seeds.

Members of the purslane family are often grown as ornamentals and include portulaca (*Portulaca* spp.), spring-beauty (*Claytonia* spp.), pygmyroot (*Lewisia* spp.) and miner's-lettuce (*Montia* spp.).

Eastern Spring-beauty
Claytonia virginica

Habitat: Moist woods and forest seeps. **General:** Stem 5–40 cm tall; tubers globe-shaped, 1–2 cm across. **Leaves:** Basal, linear, 3–14 cm long, 0.5–1.3 cm wide, stalks 3–6 cm long; stem leaves 1–10 cm long, stalkless. **Flower Cluster:** Raceme 2–10-flowered, nodding; bract 1, scale-like, about 5 mm long.

Flowers: White to pink or candy-striped, 5–12 mm across; petals 5; sepals 2, 5–7 mm long; stamens 5; pistil 1. **Fruit:** Capsule globe-shaped, enclosed by the 2 persistent sepals; seeds 6, 2–3 mm across, shiny.

Notes: The genus name commemorates John Clayton (1686–1773), an American physician who collected several plant species in Virginia. The species name means "of Virginia."

Purslane · *Portulaca oleracea*

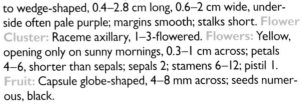

Habitat: Gardens, waste areas and dry, sandy roadsides; introduced garden vegetable from Europe and North Africa. **General:** Stem trailing, reddish green, 10–60 cm long, succulent, hairless, often rooting at nodes and forming mats; taproot 2–10 cm long. **Leaves:** Alternate but appearing opposite, succulent, oval to wedge-shaped, 0.4–2.8 cm long, 0.6–2 cm wide, underside often pale purple; margins smooth; stalks short. **Flower Cluster:** Raceme axillary, 1–3-flowered. **Flowers:** Yellow, opening only on sunny mornings, 0.3–1 cm across; petals 4–6, shorter than sepals; sepals 2; stamens 6–12; pistil 1. **Fruit:** Capsule globe-shaped, 4–8 mm across; seeds numerous, black.

Notes: *Portulaca* is the Latin name for this species. The species name *oleracea* means "vegetable-like," a reference to the plant being cultivated as a vegetable in several countries.

PINK FAMILY
Caryophyllaceae

Members of the pink family are primarily herbaceous plants. Important characteristics of this family are swollen stem nodes and opposite leaves. The simple, stalkless leaves have smooth margins. Stipules may be present or absent. The flowers, usually borne in cymes, may be perfect or unisexual. The regular flowers have 4–5 free petals, 4–5 fused or free sepals, 3–10 stamens and 1 pistil with 2–5 styles. The fruit is a capsule or achene.

Ornamental species include garden carnations (*Dianthus* spp.), baby's-breath (*Gypsophila paniculata*) and Maltese-cross (*Lychnis chalcedonica*). Several species in this family are noxious weeds.

Mouse-ear Chickweed

Cerastium arvense

Habitat: Dry, open meadows and sandy hillsides.
General: Stems 5–30 cm tall, erect to ascending, branched, hairy, often rooting at lower nodes.
Leaves: Opposite, elliptic to linear or lance-shaped, 1–7 cm long, 1–15 mm wide, upper surface often hairy; stalkless. **Flower Cluster:** Cyme 2–20-flowered; flower stalks 0.5–3 cm long. **Flowers:** White, 1.5–2 cm across; petals 5, deeply notched, longer than sepals; sepals 5; stamens 10; pistil 1, styles 5. **Fruit:** Capsule 0.8–1.5 cm long, opening by 10 valves.

Notes: The genus name *Cerastium* comes from the Greek word *keras*, "horn," a reference to the horn-like capsules. The species name *arvense* means "of cultivated fields."

Baby's-breath · *Gypsophila paniculata*

Habitat: Roadsides and waste areas; introduced ornamental from Europe and Asia.
General: Stems 40–100 cm tall, profusely branched, hairless, covered in a white, powdery film; rootstalk woody, 1–4 m deep, with sufficient reserves to survive 2 years of drought.
Leaves: Opposite, linear to lance-shaped, 2–10 cm long, 0.2–1 cm wide; margins smooth; often withering or absent when plant is in flower. **Flower Cluster:** Panicle of cymes; flower stalks 1–20 mm long. **Flowers:** White to pinkish, numerous, sweet-scented, 3–6 mm wide; petals 5; sepals 5; stamens 10; pistil 1.
Fruit: Capsule egg-shaped; seeds 2–5.

Notes: The genus name *Gypsophila* comes from the Greek words *gypsis* and *philein*, meaning "chalk lover," a reference to the habitat of some species. The species name *paniculata* refers to the type of flower cluster.

Soapwort · *Saponaria officinalis*

Habitat: Waste areas, roadsides and open, sandy sites; introduced ornamental from Europe and Asia. **General:** Stems 30–90 cm tall, leafy, hairless; rhizome thick, creeping, forming large colonies.
Leaves: Opposite, elliptic to oval or lance-shaped, 7–10 cm long, 2–4 cm wide, 3–5-nerved, hairless; margins smooth; stalks 1–15 mm long. **Flower Cluster:** Cyme terminal, head-like, to 15 cm long; lower bracts leaf-like; flower stalks 1–5 mm long.
Flowers: Pale pink, fragrant, 2–2.7 cm across, showy; petals 5, 0.8–1.5 cm long; sepals 5, fused, 1.5–2.5 cm long, prominently 20-veined; stamens 10; pistil 1, styles 2. **Fruit:** Capsule 1.5–2 cm long, opening by 4 valves, enclosed by the calyx.

Notes: The genus name comes from the Latin word *sapo*, "soap," because the leaves and flowers of this plant form a soapy lather when boiled in water. This attribute made it a prized plant by early settlers. The species name *officinalis* means "medicinal."

Knawel · *Scleranthus annuus*

Habitat: Waste areas and roadsides; introduced from Europe and
North Africa. **General:**
Stems 2.5–25 cm long, weak,
spreading, often forming mats.
Leaves: Opposite, linear to awl-
shaped, 5–24 mm long, 1–1.5 mm
wide; margins papery; stalkless.
Flower Cluster: Cyme terminal
or axillary, 1–10-flowered; bracts
longer than flowers. **Flowers:** Green, 4–5 mm
across; petals absent; sepals 5, 3–4 mm long; stamens 8–10, about twice as long as sepals; pistil 1,
styles 2. **Fruit:** Capsule oval, inflated, 3.2–5 mm
long, surrounded by the hardened calyx; seed 1.

Notes: The genus name comes from the Greek
words *skleros*, "hard," and *anthos*, "flower," a reference to the hardened calyx. The species name
annuus refers to this plant's annual life cycle.

White Cockle · *Silene latifolia* ssp. *alba*

Habitat: Roadsides,
waste areas and
fields; introduced
from Europe.
General: Stems
40–100 cm tall, cov-
ered with sticky
hairs; taproot
woody; plants male
or female. **Leaves:**
Opposite, oblong to lance-shaped or elliptic, 2–12 cm
long, 0.6–3 cm wide; margins smooth; stalks 0.5–1.2 cm
long, upper leaves stalkless. **Flower Cluster:** Cyme
3–10-flowered; bracts scale-like; flower stalks 1–5 cm
long. **Flowers:** White, 2.5–3.5 cm across, fragrant,
male or female; petals 5, deeply 2-lobed; sepals 5, fused;
stamens 10; pistil 1; styles 4–5; male calyx with 10
prominent veins, female calyx with 20 veins; opening in
the evening and closing before noon the following day. **Fruit:** Capsule 1–1.5 cm
long, surrounded by the calyx, produced by female plants only.

Notes: The genus name comes from the Greek word *sialon*, "saliva," a reference
to the sticky stem. The species name *latifolia* means "broad-leaved."

Night-flowering Catchfly · *Silene noctiflora*

Habitat: Roadsides and waste areas; introduced from Europe. **General:** Plant covered in sticky hairs throughout; stems 20–80 cm tall. **Leaves:** Opposite, oval to oblong or lance-shaped, 4–12 cm long, 2–4.5 cm wide, reduced in size upward; lower leaves stalked, upper leaves stalkless. **Flower Cluster:** Cyme 3–15-flowered; bracts leaf-like, 1–5 cm long. **Flowers:** White to pale pink, 2–2.5 cm across, fragrant; perfect or unisexual; petals 5, deeply notched; sepals 5, fused, 1–1.5 cm long, dark green, prominently 10-veined; stamens 10; pistil 1, styles 3; opening in the evening and closing before noon the following day. **Fruit:** Capsule 2.5–4 cm long, surrounded by the inflated calyx.

Notes: The species name *noctiflora* means "night-flowering." • The sticky hairs trap small insects, hence the common name "catchfly."

Bladder Campion

Silene vulgaris

Habitat: Roadsides and waste areas; introduced from Europe and Asia **General:** Stems 15–100 cm tall, hairless, often branched at the base. **Leaves:** Opposite, oval to lance-shaped, 2–8 cm long, 0.5–3 cm wide, covered with a powdery white film; stalkless. **Flower Cluster:** Cyme 5–40-flowered; bracts small; flower stalks 0.5–3 cm long. **Flowers:** White, 1–2 cm across; female or perfect; petals 5, deeply notched; sepals 5, 0.9–1.2 cm long, fused, inflated and bladder-like, with 20 pinkish white veins; stamens 10, slightly longer than sepals; pistil 1, styles 3. **Fruit:** Capsule round, 1.2–1.8 cm long, enclosed by the inflated calyx; seeds numerous.

Notes: The species name *vulgaris* means "common."

Saltmarsh Sand Spurrey
Spergularia salina

Habitat: Alkaline areas, roadsides and edges of parking lots; introduced from eastern North America. **General:** Stems 8–30 cm long, delicate, ascending to prostrate. **Leaves:** Opposite, linear, somewhat fleshy, 1–5.4 cm long; stipules 2 per node, triangular, white, 1.2–3.5 mm long. **Flower Cluster:** Cyme terminal, simple to 8-times compound. **Flowers:** Pink to rose or white, 2–6 mm wide; petals 5; sepals 5; stamens 2–3; pistil 1. **Fruit:** Capsule greenish brown, 3–6.5 mm long.

Notes: The species name *salina* means "saline." • This species is believed to be native to coastal areas and inland sites where saline soils are found. In Ontario, it is confined to areas with salt runoff from sidewalks and roadsides.

Long-leaved Starwort
Stellaria longifolia

Habitat: Moist woods, marshes and roadsides. **General:** Stems 10–50 cm tall, weak, straggling, sharply 4-angled; rhizome creeping. **Leaves:** Opposite, linear to narrowly lance-shaped, 2–6 cm long, 1–3 mm wide, firm, hairless; stalkless. **Flower Cluster:** Cyme open, terminal, 2–10-flowered; bracts 1–5 mm long, papery; flower stalks 3–30 mm long. **Flowers:** White, 5–9 mm across; petals 5, notched; sepals 5; stamens 5–10; pistil 1, styles 3. **Fruit:** Capsule blackish purple to tan, 3–6 mm long, opening by 6 teeth.

Notes: The genus name comes from the Latin word *stella*, "star," a reference to the flower shape. The species name *longifolia* means "long-leaved."

Common Chickweed
Stellaria media

Habitat: Gardens, moist waste areas and disturbed, shady sites; introduced from Europe before 1672; now found on all continents including Antarctica. **General:** Stems 5–80 cm long, weak, trailing, branched, leafy; each internode with a single line of hairs; shade tolerant; large plants producing up to 15,000 seeds. **Leaves:** Opposite, oval, 0.5–4 cm long, 0.2–2 cm wide; margins smooth; stalks 0.5–2 cm long with

a single line of hairs. **Flower Cluster:** Cyme 5- to many-flowered; bracts 0.1–4 cm long; flower stalks 0.3–4 cm long. **Flowers:** White, 2–6 mm across; petals 5 (occasionally absent), 3–5 mm long, deeply notched; sepals 5, hairy, 4–6 mm long; stamens 3–10, anthers reddish purple; pistil 1, styles 3. **Fruit:** Capsule 3–7 mm long, opening by 6 teeth; seeds 8–10, reddish brown.

Notes: Flowering often commences when the air temperature reaches 2°C.

Cow Cockle
Vaccaria hispanica

Habitat: Roadsides and waste areas; introduced from Europe and Asia. **General:** Stems 20–100 cm tall, profusely branched, hairless, bluish green owing to a white powdery film. **Leaves:** Opposite, lance-shaped, bluish green, somewhat fleshy, 2–10 cm long, 2–4 cm wide, hairless, clasping the stem. **Flower Cluster:** Panicle of numerous cymes, 15–100-flowered; flower stalks 1–3 cm long. **Flowers:** Bright pink, 1–1.3 cm across; petals 5, 1.7–2.3 cm long; sepals 5, fused, 1–1.5 cm long, flask-like, prominently 5-ribbed; stamens 10; pistil 1, styles 2. **Fruit:** Capsule round, 6–8 mm across, enclosed by the enlarged calyx; seeds dull black.

Notes: *Vaccaria* is the Latin word for "cow pasture," referring to the plant being used as livestock forage in Europe. The species name *hispanica* means "Spanish." • The seeds of cow cockle contain saponin, a substance reported to be toxic to livestock.

CROWFOOT OR BUTTERCUP FAMILY
Ranunculaceae

The crowfoot or buttercup family is composed of herbs and climbing vines with simple to compound leaves that may be alternate, opposite or whorled. Stipules are absent. Flowers are extremely variable within this family and are perfect except for a few species in *Clematis* and *Thalictrum*. The flowers are composed of 0–5 petals, 3–15 often petal-like sepals, 5 to many stamens and 5 to many styles. Members of this family are often confused with roses (Rosaceae). The technical difference between the 2 families is the absence or presence of a hypanthium, the fusion of sepals, petals and stamens; buttercups do not have a hypanthium, whereas roses do. The fruit is a cluster of follicles, achenes or berries.

Several members of this family are grown as ornamentals, including anemones (*Anemone* spp.), clematis (*Clematis* spp.) and columbines (*Aquilegia* spp.). Several species in this family are poisonous.

Doll's-eyes
Actaea pachypoda

Habitat: Moist woods and forests. **General:** Stem 40–80 cm tall; roots fibrous. **Leaves:** Alternate, compound, 2–3 times divided; leaflets oval, to 10 cm long, hairless; margins 3-lobed to coarsely toothed. **Flower Cluster:** Raceme 2–17 cm long, with 25 or more flowers, long-stalked. **Flowers:** White, 5–7 mm wide, sepals and petals falling off as flower opens; petals 4–10, white; sepals 3–5, whitish green; stamens 15–50; pistil 1. **Fruit:** Berries white, shiny, 6–11 mm across, prominently black-dotted at apex; fruiting stalks bright red, 2–3 mm thick; poisonous.

Notes: The genus name comes from the Greek word *aktea*, "elder," a reference to the leaves, which resemble those of elderberry (*Sambucus* spp.). The species name *pachypoda* means "thick-footed." • The berries resemble the white eyes of china dolls, hence the common name.

Baneberry · *Actaea rubra*

Habitat: Moist woods and forests.
General: Stems 30–100 cm tall, hairless.
Leaves: Alternate, 1–5 per stem, compound, 2–3 times divided into 3 leaflets; leaflets oval, 3–7 cm long; margins sharply toothed or lobed. **Flower Cluster:** Raceme thimble-shaped, 1–5 cm long, 25 or more flowers, long-stalked. **Flowers:** White, 2–4 mm across, sepals and petals falling off as flower opens; petals 4–10; sepals 3–5, petal-like; stamens 15–50; pistil 1. **Fruit:** Berry red or white, showy, 0.6–1 cm across, black dot at apex; fruiting stalks dull green or brown; poisonous.

Notes: The species name *rubra* means "red," a reference to the colour of the fruit.

Sharp-lobed Hepatica

Anemone acutiloba

Habitat: Dry to moist, deciduous forests.
General: Flowering stem 5–19 cm tall; rhizome long.
Leaves: Basal, 3–15, evergreen, 1.3–8 cm long, 1.8–11.5 cm wide, 3-lobed, underside often purplish; stalks hairy.
Flower Cluster: Solitary; bracts 3, 5–18 mm long, tips pointed. **Flowers:** White, pink or blue, showy, 1.2–2.5 cm across; petals absent; sepals 5–12, petal-like, 0.6–1.5 cm long; stamens 10–30; pistils numerous. **Fruit:** Achene 3.5–5 mm long; in globe-shaped clusters.

Notes: The species name *acutiloba* means "sharp-lobed." • The common name "hepatica" means "liver," referring to the lobed leaves, whose shape is reminiscent of the liver.

Round-lobed Hepatica

Anemone americana

Habitat: Dry to moist, mixed forests. **General:** Flowering stems 5–18 cm tall; rhizomes horizontal. **Leaves:** Basal, 3–15, 1.5–7 cm long, 2–10 cm wide, 3-lobed, underside often purplish; margins smooth; stalks 5–20 cm long. **Flower Cluster:** Solitary; bracts 3, 0.7–1.8 cm long, tips blunt. **Flowers:** White to pink or bluish, 1–2.5 cm wide; petals absent; sepals 5–12, 0.8–1.5 cm long; stamens 10–30; pistils numerous. **Fruit:** Achene 3.5–5 mm long; in globe-shaped clusters.

Notes: The species name means "of America."

Canada Anemone

Anemone canadensis

Habitat: Moist, open areas and roadsides. **General:** Stems 20–70 cm tall; root bulb-like. **Leaves:** Basal leaves 1–5, 4–10 cm long, 4–7 cm wide, palmately divided into 3–7 wedge-shaped segments, each 3-cleft, prominently veined, margins sharply toothed, stalks 8–22 cm long; stem leaves similar, in whorls of 2–3, deeply 2–3-lobed, margins coarsely toothed, stalkless. **Flower Cluster:** Solitary; stalks long, hairy; whorl of leaf-like bracts at the base. **Flowers:** White, 2.5–4 cm across; petals absent; sepals 4–6, petal-like, 1–2 cm long; stamens 80–100; pistils numerous. **Fruit:** Achene 3–6 mm long; fruiting head globe-shaped, on stalk 7.5–11.5 cm long.

Notes: The genus name is from the Greek word *anemos*, meaning "wind."
• Anemones are often called "windflowers," a reference to the fruiting heads, which are on long stalks that blow in the wind.

Long-fruited Anemone, Thimbleweed
Anemone cylindrica

Habitat: Dry, open woods, roadsides and pastures. General: Stems 15–80 cm tall, covered with grey, woolly hairs; base somewhat woody. Leaves: Basal leaves 5–10, palmately divided into 5 diamond-shaped segments, terminal segment 2.5–6 cm long and 3–10 cm wide, margins sharply toothed, stalks 9–21 cm long; stem leaves similar, whorled, long-stalked. Flower Cluster: Cyme umbel-like, 2–8-flowered, stalks 10–30 cm long, hairy; whorl of 4–12 bracts at the base. Flowers: Creamy to greenish white, 1.5–2.2 cm across; petals absent; sepals 4–6, petal-like, silky-haired on outer surface; stamens 50–75; pistils numerous. Fruit: Achene 2–3 mm long, densely woolly; fruiting head cylindric, 2–4 cm long, on stalk 10–30 cm long.

Notes: The species name refers to the cylindric or thimble-shaped fruiting head.

Wood Anemone
Anemone quinquefolia

Habitat: Moist, wooded areas. General: Stems 5–30 cm tall, delicate; rhizome slender. Leaves: Basal leaf 1, divided into 3–5 coarsely toothed, wedge-shaped leaflets, terminal leaflet 1–4.5 cm long and 1.4–3.7 cm wide, stalk 4–25 cm long; stem leaves similar, whorled, short-stalked, terminal leaflet 0.4–1.5 cm wide. Flower Cluster: Solitary, long-stalked; whorl of 3 bracts at the base. Flowers: White to pinkish white, 2.2–2.7 cm across; petals absent; sepals 5, petal-like, 0.6–2.5 cm long; stamens 30–60; pistils numerous. Fruit: Achene 2.5–4.5 mm long, stiff-haired; fruiting head globe-shaped, on stalk 2–6 cm long.

Notes: The species name *quinquefolia* means "5-leaved."

Blue Columbine
Aquilegia brevistyla

Habitat: Moist, open forests and meadows. General: Stems 20–80 cm tall, often hairy above; rhizome short, woody. Leaves: Basal leaves 5–30 cm long, compound, 2–3 times divided into 3-lobed leaflets, the leaflets 1.2–4.4 cm long, long-stalked; stem leaves alternate, simple, 3-lobed, stalk-less. Flower Cluster: Solitary, on long stalks. Flowers: Blue to purple, nodding, 1.5–2.5 cm long; petals 5, yellowish white; sepals 5, petal-like; stamens numerous;

pistils 5; spur bluish, 0.5–1 cm long, hooked. Fruit: Follicle 1.5–2.5 cm long, in clusters of 5.

Notes: The species name *brevistyla* means "short-styled."

Red Columbine, Wild Columbine, Rock-bells
Aquilegia canadensis

Habitat: Rocky, open woods. General: Stems 30–200 cm tall, woody at the base. Leaves: Basal leaves 7–30 cm long, 10–15 cm wide, compound, 2 times divided, the leaflets 9–27, 1.7–5.2 cm long, 3-lobed; stem leaves alternate, reduced in size upward. Flower Cluster: Solitary, on long stalks. Flowers: Red, nodding, showy, 3–5 cm long; petals 5, red, spurred; sepals 5, red; stamens numerous, 1.5–2.3 cm long; pistils 5; spurs red, straight, 1.3–2.5 cm long. Fruit: Follicle 1.5–3.1 cm long, in clusters of 5.

Notes: The genus name may be derived from the Latin word *aquila*, "eagle," a possible reference to the spurs. The species name means "of Canada."

Marsh-marigold
Caltha palustris

Habitat: Streambanks, road-side ditches and pond edges.
General: Stems 20–60 cm tall, succulent, hollow and branched above; roots fibrous; poisonous. **Leaves:** Basal leaves heart to kidney-shaped, 2–12.5 cm long, 2–18 cm across, long-stalked, margins smooth to toothed; stem leaves alternate, stalkless.
Flower Cluster: Solitary or 2–7-flowered cyme, long-stalked, in leaf axils. **Flowers:** Yellow, 1–4.5 cm across; petals absent; sepals 5–9, petal-like; stamens numerous; pistils 4–15. **Fruit:** Follicle 4–15 mm long, in clusters of 4–15.

Notes: The species name *palustris* means "marsh-loving." • Marsh-marigold is one of the first flowers to appear in spring.

Goldthread
Coptis trifolia

Habitat: Wet to moist, coniferous to mixedwood forests, often growing in moss. **General:** Flowering stems 3–17 cm tall; rhizome bright yellow to orange. **Leaves:** Basal, shiny, evergreen, compound with 3 oval to triangular leaflets; leaflets 0.8–1.5 cm long; margins toothed; stalks long.
Flower Cluster: Solitary; on stalks 3–17 cm long arising from the base.
Flowers: White, 0.7–1.3 cm across; petals 5–7, yellowish green, club-shaped, shorter than stamens; sepals 5–7, white, falling off early; stamens 30–60, white; pistils 4–7. **Fruit:** Follicle 4–7 mm long; in clusters of 4–7 on stalks 4–10 mm long.

Notes: The genus name *Coptis* comes from the Greek word *kopto*, "to cut," a reference to the toothed leaf margins. The species name *trifolia* means "3-leaved."

Small-flowered Crowfoot
Ranunculus abortivus

Habitat: Rich, open woods and clearings. General: Stems 10–60 cm tall, hairless; roots thread-like. Leaves: Basal leaves round to kidney-shaped, simple to 3-parted, 1.4–4.2 cm long, 2–5.2 cm wide, margins with rounded teeth; stem leaves 3–5-parted, stalkless. Flower Cluster: Cyme 3–50-flowered; flower stalks hairless. Flowers: Yellow, 4–6 mm across; petals 5, 1.5–3.5 mm long; sepals 5, 2.5–4 mm long; stamens and pistils numerous. Fruit: Achene 3–6 mm long; in ovoid heads.

Notes: The species name *abortivus* means "aborted," a reference to the petals, which are shorter than the sepals.

Tall Buttercup
Ranunculus acris

Habitat: Moist, open areas; introduced from Europe. General: Stems 20–100 cm tall, slightly hairy, leafy near the base, branched above; roots fibrous. Leaves: Basal leaves 2–9 cm long, 1.8–5.2 cm wide, long-stalked, deeply divided into 3–7 segments, divisions cleft into narrow-toothed segments; stem leaves alternate, 3-lobed, short-stalked. Flower Cluster: Cyme few-flowered, long-stalked. Flowers: Yellow, 2–3.5 cm across; petals 5; sepals 5, hairy; stamens and pistils numerous. Fruit: Achene 2–3 mm long; fruiting head globe-shaped, 0.5–1 cm across.

Notes: The genus name *Ranunculus* comes from the Greek word *rana*, "frog," alluding to the marshy habitat of many species. The species name *acris* means "acrid" or "bitter," a reference to the plant juices.

Lesser Celandine, Pilewort
Ranunculus ficaria

Habitat: Moist woods, streambanks and disturbed sites; introduced from Europe. **General:** Stems erect, 5–30 cm tall, hairless; roots tuberous; plants disappearing by midsummer, leaving bare, open patches where the colony grew. **Leaves:** Appearing basal but technically alternate or opposite, round to triangular or heart-shaped, 1.8–3.7 cm long, 2–4 cm wide, base heart-shaped; margins smooth. **Flower Cluster:** Solitary, on hairless stalks. **Flowers:** Yellow, occasionally white-splotched, 2–3.5 cm wide; petals 8–12; sepals 3–4, green; stamens and pistils numerous. **Fruit:** Achene 2.5–3 mm long; in dome-shaped heads.

Notes: An aggressive weed, lesser celandine outcompetes early blooming native plants, leaving large patches of bare ground where the colony was established. Eradication is difficult because the small tubers are easily spread while being removed.

Cursed Crowfoot
Ranunculus sceleratus

Habitat: Marshes, ditches and pond edges. **General:** Stems 20–60 cm tall, hollow, branched, hairless. **Leaves:** Basal leaves 1–5 cm long, 1–7 cm across, round to kidney-shaped, deeply 3-lobed, the lobes further divided, stalks long; stem leaves alternate, narrowly 3-lobed or unlobed, stalkless. **Flower Cluster:** Solitary, in upper leaf axils. **Flowers:** Yellow, 6–8 mm across; petals 3–5; sepals 3–5, reflexed; stamens and pistils numerous. **Fruit:** Achene 1–1.5 mm long; fruiting head cylindric, 5–13 mm long.

Notes: The species name *sceleratus* means "cursed" or "wicked," a reference to the sap, which can blister skin.

Tall Meadow-rue
Thalictrum dasycarpum

Habitat: Moist woods and wet meadows.
General: Stems 40–150 cm tall, hairless; plants male or female. Leaves: Alternate, brownish green to dark green, compound, 3–5 times divided; leaflets oval to wedge-shaped, 1.5–6 cm long, 0.8–4.5 cm wide, tips 2–5-lobed; lower leaves stalked, upper leaves stalkless; margins rolled under. Flower Cluster: Panicle pyramid-shaped, 15–25 cm long, many-flowered. Flowers: Greenish white, 3–8 mm across, male or female; petals absent; sepals 4–5; stamens 7–30, filaments white to purplish; pistils 7–13. Fruit: Achene 2–4.6 mm long, prominently veined, hairy to glandular.

Notes: The species name *dasycarpum* means "hairy-fruited."

BARBERRY FAMILY · Berberidaceae
Family description on p. 92.

Blue Cohosh
Caulophyllum thalictroides

Habitat: Rich, wooded areas. General: Stems 20–90 cm tall, hairless; emerging plants bluish purple or yellowish green; easily recognizable in spring by its colour. Leaves: Alternate, 2, stalkless, purple or yellowish green, covered in bluish white film, compound; lowest leaf 3-parted with 24–30 leaflets; upper leaf with 9–12 leaflets; leaflets 2.5–8 cm long, 2–10 cm wide, 2–5-lobed. Flower Cluster: Panicle 5–70-flowered, 3–6 cm long; bracts 1–3 mm long. Flowers: Purple or yellowish green, 1–1.5 cm across; 6 petals (reduced to small glands at the base of sepals); 6 sepals; 6 stamens; 1 pistil. Fruit: Berry-like, blue, 5–8 mm across; technically an achene as the ovary wall ruptures during seed development, exposing the blue "berry."

Notes: *Caulophyllum* from the Greek words *kaulos*, "stem," and *phyllon*, "leaf." The species name *thalictroides* alludes to the fact that the leaves of blue cohosh resemble those of meadow-rue (*Thalictrum* spp.), above.

Twinleaf, Rheumatism-root
Jeffersonia diphylla

Habitat: Moist, deciduous forests. **General:** Flowering stems 12–33 cm tall, leafless, 30–45 cm tall at maturity; rhizome short, producing 4–8 leaves per year. **Leaves:** Basal, stalks 9–25 cm long at flowering, 18–43 cm long at fruiting, deeply divided into 2 irregular lobes or leaflets resembling butterfly wings, 1–4 cm long at flowering, 6–15 cm long at fruiting; young leaves bluish green turning dark green with age. **Flower Cluster:** Solitary. **Flowers:** White, showy, 2–3 cm wide; petals 8; sepals 4, falling as flower opens; stamens 8 or 16; pistil 1. **Fruit:** Capsule pear-shaped, 1.8–3.8 cm long, leathery, opening by a lid to release numerous seeds.

Notes: The genus name *Jeffersonia* commemorates Thomas Jefferson (1743–1826), the third president of the United States. The species name *diphylla* means "2-leaved." • First Nations peoples boiled the roots to treat rheumatism, hence the common name "rheumatism-root."

Mayapple, Wild Mandrake
Podophyllum peltatum

Habitat: Rich woods. **General:** Stems 30–60 cm tall; rhizome creeping, often forming large colonies. **Leaves:** 1–2, shield-like, 18–40 cm across, deeply 5–9-lobed, the lobes variously toothed or lobed; margins smooth to coarsely toothed; sterile plants with 1 leaf; fertile plants with 2 opposite leaves, stalks 5–15 cm long. **Flower Cluster:** Solitary; stalk 1.5–6 cm long, originating in the angle between the 2 leaves. **Flowers:** Creamy white, nodding, showy, 3–5 cm across; petals 6–9, waxy; sepals 6, falling as flower opens; stamens 12–18; pistil 1. **Fruit:** Berry 2–5 cm long, fleshy; mature fruit edible; seeds 30–50, extremely poisonous.

Notes: The genus name comes from the Greek words *podos*, "foot," and *phyllon*, "leaf," a reference to the leaf and its attachment. The species name *peltatum* means "shield-like," referring to the leaf shape. • The leaves, roots, seeds and immature fruit are poisonous.

POPPY FAMILY
Papaveraceae

The poppies are a family of herbaceous plants with milky or coloured juice. Most species have simple or compound basal leaves, but some have alternate, opposite or whorled stem leaves. The flowers are solitary or are borne in cymes or racemes. Flowers have 4 or more petals, 2–3 sepals that fall off as the flower opens, numerous stamens and 2 or more fused pistils. The fruit is a many-seeded capsule.

The fumitory family (Fumariaceae) is now included in the poppy family. The sub-family characteristics are listed below:

Papaveroideae	Fumarioideae
Flowers regular	Flowers irregular
Coloured sap	Watery sap

Members of this family are grown for ornamental purposes and include California poppy (*Eschscholzia californica*) and prickly poppy (*Argemone mexicana*).

Celandine, Greater Celandine
Chelidonium majus

Habitat: Moist, open forests; introduced from Europe and Asia as a wart treatment. **General:** Stems 20–100 cm tall, somewhat succulent and bristly, branched, leafy; sap yellowish orange; shade tolerant. **Leaves:** Alternate, 5–35 cm long, pale green, deeply 5–9-lobed, lateral segments oblong, terminal segment 3-lobed; margins wavy to irregularly toothed; stalks 2–10 cm long, bristly haired. **Flower Cluster:** Umbel few-flowered, stalk 2–10 cm long; bracts present; flower stalks 0.5–3.5 cm long. **Flowers:** Yellow, 1–2 cm wide; petals 4; sepals 2, falling as flower opens; stamens 12 to many; pistil 1. **Fruit:** Capsule, linear to oblong, erect, 2–5 cm long, hairless; opening on 2 sides from the bottom upward.

Notes: The genus name comes from the Greek word *chelidon*, "a swallow," a reference to the bloom period coinciding with the swallow migration. The species name *majus* means "greater."

Golden Corydalis
Corydalis aurea

Habitat: Open woods, clearings and waste areas. **General:** Stems 10–30 cm tall, pale green, leafy, hairless. **Leaves:** Alternate, pale green, 7–10 cm long, divided twice into narrow segments. **Flower Cluster:** Raceme few-flowered. **Flowers:** Golden yellow, irregular, 10–15 cm long; petals 4, outer pair spurred, the spur 4–5 mm long; sepals 2; stamens 6; pistil 1. **Fruit:** Capsule 1–2 cm long; seeds 2–2.5 mm wide, shiny, black.

Notes: The species name *aurea* means "golden," a reference to the flower colour.

Pink Corydalis
Corydalis sempervirens

Habitat: Open, wooded areas. **General:** Stems 30–60 cm tall, erect, branched, pale green, hairless. **Leaves:** Alternate, bluish green, deeply divided into 3 sharply toothed segments, the segments 1–2 cm long; lower leaves stalked, upper leaves stalkless. **Flower Cluster:** Raceme 1–8-flowered, in leaf axils. **Flowers:** Purplish pink with yellow tips, 1.2–2 cm long; petals 4; sepals 2; stamens 6; pistil 1; spur on outer petal 2–5 mm long. **Fruit:** Capsule 2–4 cm long, erect.

Notes: The genus name *Corydalis* comes from the Greek word *korudalis*, "crested lark," because the crested petals were thought to resemble a lark. The species name *sempervirens* means "always green," a reference to the lower leaves, which remain green under the snow.

Squirrel-corn
Dicentra canadensis

Habitat: Rich woods. General: Flowering stems 10–30 cm tall; roots fibrous, bearing yellowish white tubers about the size of corn kernels. Leaves: Basal, greyish green, 14–24 cm long, 6–14 cm wide, divided 4 times into narrow, oblong to linear leaflets; stalks 8–16 cm long. Flower Cluster: Raceme 15–27 cm long, 3–12-flowered; bracts oval, 2–5 mm long; flower stalks 3–7 mm long. Flowers: Greenish white, fragrant, heart-shaped, nodding, 8–12 mm long; petals 4; sepals 2, 2–4 mm long; stamens 6; pistil 1; spurs 2, about 1 mm long. Fruit: Capsule oval, 9–13 mm long.

Notes: The genus name *Dicentra* comes from the Greek word for "2-spurred," a reference to the flowers.

Dutchman's-breeches
Dicentra cucullaria

Habitat: Moist, rich woods. General: Flowering stem 15–40 cm tall; roots bearing white to pink, teardrop-shaped bulblets. Leaves: Basal, 14–16 cm long, 6–14 cm wide, divided 4 times into narrow leaflets; leaflets 0.5–1.5 cm long; margins smooth; stalks 8–16 cm long. Flower Cluster: Raceme 3–14-flowered; bracts minute; flower stalks 4–7 mm long. Flowers: White, pantaloon-shaped, nodding, fragrance absent, 1.4–2 cm long; petals 4, often pinkish with yellow tips; sepals 2, 2–5 mm long; stamens 6; pistil 1; spur 1–3 mm long. Fruit: Capsule oval, 0.9–1.3 cm long.

Notes: The species name *cucullaria* means "hood-like," a reference to the flowers.

Common Fumitory
Fumaria officinalis

Habitat: Waste areas, roadsides and disturbed sites: introduced from southern Europe and North Africa. **General:** Stems 10–70 cm tall, sprawling, green, hairless, somewhat succulent; taproot slender. **Leaves:** Alternate, greyish green, 2–8 cm long, divided 3–4 times into narrow segments, each 1–2 mm wide; margins smooth; stalks long. **Flower Cluster:** Raceme 1–7.5 cm long, 10–40-flowered (commonly about 20), opposite the leaves; bracts 1.5–3 mm long; flower stalks 2–4 mm long. **Flowers:** Purplish pink to white at the base, reddish purple to maroon at tips, irregular, 0.6–1 cm long; petals 4, the upper spurred; sepals 2, very small, toothed; stamens 6, in groups of 3; pistil 1; spur about 2.5 mm long. **Fruit:** Capsule globe-shaped, 1–2 mm across, warty; seed 1.

Notes: The genus name comes from the Latin word *fumus*, "smoke," presumably alluding to the odour of the roots. The species name *officinalis* means "medicinal."

Bloodroot, Puccoon-root
Sanguinaria canadensis

Habitat: Rich woods. **General:** Flowering stems 10–15 cm tall, to 40 cm at fruiting; sap blood red; rhizome branched. **Leaves:** Basal, 1, round to heart- or kidney-shaped, 10–25 cm wide, bluish green, palmately veined, 3–9-lobed; margins wavy to rounded-toothed; stalks 5–15 cm long; leaf often appears to surround flowering stalk. **Flower Cluster:** Solitary, on a leafless stem. **Flowers:** White, showy, 2–5 cm wide; petals 8–16; sepals 2, falling as flower opens; stamens numerous; pistil 1. **Fruit:** Capsule 3–6 cm long; seeds black to reddish orange.

Notes: The genus name, which is from the Latin word *sanguis*, meaning "blood," and the common name "bloodroot" both refer to the blood red sap exuded by the stem when broken.

Wood Poppy, Celandine Poppy

Stylophorum diphyllum

Habitat: Moist, deciduous woods. **General:** Stems 30–50 cm tall, greenish purple at the base; sap yellow to orange; rhizome stout. **Leaves:** Basal leaves oval, deeply lobed, 20–50 cm long, stalks 10–20 cm long; stem leaves 2–3, subopposite, stalkless, 5–7-lobed; margins irregularly toothed. **Flower Cluster:** Umbels of 1–5; stalks 2.5–8 cm long; bracts 0.7–1.2 cm long. **Flowers:** Yellow, 3–7 cm across; petals 4; sepals 2, about 1.5 cm long; stamens many; pistil 1, stigma 3–5-lobed. **Fruit:** Capsule 2–3.5 cm long, hairy, nodding; seeds pale brown.

Notes: The genus name comes from the Greek words *stylos*, "style," and *phoros*, "bearing," a reference to the prominent style. The species name *diphyllum* means "2-leaved," referring to the 2 stem leaves. • Wood poppy is rare in Ontario and has been classified as endangered in Canada.

SPIDERFLOWER FAMILY · Cleomaceae

The spiderflowers are a small family of flowering plants with about 300 species worldwide. Formerly included in the caper family (Capparaceae), genetic research has shown a closer relationship to the mustard family (Brassicaceae). These herbaceous plants have alternate, palmately compound leaves with 3–11 leaflets. The flowers, borne in racemes or corymbs, are regular but occasionally irregular. The perfect flowers have 4 petals, 4 sepals, 6–32 stamens and 1 pistil. The fruit is an elongated capsule with 1–40 seeds.

Several species are grown for their ornamental value, including spiderflower (*Tarenaya hassleriana*). A few species are designated weeds.

Clammyweed
Polanisia dodecandra

Habitat: Dry, sandy soil and disturbed sites. **General:** Stems 10–50 cm tall, branched, covered in sticky hairs, unpleasant odour. **Leaves:** Alternate, 2–5 cm long, 1–1.5 cm wide, compound, divided into 3 leaflets; leaflets oval to oblong or lance-shaped, 1–4 cm long; margins smooth; stalks 10–40 cm long. **Flower Cluster:** Raceme 5–15 cm long, several-flowered. **Flowers:** Yellowish white, 0.8–1.2 cm long; petals 4, notched at tip; sepals 4, purplish green; stamens 5–16, purple, unequal in length; pistil 1. **Fruit:** Capsule 2–5 cm long, finely net-veined.

Notes: The genus name *Polanisia* comes from the Greek words *polu*, "many," and *anisos*, "unequal," a reference to the different lengths of the stamens. The species name *dodecandra* means "12 stamens."

Spiderflower, Cleome
Tarenaya hassleriana

Habitat: Waste areas, roadsides and disturbed sites; introduced ornamental from tropical North America. **General:** Stems 20–100 cm tall, covered in sticky hairs. **Leaves:** Alternate, palmately divided into 5–7 leaflets; leaflets oblong to lance-shaped, 6–10 cm long, finely haired; margins finely toothed; stalks long, often with 2 spines at the base. **Flower Cluster:** Raceme many-flowered; bracts simple. **Flowers:** White to pink; petals 4; sepals 4; stamens 6, filaments long; pistil 1. **Fruit:** Capsule 5–10 cm long, conspicuously veined; fruiting stalks 5–10 cm long.

Notes: The species name honours Swiss physician and botanist Emile Hassler (1894–1937), who is known for his collections of flora in Paraguay.

MUSTARD FAMILY
Brassicaceae

Formerly known as Cruciferae, the mustard family includes over 3000 species of shrubs and herbaceous plants. The mustards have alternate, simple to pinnately compound leaves. The leaves and stems are often covered in variously shaped hairs that can assist in identifying the plant. The flowers have 4 petals, 4 sepals, 6 stamens and 1 pistil. The stamens are of 2 lengths, 4 long-stalked and 2 short-stalked, the 2 short stamens located outside the inner 4 longer ones. From above, the flowers resemble a cross or crucifix, hence the old family name Cruciferae, which means "cross-bearing." The pod-like fruits are called silicles or siliques. A silicle is long and slender, whereas a silique is shorter and somewhat wider.

Several members of the mustard family are economically important food crops. One species, *Brassica oleracea*, has through hybridization given rise to collard greens, broccoli, cauliflower, cabbage, Brussels sprouts and kohlrabi. Other species of importance are radish (*Raphanus sativus*), canola (*Brassica rapa* and *B. napus*) and turnip (*Brassica rapa* var. *rapa*). Several species in this family are noxious weeds.

Garlic Mustard · *Alliaria petiolata*

Habitat: Waste areas and woods; introduced from Europe as a medicinal herb and salad green in 1879. **General:** Stems 10–100 cm tall, rarely branched, strong garlic odour; somewhat shade tolerant. **Leaves:** Basal leaves dark green, kidney-shaped, 3–12 cm across, margins with rounded teeth, stalks 1–5 cm long, hairy; stem leaves alternate, triangular, 3–8 cm long, stalkless, margins coarsely toothed, reduced in size upward; young leaves with distinctive garlic odour when crushed. **Flower Cluster:** Raceme elongating to 25 cm long in fruit. **Flowers:** White, 6–10 mm wide; petals 4, 3–6 mm long; sepals 4, green; stamens 6; pistil 1. **Fruit:** Silique pod-like, 2.5–6 cm long, 1–2 mm wide; fruiting stalks 4–5 mm long, spreading; seeds 6–22.

Notes: The genus name comes from the Latin word *allium*, "onion," a reference to the plant's odour. • Garlic mustard's ability to tolerate semi-shade has allowed it to spread throughout forests, where it outcompetes many native spring wildflowers.

Yellow Rocket
Barbarea vulgaris

Habitat: Roadsides, ditches and waste areas; introduced from Europe and Asia about 1800; designated weed in Ontario. General: Stems 30–90 cm tall, hairless, leafy, branched near the top; roots fibrous. Leaves: Basal leaves dark green, 5–25 cm long, stalks 2–8 cm long, margins with 2–8 lateral lobes and 1 large terminal lobe; stem leaves alternate, 1–2.5 cm long, 1–4 pairs per stem, reduced in size upward, margins wavy to lobed, stalkless and partially clasping the stem, lower leaves long-stalked. Flower Cluster: Raceme pyramid-shaped, elongating in fruit. Flowers: Yellow, 1–1.6 cm across; petals 4, 5–8 mm long; sepals 4; stamens 4; pistil 1. Fruit: Silicle pod-like, 1.5–3.8 cm long, erect, on stalks 3–6 mm long; seeds 5–20.

Notes: The genus name commemorates St. Barbara, a 4th-century saint. The species name *vulgaris* mean "common," a reference to this plant's wide distribution.

Hoary Alyssum
Berteroa incana

Habitat: Fields, roadsides and waste areas; introduced from the Lake Baikal region of Russia. General: Stems 20–90 cm tall, leafy, often purplish and branched above, covered in grey, star-shaped hairs. Leaves: Basal leaves oblong to lance-shaped, long-stalked, margins smooth; stems leaves alternate, 1–4 cm long, 0.7–1.2 cm wide, stalkless, not clasping; all leaves covered in whitish grey, star-shaped hairs. Flower Cluster: Raceme 5–25 cm long, elongating in fruit; flower stalks 7–8 mm long. Flowers: White, numerous, 2–3 mm across; petals 4, deeply notched; sepals 4, densely haired; stamens 6; pistil 1. Fruit: Silique pod-like, oval, 5–8 mm long, 3–4 mm wide, erect, covered with grey, star-shaped hairs; seeds 4–12, dark reddish brown.

Notes: The genus name commemorates Italian botanist and physician Guiseppe Bertero (1789–1831). The species name *incana* means "hoary" or "light grey," a reference to the plant's greyish green colour.

Field Mustard

Brassica rapa var. *rapa*

Habitat: Waste areas and roadsides; introduced from Europe.
General: Stems 20–120 cm tall,
bluish green, branched from the
base. Leaves: Alternate; lower
leaves 15–20 cm long, 5–7 cm wide,
stalked, wavy-toothed to deeply
2–4-lobed; upper leaves oblong to
lance-shaped, stalkless and clasping
the stem. Flower Cluster: Raceme
5–50-flowered, stalks 2–3.5 cm long.
Flowers: Yellow, 8–15 mm across;
petals 4, 6–11 mm long; sepals 4;
stamens 6; pistil 1. Fruit: Silique
pod-like, 3–8 cm long; on ascending
stalks 0.7–2.5 cm long; seeds dark
brown to black.

Notes: This species is very closely
related to cultivated canola (*B. napus*),
which is grown for its seed oil.

American Sea Rocket

Cakile edentula

Habitat: Sandy shorelines. General:
Stems spreading, leafy, profusely
branched, 10–30 cm tall, somewhat
fleshy; taproot shallow. Leaves: Alternate, fleshy, oval to lance-shaped,
7–13 cm long; margins wavy or toothed;
stalks short. Flower Cluster: Raceme
5–15 cm long, 2–15-flowered; flower
stalks 2–5 mm long. Flowers: Pale
purple, 5–7 mm wide; petals 4; sepals 4;
stamens 6; pistil 1. Fruit: Silique pod-
like, jointed, 0.8–2.8 cm long; terminal
segment 0.8–2 cm long, lower segment
4–8 mm long; seeds 1–2.

Notes: The species name *edentula*
means "small-toothed," a reference to
the leaf margins. • The common name
refers to the plant's habitat and its
resemblance to rocket, another
cruciferous plant.

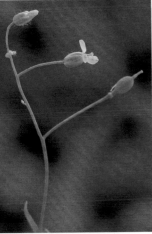

Small-seeded False Flax
Camelina microcarpa

Habitat: Fields and waste areas; introduced from Europe. **General:** Stems 20–90 cm tall, leafy, greyish green, covered with short, simple or star-shaped hairs. **Leaves:** Basal leaves spatula-shaped with a tapered stalk; stem leaves alternate, 0.9–1.5 cm long, 4–6 mm wide, clasping the stem, base arrowhead-shaped; lower leaves covered in grey, star-shaped hairs about 2 mm long. **Flower Cluster:** Raceme compact, elongating to over 20 cm fruit. **Flowers:** Pale yellow, 3–6 mm across; petals 4; sepals 4; stamens 6; pistil 1. **Fruit:** Silicle pod-like, pear-shaped, 4–6 mm long; seeds 10 or more; fruiting stalks 0.6–2.5 cm long.

Notes: The genus name *Camelina* comes from the Greek words *chamai*, "on the ground," and *linon*, "flax," alluding to the plant inhabiting flax fields. The species name *microcarpa* means "small-seeded."

Shepherd's-purse
Capsella bursa-pastoris

Habitat: Fields, waste areas and roadsides; introduced from Europe. **General:** Stems 10–80 cm tall, smooth to somewhat hairy. **Leaves:** Basal leaves oblong, 3–15 cm long, to 4 cm wide, margins wavy to deeply lobed; stem leaves alternate, clasping the stem, reduced in size upward, margins smooth to lobed; leaves highly variable from plant to plant. **Flower Cluster:** Raceme 10–40 cm long, elongating in fruit. **Flowers:** White, 3–8 mm across; petals 4; sepals 4; stamens 6; pistil 1. **Fruit:** Silicle pod-like, triangular to heart-shaped, 5–8 mm long; seeds orange.

Notes: *Capsella* means "little box." The species name *bursa-pastoris* means "purse of the shepherd." • The triangular fruit resembles the purse carried by early shepherds, hence the species and common names.

Cut-leaved Toothwort
Cardamine concatenata

Habitat: Wet woods and swampy areas. **General:** Stems 20–40 cm tall; rhizome constricted at joints, the segments 2–3 cm long. **Leaves:** Whorl of 3 leaves at or above mid-stem; stem leaves 5–13 cm across, compound, with 3 leaflets; leaflets deeply divided into 3 segments, the segments linear to lance-shaped and coarsely toothed; basal leaves similar, usually absent at time of flowering. **Flower Cluster:** Raceme 3–10-flowered; flower stalks 5–12 mm long. **Flowers:** White, occasionally tinged pink or pale purple, 1–2 cm across, fragrant; petals 4; sepals 4, greenish purple; stamens 6; pistil 1. **Fruit:** Silique pod-like, 2–5 cm long, hairless.

Notes: The species name *concatenata* means "joined" or "linked together." • This species flowers in early spring before the leaves of trees appear. The fruit and plant have often disappeared by the time the leaves have fully emerged.

Two-leaved Toothwort, Broad-leaved Toothwort
Cardamine diphylla

Habitat: Moist woods and swampy areas. **General:** Stems 10–40 cm tall; rhizome short, pepper-flavoured. **Leaves:** Basal leaves 4–10 cm long, 2–6.5 cm wide, long-stalked, compound with 3 leaflets, the leaflets elliptic to oval, coarsely toothed; stem leaves opposite, 2, similar to basal leaves, margins toothed. **Flower Cluster:** Raceme 3–10-flowered; flower stalks 4–12 mm long. **Flowers:** White to pale blue, 1.6–2.5 cm across; petals 4; sepals 4; stamens 6; pistil 1. **Fruit:** Silique pod-like, flattened, 2–4 cm long.

Notes: The species name *diphylla* means "2-leaved," a reference to the number of stem leaves.

Purple Cress
Cardamine douglasii

Habitat: Moist to wet, wooded areas. **General:** Stems 10–40 cm tall, erect; root thick, tuber-like. **Leaves:** Alternate, 3–5 per stem, narrowly oblong to oval, reduced in size upward; margins smooth to wavy-toothed; lower leaves stalked, upper leaves stalkless. **Flower Cluster:** Raceme compact, elongating with age. **Flowers:** Pink to purple, 0.5–1.5 cm across; petals 4; sepals 4; stamens 6; pistil 1. **Fruit:** Silique pod-like, 1.5–2.5 cm long.

Notes: The species name commemorates David Douglas (1798–1834), a Scottish botanist who collected and documented several plant species in North America.

Bitter Cress
Cardamine pensylvanica

Habitat: Wet ground, marshes and streambanks. **General:** Stems 10–50 cm tall, branched, arising from basal rosette of leaves. **Leaves:** Basal leaves 2–8 cm long, pinnately compound with 1–8 pairs of rounded leaflets, terminal segment largest; stem leaves alternate, oval to linear or lance-shaped, reduced in size upward. **Flower Cluster:** Raceme compact, elongating in fruit. **Flowers:** White, 3–10 mm across; petals 4; sepals 4; stamens 6; pistil 1. **Fruit:** Silique pod-like, 1–3 cm long, stalks ascending.

Notes: *Cardamine* comes from the ancient Greek word *kardamon*, the name for an unknown mustard plant. The species name means "of Pennsylvania."

Hare's-ear Mustard,
Treacle · *Conringia orientalis*

Habitat: Waste areas, roadsides and disturbed sites; introduced from the eastern Mediterranean region of Asia. General: Stems 30–100 cm tall, somewhat succulent, simple to branched, hairless. Leaves: Bluish green, hairless, somewhat fleshy; basal leaves oval, 5–13 cm long, margins smooth, stalks short; stem leaves alternate, base heart-shaped, stalkless and clasping. Flower Cluster: Raceme compact, 10–25-flowered, elongating to 20 cm at maturity. Flowers: Yellowish to greenish white, 0.5–1.1 cm across; petals 4, 0.8–1.2 cm long; sepals 4; stamens 6; pistil 1. Fruit: Silique pod-like, 4-angled, 4–15 cm long, 1–2 mm wide, slightly twisted or curved; stalks erect, 1–1.5 cm long; seeds oblong, dark reddish brown.

Notes: The genus name commemorates German physician Hermann Conring (1606–81). The species name *orientalis* means "eastern."

Flixweed

Descurainia sophia

Habitat: Roadsides and waste areas; introduced from Europe and Asia before 1821. General: Stems 30–100 cm tall, branched, greyish green owing to star-shaped hairs. Leaves: Alternate, 2–10 cm long, divided 2–3 times into narrow segments, covered in fine, grey, star-shaped hairs; stalks 0.5–2 cm long. Flower Cluster: Raceme compact, elongating in fruit. Flowers: Pale yellow, 2–4 mm across; petals 4; sepals 4; stamens 6; pistil 1. Fruit: Silique pod-like, 1.5–3 cm long, on spreading stalks 0.7–1.2 cm long; seeds 20–40.

Notes: The genus name honours French apothecary François Déscurain (1658–1740). The species name comes from the Greek word *sophia*, meaning "skill, wisdom or knowledge."

Sand Rocket, Stinking Wall-rocket

Diplotaxis muralis

Habitat: Sandy to gravelly soils along roadsides and in waste areas; introduced from Europe. **General:** Stems 20–50 cm tall, branched at base, hairless. **Leaves:** Basal, oblong to lance-shaped, 5–10 cm long, 1–3 cm wide; margins coarsely toothed to deeply lobed. **Flower Cluster:** Raceme compact, elongating in fruit. **Flowers:** Bright yellow, 1–2 cm across; petals 4; sepals 4; stamens 6; pistil 1. **Fruit:** Silique pod-like, flat, 2–3 cm long, 2 mm wide, on ascending stalks 0.8–2 cm long.

Notes: *Diplotaxis* comes from the Greek words *diplos* and *taxis*, meaning "double arrangement," referring to the fruit, which has 2 rows of seeds. The species name *muralis* means "wall."

Dog Mustard

Erucastrum gallicum

Habitat: Roadsides, railway grades and waste areas; introduced from Europe in the early 1900s. **General:** Stems 20–100 cm tall, branched; lower stem with bristly, downward-pointing hairs. **Leaves:** Alternate, oblong to lance-shaped, 3–25 cm long, 1–6 cm wide, covered with stiff, simple hairs, irregularly toothed and deeply lobed often to the midrib; short-stalked to stalkless. **Flower Cluster:** Raceme 10–30 cm long, elongating in fruit. **Flowers:** Pale yellow, 5–7 mm across; petals 4, with greenish veins; sepals 4; stamens 6; pistil 1. **Fruit:** Silique pod-like, 4-sided, 2–5 cm long; stalks upward-curved, 6–10 mm long.

Notes: The genus name *Erucastrum* comes from the Greek words *eruca* and *astrum*, referring to this plant's resemblance to species in the genus *Eruca*. The species name *gallicum* refers to Gaul, an ancient region of present-day western Europe.

Wormseed Mustard
Erysimum cheiranthoides

Habitat: Roadsides and waste areas.
General: Stems 20–50 cm tall, simple to branched, often purplish green near the base. **Leaves:** Alternate, oblong to lance-shaped, 2–10 cm long, 0.2–2 cm wide, cov-ered with branched hairs; margins smooth to finely toothed; short-stalked to stalk-less. **Flower Cluster:** Raceme compact, elongating to 20 cm long in fruit. **Flowers:** Yellow, 0.5–1 cm across; petals 4; sepals 4; stamens 6; pistil 1. **Fruit:** Silique pod-like, 4-sided, 1–3 cm long; stalks 0.6–1.2 cm long, held close to stem; seeds oblong.

Notes: The genus name *Erysimum* comes from the Greek word *erusimon*, a name given to an unknown species of cultivated mustard. • The common name refers to the resemblance of the seeds to small worms.

Dame's-rocket
Hesperis matronalis

Habitat: Roadsides, open forests and moist woods; introduced ornamental from Europe. **General:** Stems 30–100 cm tall, branched, covered in simple to forked hairs. **Leaves:** Alternate, oblong to lance-shaped, 5–15 cm long, 1–4 cm wide, hairy on both surfaces; margins shallowly toothed; stalkless. **Flower Cluster:** Raceme terminal, many-flowered. **Flowers:** Pinkish purple to white, fra-grant, 1.4–2.5 cm across; petals 4; sepals 4; stamens 6; pistil 1. **Fruit:** Silique pod-like, cylindric, 2.5–14 cm long, somewhat constricted between seeds; fruiting stalks ascending to spreading, 0.7–1.7 cm long.

Notes: The genus name come from the Greek word *herpera*, "evening," a refer-ence to the flowers, which are fragrant at night. The species name *matronalis* com-memorates the Roman festival of matrons, which occurs on March 1.

Common Peppergrass
Lepidium densiflorum

Habitat: Roadsides and dry, open areas; native species with a weedy nature. **General:** Stems 20–60 cm tall, profusely branched, greyish green owing to short hairs. **Leaves:** Basal leaves 3–10 cm long, 1–2 cm wide, margins deeply toothed to pinnately lobed, stalked; stem leaves alternate, margins smooth to slightly toothed, stalkless. **Flower Cluster:** Raceme densely flowered, 5–15 cm long. **Flowers:** Pinkish white, 2–3.5 mm long; petals 4, very small or absent; sepals 4; stamens 6; pistil 1. **Fruit:** Silicle pod-like, heart-shaped, 2–3 mm wide; seeds 2.

Notes: The species name *densiflorum* means "densely flowered."

Ball Mustard
Neslia paniculata

Habitat: Cultivated fields, roadsides and waste areas; introduced from Europe. **General:** Stems 20–100 cm tall, erect, yellowish green owing to numerous star-shaped hairs. **Leaves:** Basal leaves lance-shaped, short-stalked; stem leaves alternate, lance-shaped, 1–6 cm long, 0.2–2.5 cm wide, base arrowhead-shaped and clasping the stem. **Flower Cluster:** Raceme terminal, elongating in fruit. **Flowers:** Bright yellow, 1.5–3 mm wide; petals 4; sepals 4; stamens 6; pistil 1. **Fruit:** Silicle pod-like, round, 2–3 mm across, remaining attached to stem at maturity.

Notes: The genus name commemorates early French botanist Jacques Denesle (1735–1819).

Wild Radish
Raphanus raphanistrum

Habitat: Roadsides and waste areas; introduced from Europe and Asia.
General: Stems 30–90 cm tall, branched, somewhat bristly near the base. **Leaves:** Basal leaves 5–20 cm long, deeply divided into 5–15 oblong lobes, surfaces with bristly hairs; stem leaves alternate, oblong to oval, 2–7.5 cm long, less than 2 cm wide, margins wavy or with 2–5 irregularly toothed lobes. **Flower Cluster:** Raceme terminal; flower stalks 1–2.5 cm long.
Flowers: Bright to pale yellow tinged purple or purple-veined, 1–2 cm across; petals 4; sepals 4; stamens 6; pistil 1. **Fruit:** Silique pod-like, 3–7.5 cm long, strongly ribbed, constricted between seeds, breaking into barrel-shaped segments at maturity; seeds 4–10.

Notes: The genus and species names are derived from the Greek word *raphanis*, the name for radish.

Bog Yellow Cress
Rorippa palustris

Habitat: Marshy ground along lakes and streams. **General:** Stems 20–60 cm tall, branched, hairless to slightly bristly.
Leaves: Alternate, 6–15 cm long, 1.5–5 cm wide, lyre-shaped with deep to shallow lobes; lower leaves stalked, upper leaves stalkless and clasping the stem. **Flower Cluster:** Raceme terminal; bracts absent.
Flowers: Yellow, less than 6 mm across; petals 4; sepals 4; stamens 6; pistil 1. **Fruit:** Silique pod-like, curved, 3–7 mm long; stalks 4–8 mm long, spreading.

Notes: The species name *palustris* means "of the marsh," a reference to this plant's habitat.

White Mustard
Sinapis alba

Habitat: Roadsides, waste areas and cultivated fields; introduced from Europe and Asia. General: Stems 25–100 cm tall, leafy, bristly haired. Leaves: Alternate, 3.5–16 cm long, 2–8 cm wide, fiddle-shaped to deeply lobed, reduced in size upward; lower leaf stalks 1–3 cm long, upper stalks shorter. Flower Cluster: Raceme terminal. Flowers: Yellow, 1.6–3 cm across; petals 4; sepals 4; stamens 6; pistil 1. Fruit: Silique pod-like, flat, 2–4.2 cm long, covered with bristly hairs; fruiting stalks 0.6–1.2 cm long, spreading.

Notes: The species name *alba* means "white."

Wild Mustard
Sinapis arvensis

Habitat: Cultivated crops, roadsides and waste areas; introduced from Europe and Asia. General: Stems 20–180 cm tall, bristly haired at the base, hairless near the top; junction of stem and side branches often purplish green. Leaves: Alternate, 4–15 cm long, 1–10 cm wide; lower leaves with small lateral lobes and large terminal lobe, stalked; upper leaves coarsely lobed, stalkless but not clasping the stem. Flower Cluster: Raceme terminal. Flowers: Yellow 1–1.5 cm across; petals 4; sepals 4; stamens 6; pistil 1. Fruit: Silique pod-like, purplish green, 2–5 cm long, prominently ribbed; seeds 10–18.

Notes: The species name *arvensis* means "of the field," a reference to the plant's habitat.

Tall Tumble Mustard

Sisymbrium altissimum

Habitat: Roadsides and waste areas; introduced from Europe and Asia. **General:** Stems 60–120 cm tall, pale green, branched. **Leaves:** Basal leaves 4–30 cm long, with 5–8 pairs of broad, oblong to triangular lobes; stem leaves alternate, divided into linear segments, upper leaves with 2–5 thread-like segments. **Flower Cluster:** Raceme compact, elongating in fruit. **Flowers:** Pale yellow, 0.6–1.4 cm wide; petals 4; sepals 4; stamens 6; pistil 1. **Fruit:** Silique pod-like, 5–10 cm long, resembling leaf segments; fruiting stalks 0.6–1 cm long.

Notes: The species name *altissimum* means "tallest." • Mature plants become woody and brittle, breaking off at ground level to form tumbleweeds. The tumbleweeds blow across the ground in the wind, spreading seeds in their path.

Stinkweed, Field Pennycress

Thlaspi arvense

Habitat: Gardens, roadsides and waste areas; introduced from Europe and Asia. **General:** Stems 2–80 cm tall, pale green, hairless, turning yellow at maturity; strong turnip odour when crushed. **Leaves:** Basal leaves oblong to lance-shaped, 2–10 cm long, 0.4–2.3 cm wide, stalked, withered at flowering time; stem leaves alternate, stalkless and clasping the stem. **Flower Cluster:** Raceme compact, elongating in fruit. **Flowers:** White, 1–3 mm across; petals 4; sepals 4; stamens 6; pistil 1. **Fruit:** Silicle pod-like, flat, heart-shaped, 1–2 cm long; fruiting stalks 0.9–1.3 cm long; seeds 4–16.

Notes: The species name *arvense* means "of cultivated fields," a reference to this plant's weedy nature.

PITCHER-PLANT FAMILY
Sarraceniaceae

Pitcher-plants are insectivorous plants found in acidic bogs throughout the world. These herbs have hollow, alternate and basal leaves that are often filled with water. Stem leaves, when present, are reduced to small bracts. The flowers are composed of 4–6 petals, 4–6 sepals, 12 to many stamens and 1 pistil. The fruit is a capsule.

Pitcher-plant
Sarracenia purpurea

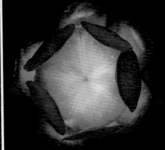

Habitat: Peat bogs and muskeg. **General:** Flowering stem 25–45 cm tall, hairless; insectivorous. **Leaves:** Basal, evergreen, pitcher-shaped, 7.5–30 cm long, green with reddish purple veins, winged on outer lower side, often filled with water. **Flower Cluster:** Solitary, atop a leafless stalk. **Flowers:** Purplish red, nodding, 5–7 cm wide; petals 5, purplish red, incurved; sepals 5, reddish green; stamens numerous; pistil 1, style umbrella-shaped. **Fruit:** Capsule 5-chambered; seeds winged.

Notes: The inside of the pitcher is lined with small, downward-pointing bristles. An insect, upon entering the pitcher, is forced downward by the bristly hairs. Unable to turn around, it drowns in the water-filled pitcher and nutrients from the insect's body are absorbed by the plant. • The pitcher-plant is the floral emblem of Newfoundland.

SUNDEW FAMILY
Droseraceae

Sundews are insectivorous plants found in bogs. The stalked, basal leaves are covered with sticky, red hairs that trap insects. These plants usually have 3–10 flowers composed of 4–8 petals, 4–8 sepals, 4–8 stamens and 1 pistil with 3–5 styles. The fruit is a capsule.

Linear-leaved Sundew
Drosera linearis

Habitat: Bogs and sandy shorelines. **General:** Flowering stems 6–13 cm tall, leafless; insectivorous. **Leaves:** Basal, linear, 2–5 cm long, about 2 mm wide, covered in sticky, red hairs; stalks 3–7 cm long, hairless. **Flower Cluster:** Raceme 1–4-flowered. **Flowers:** White, 6–8 mm wide; petals 5, about 6 mm long; sepals 5, 4–5 mm long; stamens 4–8; pistil 1. **Fruit:** Capsule 3–5 mm long; seeds black.

Notes: The species name *linearis* refers to the leaf shape.

Round-leaved Sundew
Drosera rotundifolia

Habitat: Peat bogs; often over-looked because the leaves blend in with the sphagnum moss. **General:** Flowering stems 10–20 cm tall; insectivorous. **Leaves:** Basal, round, 3–10 cm long, 0.6–1 cm across, covered with sticky, red hairs; in a rosette of 4–12 leaves. **Flower Cluster:** Raceme 3–10-flowered, often 1-sided. **Flowers:** White, 3–5 mm across; petals 5; sepals 5; stamens 4–8; pistil 1. **Fruit:** Capsule 3–5 mm long.

Notes: The genus name *Drosera* comes from the Greek word *droseros*, meaning "dewy," a reference to the dew-like glands on the leaves. The species name *rotundifolia* means "round-leaved."

DITCH STONE-CROP FAMILY
Penthoraceae

The ditch stonecrop family is represented by 2 species worldwide. Plants in this family are fleshy, rhizomatous herbs with alternate leaves. Stipules are absent. Flowers are borne in terminal, scorpioid racemes. The perfect flowers have 1–8 inconspicuous petals (sometimes absent), 5–8 basally fused sepals, 10 stamens and 1 pistil. The ovary is superior to partly inferior. The fruit, a capsule, contains numerous seeds.

The taxonomy of this family is greatly debated. It is sometimes included in the stonecrop family (Crassulaceae), the saxifrage family (Saxifragaceae) or, most recently, the water-milfoil family (Haloragaceae).

Ditch Stonecrop
Penthorum sedoides

Habitat: Ditches, marsh edges and wet soils. **General:** Stems erect, 15–60 cm tall; roots fibrous, producing stolons. **Leaves:** Alternate, elliptic to lance-shaped, 2–12 cm long, 1–4 cm wide; margins finely toothed; short-stalked to stalkless. **Flower Cluster:** Cyme 2–6-branched, 2–8 cm across; flowers nodding on upper side of branch, on stalks 1–3 mm long. **Flowers:** Yellowish to whitish green, 4–7 mm long; petals absent (if present, 1–8, inconspicuous); sepals 5; stamens 10, anthers yellowish pink; pistils 5, united below the middle. **Fruit:** Follicle 4–7 mm long, greenish brown to reddish; sepals persistent.

Notes: The genus name comes from the Greek words *pente*, "five," and *oros*, "a mark," a reference to the 5-parted flowers. The species name *sedoides* means "like *Sedum*," referring to this plant's resemblance to those in the genus *Sedum*.

STONECROP FAMILY
Crassulaceae

The stonecrops are succulent herbs and shrubs with alternate or opposite, simple leaves. The flowers, borne in cymes, have 4–5 petals, 4–5 sepals, 4–10 stamens and 4–5 pistils. Nectaries are often present at the base of the ovary. The fruit is a cluster of follicles.

Ornamental members of this family include jade plant (*Crassula ovata*) and hen-and-chicks (*Sempervivum* spp.).

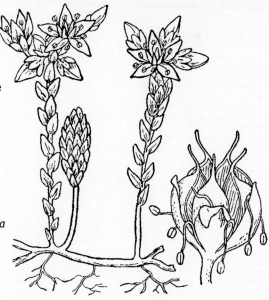

Mossy Stonecrop
Sedum acre

Habitat: Sandy to rocky soils; introduced ornamental from Europe and Asia.
General: Stems 2–12 cm tall, succulent, creeping, hairless; sterile and fertile branches present. **Leaves:** Alternate, crowded, overlapping, 2–6 mm long, oval in cross-section.
Flower Cluster: Cyme few-flowered, often 3-branched. **Flowers:** Yellow, 8–15 mm wide; petals 4–5; sepals 4–5; stamens 8 or 10; pistils 5. **Fruit:** Follicle 3–10 mm long, in clusters of 5.

Notes: The genus name *Sedum* comes from a Latin word for "succulent." The species name *acre* means "acrid" or "bitter-tasting." • Stonecrops thrive in drought conditions because the fleshy leaves retain water.

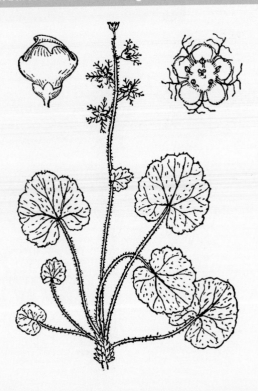

SAXIFRAGE FAMILY
Saxifragaceae

Members of the saxifrage family have simple, alternate, opposite or basal leaves. Stipules are absent. Flowers are borne in spikes, racemes or panicles. The perfect, regular flowers have 5 distinct petals, 4–5 fused sepals, 4–10 stamens and 3–5 pistils. The pistils are often fused at the base. The fruit is a capsule, berry or follicle.

Several species of this family are grown horticulturally. Including false spiraea (*Astilbe* spp.), bergenia (*Bergenia* spp.) and alumroot (*Heuchera* spp.)

Golden Saxifrage

Chrysosplenium alternifolium
ssp. *iowense*

Habitat: Mossy streambanks and marshy ground. General: Stems 3–15 cm tall, pale green to yellowish green; stolons leafy. Leaves: Alternate or opposite, round to kidney-shaped, 0.5–2 cm across, 0.4–2.5 cm long; margins wavy-toothed; stalks 1–3.5 cm long. Flower Cluster: Cyme terminal, 5–12-flowered; bracts yellow. Flowers: Golden yellow, 1.5–5 mm across, bell-shaped; petals absent; sepals 4; stamens 2–8; pistils 2. Fruit: Capsule somewhat 2-lobed; seeds several, light brown.

Notes: *Chrysosplenium* means "shining gold," a reference to the colour of the plant. The species name *alternifolium* means "alternate-leaved."

Two-leaved Mitrewort

Mitella diphylla

Habitat: Moist, rocky woods. General: Stems 10–40 cm tall. Leaves: Basal leaves oval to heart-shaped, 2–7.5 cm long, 1–3 cm wide, resembling maple leaves, hairy, long-stalked; stem leaves 2, opposite, heart-shaped, 3–5-lobed, margins toothed, stalkless, located at midstem. Flower Cluster: Raceme 5–20-flowered. Flowers: White, 3–6 mm wide, snowflake-like; petals 5, fringed; sepals 5; stamens 10; pistil 1; styles 2. Fruit: Capsule mitre-shaped, 2–4 mm long.

Notes: The species name *diphylla* means "2-leaved." • The common name "mitrewort" refers to the fruit, which is shaped like a bishop's mitre.

Bishop's-cap
Mitella nuda

Habitat: Cool, mossy woods. **General:** Flowering stems 5–20 cm tall; rootstock creeping. **Leaves:** Basal leaves heart-shaped, 1.5–6 cm long, 2–5 cm across, both surfaces with stiff hairs; stem leaf, when present, oval and located at midstem. **Flower Cluster:** Raceme 3–10-flowered, 2–10 cm long. **Flowers:** Greenish white, 0.5–1 cm across; petals 5, feather-like; sepals 4; stamens 10; pistils 2. **Fruit:** Capsule 2-valved at tip; seeds black, shiny.

Notes: The genus name *Mitella* comes from the Greek word *mitra*, "cap," a reference to the shape of the fruit. The species name *nuda*, "naked," refers to the flowering stem.

Early Saxifrage
Saxifraga virginiensis

Habitat: Dry to moist, open woods. **General:** Flowering stem 5–10 cm tall, covered in sticky hairs; fruiting stems 20–40 cm tall; roots fibrous. **Leaves:** Basal, oblong to oval, 2–7.5 cm long, base wedge-shaped; margins toothed; stalks short, winged. **Flower Cluster:** Cyme panicle-like, compact, expanding with age. **Flowers:** White, fragrant, 5–12 mm wide; petals 5; sepals 5; stamens 10; pistils 2; appearing before leaves of trees emerge. **Fruit:** Capsule 2-beaked; seeds numerous.

Notes: The genus name comes from the Greek words *saxum*, "rock," and *frango*, "to break," referring to the fact that many saxifrages live in the cracks of rocks. The species name means "of Virginia."

Heartleaf Foamflower
Tiarella cordifolia

Habitat: Open to shady, dry to swampy, deciduous woods. General: Stems 15–40 cm tall. Leaves: Basal (may have 1 stem leaf), 3–13 cm long, 3–10 cm wide, shallowly palmately 3–5-lobed; margins toothed; stalks 2–20 cm long. Flower Cluster: Raceme 15–50-flowered, 4–40 cm long, 1–2 cm wide; flower stalks with dissected bracts. Flowers: White, bell-shaped, 3–6 mm long, 3–8 mm wide; petals 5, 3-toothed at tip; sepals 5; stamens 10; pistil 1. Fruit: Capsule 5–11 mm long; seeds 4–15.

Notes: *Tiarella* comes from the Latin words *tiara*, "turban," and *ella*, a suffix meaning "small," a reference to the shape of the capsule. The species name *cordifolia* means "heart-leaved."

ROSE FAMILY · Rosaceae

Family description on p. 49.

Agrimony
Agrimonia striata

Habitat: Open woods. General: Stems 50–120 cm tall, erect, covered in short, stiff, brown hairs. Leaves: Alternate, compound with 5–9 main leaflets interspersed with several smaller leaflets; leaflets 2–7 cm long, 1–4.4 cm wide, coarsely veined, covered with soft hairs, underside glandular-dotted. Flower Cluster: Raceme spike-like, 5–20 cm long, many-flowered. Flowers: Yellow, 0.6–1.2 cm across; petals 5; sepals 5; stamens 5–15; pistils 2.

Fruit: Achenes, in pairs; persistent calyx covered with hooked bristles.

Notes: The species name *striata* means "striped."

Silverweed
Argentina anserina

Habitat: Moist, open areas, often on sandy shores and streambanks. **General:** Stems 50–200 cm long, tufted, trailing, producing reddish green runners (stolons) that root and form new plants. **Leaves:** Basal, 10–45 cm long, 2–8 cm wide, pinnately compound with 7–25 leaflets interspersed with a few smaller leaflets; leaflets oblong to elliptic, to 4 cm long, green above, silvery white and hairy below; margins sharply toothed. **Flower Cluster:** Solitary, on leafless stalks at stolon nodes; bracts 5, small, alternating with sepals. **Flowers:** Bright yellow, 1.6–2.5 cm across; petals 5; sepals 5; stamens 20–25; pistils numerous. **Fruit:** Achene; in head-like clusters.

Notes: The species name *anserina* means "belonging to geese," a reference to this plant's habitat of moist, sandy shorelines and areas frequented by wild geese.

Marsh Cinquefoil
Comarum palustre

Habitat: Marshes, swamps and shorelines. **General:** Stems 20–60 cm tall, reddish brown; rhizome reddish brown, creeping, 20–60 cm long, woody. **Leaves:** Alternate, purplish green, long-stalked, pinnately compound with 5–7 leaflets; leaflets oblong to lance-shaped, 2–10 cm long, 1–3 cm wide, margins sharply toothed; upper stem leaves short-stalked, leaflets 5 (3 large, 2 small); stipules sheathing. **Flower Cluster:** Cyme 1–8-flowered, leafy; bracts alternating with and shorter than sepals. **Flowers:** Reddish purple to maroon, 1.5–3.5 cm across; petals 5, reddish; sepals 5, purple; stamens 2–25; pistils several. **Fruit:** Achene smooth, brown.

Notes: The species name *palustris* means "of the marsh," referring to this plant's habitat.

Woodland Strawberry

Fragaria vesca

Habitat: Open woods. **General:** Flowering stems 5–25 cm tall; rootstock short; runners (stolons) producing new plants at nodes. **Leaves:** Basal, 3–20 cm long, compound with 3 leaflets; leaflets oblong to oval, 1–3 cm long, 1–2 cm wide, stalkless; terminal leaflet with 1 terminal tooth longer than the adjacent teeth; margins toothed above the middle. **Flower Cluster:** Raceme or panicle 3–15-flowered, rising above leaves. **Flowers:** White, 1.5–2 cm across; petals 5; sepals 5; stamens and pistils numerous. **Fruit:** Achenes borne on the surface of a red, berry-like fruit (strawberry); edible.

Notes: *Fragaria* comes from the Latin word *fraga*, a reference to the fragrant scent of the fruit. The species name *vesca* means "little." • The fleshy, red part of the strawberry is actually the enlarged tip of the flowering stem. This type of fruit is called an "accessory." The seeds are actually single-seeded fruits called achenes.

Yellow Avens

Geum aleppicum

Habitat: Moist woods and meadows. **General:** Stems 40–100 cm tall, bristly haired; rootstock stout. **Leaves:** Basal leaves numerous, stalked, 15–40 cm long, compound with 5–7 toothed leaflets, the terminal leaflet 5–10 cm long, deeply 3-lobed and wedge-shaped at the base; stem leaves alternate, compound with 3–5 leaflets, reduced in size upward, stalkless, margins toothed; stipules large. **Flower Cluster:** Panicle; flower stalks bristly haired; bracts 5, alternating with sepals. **Flowers:** Bright yellow, 1–2.5 cm across; petals 5; sepals 5; stamens 10 to many; pistils numerous. **Fruit:** Achene with hooked beak, in head-like clusters 1–1.5 cm across.

Notes: *Geum* is the ancient Latin name for a plant in this genus. The species name means "of Aleppo," a city in Syria.

White Avens, Canada Avens
Geum canadense

Habitat: Dry to moist, open woods. **General:** Stems 40–100 cm tall; covered with short and long hairs; rhizome stout. **Leaves:** Basal leaves 2–15 cm long, long-stalked, compound with 3–5 leaflets, the leaflets oblong to lance- or diamond-shaped, margins coarsely toothed; stem leaves alternate, short-stalked to stalkless, leaflets 3, margins coarsely toothed; stipules leaf-like, to 2 cm long and 1 cm wide. **Flower Cluster:** Corymb few-flowered; bracts 5, alternating with sepals. **Flowers:** White, 1–1.4 cm across; petals 5; sepals 5, pointed; stamens and pistils numerous. **Fruit:** Achene bristly haired, in oblong clusters 1–1.5 cm long; styles persistent, hooked, 7–9 mm long.

Notes: The species name means "of Canada."

Purple Avens
Geum rivale

Habitat: Marshes and wet meadows. **General:** Stems 40–100 cm tall, surface with scattered, bristly hairs. **Leaves:** Basal leaves 5–30 cm long, fiddle-shaped, long-stalked, compound with 3–5 main leaflets and several smaller leaflets, the terminal leaflet somewhat 3-lobed and coarsely toothed; stem leaves 3-lobed, slightly hairy, margins coarsely toothed. **Flower Cluster:** Raceme 2–4-flowered, in upper leaf axils; flower stalks long; bracts 5, alternating with sepals. **Flowers:** Pinkish purple, nodding, 1.5–2 cm across; petals 5, pinkish yellow with purple veins; sepals 5, yellowish purple; stamens and pistils numerous. **Fruit:** Achene 6–8 mm long; in head-like clusters 1–2 cm across; styles elongating and becoming feathery with age.

Notes: The species name comes from the Latin word *rivalis*, "brook," a reference to this plant's wet, marshy habitat.

Three-flowered Avens
Geum triflorum

Habitat: Dry meadows and hillsides.
General: Stems 10–40 cm tall, covered in soft hairs; rootstalk woody.
Leaves: Basal leaves 10–20 cm long, pinnately compound with 7–17 leaflets, the leaflets wedge-shaped, 1–5 cm long, increasing in size toward the tip, margins coarsely toothed; stem leaves alternate, smaller, comb-like or pinnately lobed. **Flower Cluster:** Cyme 3-flowered, long-stalked; bracts 5, reddish purple, alternating with sepals.
Flowers: Purplish pink to yellow, 1.2–2 cm across; petals 5, pink or yellowish; sepals 5, purplish pink; stamens numerous; pistils numerous. **Fruit:** Achene; style feathery, 3–5 cm long, persistent; fruiting clusters erect at maturity.
Notes: The species name and the common name both refer to the 3-flowered stems.

White Cinquefoil
Potentilla arguta

Habitat: Moist, open meadows. **General:** Stems 40–100 cm tall, reddish green, covered with sticky, brown hairs. **Leaves:** Basal leaves stalked, pinnately compound with 7–11 leaflets, the leaflets oval, 1.5–4 cm long, the terminal leaflet 1–7 cm long, margins toothed; stem leaves alternate, leaflets 3–5, margins coarsely toothed. **Flower Cluster:** Cyme compact, elongating in fruit; bracts 5, alternating with sepals. **Flowers:** Creamy white, 1.2–2 cm across; petals 5; sepals 5; stamens 25–30; pistils numerous. **Fruit:** Achene light brown, about 1 mm long.

Notes: The genus name *Potentilla* comes from the Greek word *potens*, "powerful," a reference to the medicinal properties of some members of this genus. The species name *arguta* means "sharp-toothed."

Rough Cinquefoil · *Potentilla norvegica*

Habitat: Moist meadows, waste areas and road-sides; native but often weedy. **General:** Stems 20–90 cm tall, often reddish green, branched, bristly haired. **Leaves:** Alternate, long-stalked, compound with 3 leaflets; leaflets elliptic to oblong or oval, 2–10 cm long, 0.8–4 cm wide, margins coarsely toothed, surfaces with long hairs; leaves reduced in size upward; stipules 4–5-lobed, 1–2.5 cm long. **Flower Cluster:** Cyme few-flowered, leafy; bracts 5, alternating with sepals; flower stalks 3–9 mm long. **Flowers:** Yellow, 7–12 mm across; petals 5; sepals 5; stamens 15–20; pistils numerous. **Fruit:** Achene; in head-like clusters, often concealed by enlarged sepals.

Notes: The species name means "of Norway," the location where this species was first collected. • This plant is found throughout the Northern Hemisphere.

Bushy Cinquefoil
Potentilla paradoxa

Habitat: Moist, sandy areas. **General:** Stems 20–50 cm long, spreading to trailing, bristly haired to hairless. **Leaves:** Alternate, stalked, compound with 7–11 leaflets; leaflets oblong to wedge-shaped, 1–3 cm long, margins sharply toothed. **Flower Cluster:** Cyme leafy, many-flowered. **Flowers:** Yellow, 5–7 mm wide; petals 5; sepals 5; stamens 10–20; pistils numerous. **Fruit:** Achene 0.5–1 mm long; cork-like thickening on 1 side nearly as large as the achene.

Notes: The species name *paradoxa* means "strange" or "unusual," a reference to the corky thickening on the achene.

Sulphur Cinquefoil
Potentilla recta

Habitat: Waste areas and roadsides; introduced from Europe and
Asia. General: Stems 30–70 cm
tall, erect, hairy; multiple stems
originating from the same crown.
Leaves: Basal leaves long-stalked,
palmately compound with 5–9 leaflets, the leaflets oblong to lance-
shaped, 2.5–14 cm long, surfaces
with soft hairs, margins with 7–17
teeth; stem leaves alternate, com-
pound with 3 leaflets; stipules leaf-like. Flower
Cluster: Corymb flat-topped, 15–35-flowered;
bracts 5, alternating with sepals. Flowers: Pale yellow
to sulphur yellow, 1.5–2.5 cm across; petals 5; sepals 5;
stamens 25–35; pistils numerous. Fruit: Achene
1–1.3 mm long; in head-like clusters.

Notes: The genus name comes from the Greek word
potens meaning "powerful," a reference to the medicinal
properties of some members of this genus. The species
name *recta* means "erect" and refers to the stem.

Dewberry
Rubus pubescens

Habitat: Moist woods.
General: Stems 10–30 cm
tall, trailing 10–100 cm,
base becoming semi-woody;
bark reddish brown.
Leaves: Alternate, stalks
2–7 cm long, compound
with 3–5 leaflets; leaflets
oval to diamond-shaped,
2–10 cm long, 1–4.5 cm
wide, margins sharply
toothed. Flower Cluster:
Solitary or raceme with
2 or 4 flowers; borne from the crown or at nodes on runners. Flowers: Pinkish to
white, 0.8–1.2 cm across; petals 5; sepals 5, reflexed; stamens and pistils numerous.
Fruit: Aggregate of drupelets, forming a reddish purple raspberry 1–1.5 cm across;
edible.

Notes: The genus name *Rubus* is the Latin name for blackberry. The species name
pubescens means "downy."

Barren Strawberry, Dry Strawberry · *Waldsteinia fragarioides*

Habitat: Dry, rich woods. **General:** Flowering stems 5–20 cm tall, hairy, a few scale-like leaves present; rhizome short. **Leaves:** Basal, evergreen, stalked, 10–20 cm long, compound with 3 leaflets, resembling the leaves of strawberry (*Fragaria* spp.); leaflets oblong to oval, 2.5–5 cm long, base wedge-shaped, margins toothed. **Flower Cluster:** Corymb 2–8-flowered; bracts absent. **Flowers:** Yellow, 1–1.5 cm wide; petals 5; sepals 5; stamens numerous; pistils 3–6. **Fruit:** Achenes 2–10; in clusters on dry receptacles.

Notes: The genus commemorates Count Franz Adam Waldstein-Wartenberg (1759–1823), an Austrian botanist. The species name *fragarioides* means "resembling *Fragaria*." • The common names refer to the dry fruit of this strawberry-like plant.

LEGUME OR PEA FAMILY Fabaceae

LEGUME OR PEA FAMILY · Fabaceae

Family description on p. 54.

Hog Peanut · *Amphicarpa bracteata*

Habitat: Moist woods. **General:** Climbing stems 30–120 cm long/tall, twining, hairy. **Leaves:** Alternate, long-stalked, compound with 3 leaflets; leaflets oval, pointed, 2–7.5 cm long; stipules oval to lance-shaped, prominently nerved. **Flower Cluster:** Raceme densely flowered, in upper leaf axils. **Flowers:** Pale purple to white, irregular, 1.2–1.8 cm long; petals 5; sepals 5, the upper 2 fused; stamens 10; pistil 1; self-pollinating, inconspicuous flowers produced at base of plant. **Fruit:** Legume produced by showy flowers pod-like, 1.6–3.8 cm long, seeds 3–4; legume produced by inconspicuous flowers pear-shaped, fleshy, 1-seeded, usually underground.

Notes: The genus name comes from the Greek words *amphi*, "both kinds," and *carpos*, "fruit," a reference to the 2 types of fruit. • The common name refers to the fondness that hogs have for the underground fruit of this plant. • Seeds of the aboveground fruits are inedible, whereas those of the underground fruits are edible and peanut-like.

Groundnut
Apios americana

Habitat: Moist, open forests and thickets. General: Climbing stems 30–300 cm long, twining; rhizome bearing pear-shaped tubers. Leaves: Alternate, long-stalked, 10–20 cm long, pinnately compound with 5–7 leaflets; leaflets oval to lance-shaped, 4–6 cm long; stipules falling as leaves expand. Flower Cluster: Raceme densely flowered, in leaf axils; bracts small. Flowers: Purplish brown, irregular, 1–1.3 cm long; petals 5; sepals 5; stamens 10; pistil 1. Fruit: Legume pod-like, 5–10 cm long.

Notes: The genus name *Apios* is Greek for "pear," a reference to the shape of the tubers. • The Pilgrims relied upon the tubers of this plant during their first years in North America.

Canada Tick-trefoil
Desmodium canadense

Habitat: Open woods and meadows. General: Stems 40–180 cm tall, bushy, finely haired. Leaves: Alternate, stalks 0.2–2 cm long, compound with 3 leaflets; leaflets oblong to lance-shaped, 3–9 cm long, underside hairy; stipules lance-shaped,

2–8 mm long, margins hairy. Flower Cluster: Raceme densely flowered; bracts conspicuous. Flowers: Pink to purple, irregular, 1–1.3 cm long; petals 5; sepals 5; stamens 10; pistil 1. Fruit: Loment pod-like, breaking into 3–5 single-seeded segments; segments 5–7 mm long, covered in bristly, hooked hairs.

Notes: The genus name *Desmodium* comes from the Greek word *desmos*, "bond" or "connection," a possible reference to the fruit segments.

Naked-flowered Tick-trefoil
Desmodium nudiflorum

Habitat: Rich woods. **General:** Stems branched from the base; sterile branch 10–30 cm tall, leafy at the top; fertile branch 40–100 cm tall, leafless. **Leaves:** Alternate, long-stalked, compound with 3 leaflets; leaflets 4–10 cm long, 2–8 cm wide, lateral leaflets oval to oblong, terminal leaflet elliptic to oval. **Flower Cluster:** Raceme loosely flowered; flower stalks 1–2 cm long. **Flowers:** White to purplish, irregular, 6–8 mm long; petals 5; sepals 5; stamens 10; pistil 1. **Fruit:** Loment pod-like, breaking into 2–4 segments; segments 8–11 mm long, covered in bristly, hooked hairs.

Notes: The species name *nudiflorum* means "naked flower," a reference to the leafless flowering stem.

Goat's-rue · *Galega officinalis*

Habitat: Waste areas and roadsides; introduced from Europe and Asia as a forage crop in the late 1800s. **General:** Stems 50–200 cm tall, bushy with several stems arising from the same crown, hollow, branched above, hairless. **Leaves:** Alternate, compound with 11–17 leaflets; leaflets elliptic to lance-shaped, 1–5 cm long, 0.4–1.5 cm wide; stipules arrowhead-shaped, toothed or lobed. **Flower Cluster:** Raceme 7–10 cm long, 20–50-flowered. **Flowers:** Purple to white, irregular, 1–1.5 cm long; petals 5; sepals 5; stamens 10; pistil 1. **Fruit:** Legume pod-like, 2–5 cm long, 2–3 mm wide; seeds 1–9, greenish yellow to reddish brown.

Notes: The genus name *Galega* comes from the Greek word *gala*, "milk," a reference to this plant supposedly increasing milk production in goats. • The leaves contain galegin, a poisonous alkaloid, making this plant unsuitable for livestock, so its use as a forage crop was abandoned.

Everlasting Pea, Perennial Sweetpea

Lathyrus latifolius

Habitat: Roadsides, waste areas and vacant lots; introduced ornamental from Europe. General: Climbing or trailing stems 50–200 cm long/tall, weak, prominently winged; rhizome stout. Leaves: Alternate, long-stalked, pinnately compound with 2 leaflets and a branched tendril; leaflets elliptic to lance-shaped, 4–8 cm long, 1–3 cm wide; stipules lance-shaped, leaf-like, 1.5–4 cm long. Flower Cluster: Raceme 4–10-flowered. Flowers: Purple, pink or white, fragrant, irregular, 1.5–2.5 cm long; petals 5; sepals 5; stamens 10; pistil 1. Fruit: Legume pod-like, 2–10 cm long, 0.5–1.5 cm wide.

Notes: *Lathyrus* is the Latin name for a pea. The species name *latifolius* means "broad-leaved." • The common name "perennial sweetpea" refers to the life cycle of this species and the fragrant flowers that resemble garden sweetpeas.

Yellow Pea Vine

Lathyrus ochroleucus

Habitat: Moist, wooded areas. General: Climbing stems 20–100 cm long/tall, hairless. Leaves: Alternate, pinnately compound with 6–10 leaflets and a branched tendril; leaflets oval, 2.5–5 cm long, hairless; stipules oval to somewhat heart-shaped, 1–2 cm long. Flower Cluster: Raceme 5–10-flowered, in leaf axils. Flowers: Yellowish white, irregular, 1.2–1.5 cm long; petals 5; sepals 5; stamens 10; pistil 1. Fruit: Legume pod-like, 2–4 cm long, hairless; seeds 4–6, considered poisonous.

Notes: The species name *ochroleucus* means "yellowish white."

Tuberous Vetchling
Lathyrus tuberosus

Habitat: Waste areas, roadsides and hillsides; introduced from Europe and western Asia as an erosion-control agent. **General:** Climbing stems 30–100 cm tall/long, often 4-angled, hairless; roots with tubers 5–12 mm across. **Leaves:** Alternate, compound with 2 leaflets and a thread-like, 3-branched tendril; leaflets narrowly elliptic, 2–5 cm long, short-stalked; stipules pointed, 9–12 mm long. **Flower Cluster:** Raceme 2–5-flowered, 2–10 cm long, in leaf axils. **Flowers:** Bright red to pink or violet, 1.2–1.8 cm long, resembling a sweetpea; petals 5; sepals 5; stamens 10; pistil 1. **Fruit:** Legume pod-like, 2.5–3.5 cm long, 4–6 mm wide, dull brown, hairless; seeds 3–6, brown, 3–4 mm across.

Notes: The species name *tuberosus* means "tuber-bearing." • This species is designated a noxious weed in Ontario.

Bird's-foot Trefoil
Lotus corniculatus

Habitat: Roadsides, waste areas and gardens; introduced forage crop from Europe and Asia. **General:** Stems 10–60 cm long, weak to trailing, several arising from the same crown, hairless to soft-haired; taproot deep, cord-like. **Leaves:** Alternate, pinnately compound with 5 leaflets; leaflets elliptic to oblong, 5–17 mm long, 5–8 mm wide, margins finely toothed, 2 lower leaflets often mistaken for stipules; stipules very small or absent. **Flower Cluster:** Raceme umbel-like, 2–10-flowered, on stalk 3–10 cm long. **Flowers:** Bright yellow, often red-tinged, irregular, 0.8–1.5 cm long; petals 5; sepals 5; stamens 10; pistil 1. **Fruit:** Legume brown, 2–4 cm long, tip claw-like; seeds 5–20.

Notes: The species name *corniculatus* means "with small horns," a reference to the claw-like tips of the fruit. • The arrangement of the pods resembles a bird's-foot, hence the common name.

Sundial Lupine
Lupinus perennis

Habitat: Dry, open woods. **General:** Stems 20–60 cm tall, finely haired. **Leaves:** Alternate, stalks 2–6 cm long, palmately compound with 7–11 leaflets; leaflets lance-shaped, 1–5 cm long, margins smooth. **Flower Cluster:** Raceme erect, 10–20 cm long, many-flowered. **Flowers:** Blue to pink or white, irregular, 1–1.6 cm long; petals 5; sepals 5; stamens 10; pistil 1. **Fruit:** Legume pod-like, 3–5 cm long, hairy.

Notes: *Lupinus* is the Latin name for lupine. The species name *perennis* means "perennial," a reference to the plant's life cycle.

Black Medick · *Medicago lupulina*

Habitat: Waste areas, roadsides and lawns; introduced from Europe and North Africa in 1792. **General:** Stems 20–80 cm long, trailing, often forming mats; taproot thick, cord-like. **Leaves:** Alternate, compound with 3 leaflets; leaflets oval, 0.5–3 cm long, 0.3–1 cm wide, hairy, terminal leaflet stalked, lateral leaflets stalkless; stipules lance- or awl-shaped. **Flower Cluster:** Raceme globe-shaped, 1–1.2 cm across, 20–50-flowered. **Flowers:** Yellow, irregular, 3–4 mm long; petals 5; sepals 5; stamens 10; pistil 1. **Fruit:** Legume pod-like, black, kidney-shaped, 1.5–3 mm long; seeds 1.

Notes: The genus name comes from the Greek word *medice* meaning "alfalfa." The species name *lupulina* means "hops-like," a reference to the clustered fruits.

Alfalfa
Medicago sativa

Habitat: Roadsides and waste areas; introduced forage crop from Europe. **General:** Stems 20–90 cm tall; taproot deep. **Leaves:** Alternate, compound with 3 leaflets; leaflets oval, 1–3 cm long, 0.3–1 cm wide, slightly hairy. **Flower Cluster:** Raceme compact, 10–30-flowered. **Flowers:** Purple to blue, 0.5–1.1 cm long, irregular; petals 5; sepals 5; stamens 10; pistil 1. **Fruit:** Legume pod-like, hairy, 4–7 mm long, coiled 2–3 times; seeds 8–10, yellowish brown.

Notes: The species name *sativa* means "cultivated." • Alfalfa is used as a soil stabilizer to prevent erosion. • Like most legumes, alfalfa has the ability to add nitrogen to the soil with the aid of certain bacteria. This process occurs in swollen parts of the roots called nodules.

White Sweet Clover
Melilotus alba

Habitat: Waste areas and road-sides; introduced from Europe as a forage crop in the mid-1600s. **General:** Stems 0.5–2.5 m tall, sweet-scented, branched, leafy; taproot deep. **Leaves:** Alternate, 1.5–5 cm long, compound with 3 leaflets; leaflets club-shaped, 1.2–2.5 cm long, 0.3–1 cm wide, margins toothed; stip-ules 0.7–1 cm long. **Flower Cluster:** Raceme 5–20 cm long, 40–80-flowered, axillary and ter-minal. **Flowers:** White, fragrant, irregular, 4–6 mm long; petals 5; sepals 5; stamens 10; pistil 1. **Fruit:** Legume pod-like, black, smooth, papery, 3–5 mm long; seeds 1–3, yellowish brown.

Notes: The genus name comes from the Greek word *meli*, "honey," a reference to the plant's role in honey production. The species name *alba* means "white."

Crown Vetch
Securigera varia

Habitat: Waste ground, railway grades and roadsides; introduced from Europe for erosion control. **General:** Stems 20–120 cm tall, hollow, weak, ascending, branched above, hairless; rhizomes fleshy, drought tolerant. **Leaves:** Alternate, 6–15 cm long, pinnately compound with 11–29 leaflets; leaflets oval to oblong, 6–20 mm long, 5–9 mm wide; stipules awl-shaped, 1–6 mm long. **Flower Cluster:** Raceme head- or umbel-like, 10–20-flowered, on stalks 1–7 cm long from leaf axils; flower stalks 3–4 mm long. **Flowers:** White to pink or purplish, irregular, 1–1.5 cm long; petals 5; sepals 5; stamens 10; pistil 1. **Fruit:** Loment pod-like, 2–6 cm long, 4-angled, leathery, breaking into 3–7 segments; fruiting cluster crown-like.

Notes: The genus name *Securigera* means "axe-bearing." The species name means "variable."

Red Clover
Trifolium pratense

Habitat: Roadsides, waste areas and lawns; introduced forage crop from Europe. **General:** Stems 40–80 cm tall, erect to spreading, branched, hairy. **Leaves:** Alternate, long-stalked, compound with 3 leaflets; leaflets oval to elliptic, 1.3–5 cm long, 0.5–1.5 cm wide, inverted "V" on upper surface; stipules oblong with pointed tips. **Flower Cluster:** Raceme head-like, 1–4 cm long, 2–5 cm across, 30–60-flowered. **Flowers:** Pink to red, irregular, 1.2–2.2 cm long; petals 5; sepals 5; stamens 10; pistil 1. **Fruit:** Legume pod-like, 3–7 mm long, concealed by persistent dried corolla; seeds 1–6, yellow or purple.

Notes: The genus name comes from the Latin words *tres* and *folium*, meaning "3-leaved."

White Clover

Trifolium repens

Habitat: Gardens, lawns and waste areas; introduced forage crop from Europe. **General:** Stems 10–50 cm long, creeping to trailing, often rooting at nodes. **Leaves:** Alternate, stalks 5–20 cm long, compound with 3 leaflets; leaflets oval to elliptic, 0.5–2 cm long, shallowly notched at tip. **Flower Cluster:** Raceme globe-shaped, 1–2 cm across, 15–20-flowered. **Flowers:** White to pink, irregular, 7–9 mm long; petals 5; sepals 5; stamens 10; pistil 1. **Fruit:** Legume pod-like, 7–9 mm long; seeds 2–5, yellow.

Notes: The species name *repens* means "creeping," a reference to this plant's growth habit.

Tufted Vetch

Vicia cracca

Habitat: Waste areas, roadsides and gardens; introduced from Europe. **General:** Stems 30–100 cm long, climbing, angular; rhizome creeping, wiry. **Leaves:** Alternate, stalked, pinnately compound with 10–24 leaflets and a terminal tendril; leaflets linear to oblong, 1–3 cm long, bristle-tipped; stipules leaf-like. **Flower Cluster:** Raceme 1-sided, 10–40-flowered, long-stalked. **Flowers:** Bluish purple, irregular, 0.9–1.3 cm long; petals 5; sepals 5; stamens 10; pistil 1. **Fruit:** Legume pod-like, 1–3 cm long, 4–7 mm wide; seeds 6–12, black; fruit walls becoming twisted upon opening.

Notes: *Vicia* is the Latin name for vetch. The species name *cracca* is the Latin name for this plant.

FLAX FAMILY
Linaceae

Members of the flax family are herbaceous plants with simple, alternate leaves. Flowers appear at the ends of branches and usually last for only 1 day. The regular flowers are composed of 5 petals, 5 sepals, 5 stamens and 5 styles. The fruit is a 10-seeded capsule.

Several species are grown for their ornamental value. One species, common flax (*Linum usitatissimum*), is cultivated for the production of linseed oil and linen fibres.

Wild Blue Flax
Linum perenne

Habitat: Dry, open meadows and hillsides; introduced ornamental from western North America. **General:** Stems 20–70 cm tall, leafy, hairless; base woody. **Leaves:** Alternate, numerous, linear, 1–2 cm long, less than 7 mm wide; margins smooth. **Flower Cluster:** Cyme few-flowered. **Flowers:** Pale blue, showy, 2–3.8 cm across, lasting 1 day only; petals 5; sepals 5; stamens 5; pistil 1. **Fruit:** Capsule globe-shaped, 5–9 mm across; seeds 8–10.

Notes: *Linum* is the Latin name for flax. • The former species name *lewisii* commemorates American explorer Captain Meriwether Lewis (1774–1809), who discovered wild blue flax in western North America.

CALTROP FAMILY
Zygophyllaceae

The caltrop family includes temperate to tropical herbs and shrubs. The opposite leaves are pinnately compound. Flowers occur in a cyme or are solitary. The perfect, regular flowers have 4–5 petals, 4–5 sepals, 5, 10 or 15 stamens and 1 pistil. The fruit is a capsule.

Puncture-vine, Caltrop
Tribulus terrestris

Habitat: Waste areas and roadsides; native to the Mediterranean region of southern Europe and the northern fringes of the Sahara Desert in Africa. **General:** Stems prostrate, mat-forming, green to reddish brown, covered in dense, bristly hairs. **Leaves:** Opposite, 3–7 cm long, pinnately compound with 8–16 leaflets; leaflets 5–15 mm long, 4–7 mm wide, hairy; stipules 5–12 mm long. **Flower Cluster:** Solitary, in leaf axils. **Flowers:** Pale yellow, 1.5–2.5 cm across; petals 5; sepals 5; stamens 10; pistil 1. **Fruit:** Schizocarp pod-like, about 1.2 cm long, breaking into 5 triangular, sharply spined segments; seeds wedge-shaped.

Notes: The genus name *Tribulus* means "3-pointed" or "caltrop," a reference to the spiny fruits, which resemble a caltrop, a medieval military weapon composed of 4 spikes arranged so that no matter how it is thrown onto the ground, one spike always sticks up. The fruits can cause injury to livestock and humans, and damage to vehicle tires.

WOODSORREL FAMILY
Oxalidaceae

The woodsorrel family is a small family of herbaceous plants with fleshy rhizomes. The alternate leaves are usually pinnately or palmately compound. The perfect, regular flowers are borne in umbels and have 5 petals, 5 sepals, 10 stamens and 1 pistil with 5 styles. The fruit is a capsule.

A few species are grown for their ornamental value.

Yellow Woodsorrel
Oxalis stricta

Habitat: Moist woods, gardens and waste areas; native plant with a weedy nature. **General:** Stems erect to ascending, 5–50 cm tall, green to pinkish, hairy. **Leaves:** Alternate, 0.5–3 cm wide, palmately compound with 3 leaflets; leaflets heart-shaped, 1–2 cm wide, margins hairy. **Flower Cluster:** Umbel 1–5-flowered. **Flowers:** Yellow, 0.5–1.3 cm across; petals 5; sepals 5; stamens 10–15; pistil 1. **Fruit:** Capsule 1–3 cm long, densely haired, on bent stalks.

Notes: The genus name *Oxalis* is from the Greek word *oxys*, "acid," a reference to the bitter-tasting leaves. The species name *stricta* means "erect" or "upright," a reference to the stem.

269

GERANIUM FAMILY
Geraniaceae

The geranium family is a small group of herbaceous plants with opposite, compound leaves. The flowers are composed of 5 petals, 5 sepals, 5 or 10 stamens and a pistil of 5 carpels. The carpels are fused to form a central column, which appears as a single style. The fruit is a capsule or schizocarp. At maturity, the schizocarp splits from the bottom upward into 5 long-tailed mericarps. Each mericarp coils and uncoils depending on its moisture content, enabling it to work its way into the soil.

Members of this family include both wild and garden geraniums. The wild geraniums are represented by the genus *Geranium* and garden geraniums by *Pelargonium*.

Stork's-bill
Erodium cicutarium

Habitat: Waste areas and roadsides; introduced from southern Europe. **General:** Stems 10–40 cm long, spreading to trailing, hairy. **Leaves:** Basal with a few opposite stem leaves, pinnately compound with irregularly shaped leaflets; leaflets 1–2.5 cm long, covered with short, stiff hairs. **Flower Cluster:** Umbel 2–12-flowered; flower stalks 1–2 cm long. **Flowers:** Pink, 1.2–1.8 cm across; petals 5; sepals 5, bristle-tipped; stamens 10; pistil 1, styles 5. **Fruit:** Schizocarp capsule-like, 2–4 cm long, splitting from the bottom upward; seeds long-tailed.

Notes: The genus name comes from the Greek word *erodios*, "heron," referring to the persistent beak-like style on the fruit. The species name means "like that of *Cicuta*," a reference to the leaves, which resemble those of water-hemlock (pp. 300–01).

Bicknell's Geranium
Geranium bicknellii

Habitat: Gardens, roadsides and waste areas; native plant with a weedy nature. **General:** Stems 20–60 cm long, sticky-haired, spreading, branched. **Leaves:** Opposite, 2–7 cm across, palmately divided into 5 wedge-shaped lobes. **Flower Cluster:** Solitary or in pairs, in leaf axils. **Flowers:** Pink to rosy purple, 0.8–1.2 cm across; petals 5; sepals 5, sharp-pointed; stamens 10; pistil 1. **Fruit:** Schizocarp capsule-like, 1.5–2.5 cm long, splitting from the bottom upward; seeds long-tailed.

Notes: The species name commemorates Eugene Pintard Bicknell (1859–1925), a reknowned American amateur botanist.

Wild Crane's-bill, Wild Geranium · *Geranium maculatum*

Habitat: Moist meadows and forests. **General:** Stems 30–70 cm tall; rhizome thick. **Leaves:** Basal leaves long-stalked, divided into 5–7 wedge-shaped lobes, each 10–13 cm wide; stem leaves 2, opposite, stalkless, resembling basal leaves. **Flower Cluster:** Umbel 2–5-flowered; stalks hairy. **Flowers:** Pinkish purple, showy, 2.5–4 cm wide; petals 5; sepals 5; stamens 10; pistil 1, styles 5.

Fruit: Schizocarp capsule-like, splitting into 5 segments at maturity.

Notes: The genus name *Geranium* comes from the Greek word *geranion*, "crane," a reference to the persistent, beak-like style on the fruit. The species name *maculatum* means "spotted."

Herb-Robert
Geranium robertianum

Habitat: Moist, rich forests. **General:** Stems spreading, 10–60 cm long, branched, hairy, often reddish green. **Leaves:** Opposite, 2–7.5 cm across, dark green with reddish tinge, divided into 3 segments, each pinnately lobed and cleft. **Flower Cluster:** Cyme 1–2-flowered. **Flowers:** Pink, 1–1.5 cm wide; petals 5, 0.7–1 cm long; sepals 5; stamens 10; pistil 1, styles 5. **Fruit:** Schizocarp capsule-like, breaking into 5 mericarps at maturity; mericarps 2–3 mm long, tails 1–2 cm long.

Notes: The species and common names are believed to commemorate Robert of Molesme (1027–1111), an abbot who founded the Cistercian Order.

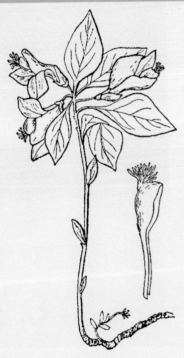

MILKWORT FAMILY
Polygalaceae

A small family of herbaceous perennials, the milkworts have simple, alternate leaves and no stipules. The flowers are borne in dense, terminal clusters. The white or purplish flowers are composed of 3 petals, 5 sepals, 8 stamens and 1 pistil. The 2 inner sepals are petal-like and larger than the other 3. The middle petal is boat-shaped and connected to the others. The fruit is a 2-seeded capsule.

Fringed Polygala
Polygala paucifolia

Habitat: Dry to moist forests. **General:** Stems 5–15 cm tall. **Leaves:** Alternate, 3–6 near the top of the stem, oval, 1.5–4 cm long, 1–2 cm wide; margins smooth; stalks short. **Flower Cluster:** Raceme 1–4-flowered, long-stalked. **Flowers:** Pink to purple, 1–2 cm long; petals 3, 1–1.8 cm long, joined at tips, the outermost with a fringed crest, 5–7 mm long; sepals 5 (outer 3 sepals 3–6 mm long, inner 2 sepals 10–17 mm long, wing-like); stamens 6–8; pistil 1. **Fruit:** Capsule round, 5–7 mm across.

Notes: The species name *paucifolia* means "few-leaved."

Seneca-root · *Polygala senega*

Habitat: Moist, open forests and grasslands. **General:** Stems 10–50 cm tall, unbranched, finely haired. **Leaves:** Alternate, oval to lance-shaped, 1–3 cm long, 1.5–3 cm wide; margins wavy; lower leaves often scale-like. **Flower Cluster:** Raceme spike-like, terminal, 2–6 cm long, many-flowered.

Flowers: Greenish white, 4–5 mm wide; petals 3; sepals 5; stamens 8; pistil 1. **Fruit:** Capsule round, 3–6 mm across; seeds hairy.

Notes: The genus name *Polygala* is derived from the Greek words *poly*, "much," and *gala*, "milk." It was believed that a diet high in milkwort would increase milk production in livestock.

SPURGE FAMILY
Euphorbiaceae

The spurge family is a large family of about 7500 species with worldwide distribution. One genus, *Euphorbia*, has over 1000 species. Members of this family include trees, shrubs, herbs and cactus-like succulents, and they often have milky juice. The simple leaves may be alternate, opposite or whorled, and can be variable in size and shape even on a single plant. Plants may be male, female or bisexual. Flowers in some genera may be produced in a specialized flower-like inflorescence called a cyathium. This highly modified flower cluster often resembles a single flower. The cup-like involucre consists of 4–5 fused bracts with a pair of nectar-secreting glands. Each cyathium contains several male flowers, each consisting of 1 stamen, often in groups of 4–5, and a single female flower. The fruit is a capsule.

The spurge family is diverse group of economically important plants. Castor oil plant (*Ricinus communis*), rubber tree (*Hevea brasiliensis*) and cassava and tapioca (*Manihot esculenta*) are produced by tropical members of this family. Other species such as crotons (*Croton* spp.), poinsettia (*Euphorbia pulcherrima*) and crown-of-thorns (*Euphorbia milii*) are grown as houseplants. Several species are designated as noxious weeds or are considered poisonous.

Three-seeded Mercury
Acalypha rhomboidea

Habitat: Moist, open areas and gardens; native species with a weedy habit. **General:** Stems 7.5–60 cm tall; milky juice absent. **Leaves:** Opposite below and alternate above, oval to lance or diamond-shaped, 2–7.5 cm long, 0.5–4 cm wide; margins with rounded teeth; stalks 1–4 cm long; turning purplish or copper-coloured in autumn. **Flower Cluster:** Spikes male, 0.4–1.5 cm long, in leaf axils; racemes female, 1–3-flowered, bracts 0.8–1.4 cm long, 9–15-lobed.

Flowers: Green, inconspicuous, 1–2 mm across; male flowers with no petals, 4 sepals and 8–16 stamens; female flowers with no petals, 3–5 sepals and 1 pistil. **Fruit:** Capsule 3-lobed; seeds 3.

Notes: *Acalypha* is the Latin name for a nettle with similar leaves. The species name means "rhomboid- or diamond-shaped," a reference to the leaf shape. • The young leaves are often copper-coloured, hence another common name, "copperleaf."

Thyme-leaved Spurge
Chamaesyce serpyllifolia

Habitat: Sandy and gravelly soils; native species with a weedy habit. General: Stems 20–60 cm across, reddish green, trailing, profusely branched, often mat-forming; exuding milky juice when broken. Leaves: Opposite, oblong to oval, 0.3–1.5 cm long, dark green with red midrib; margins finely toothed.

Flower Cluster: Cyathium 1–2 mm across, in leaf axils. Flowers: Greenish white, inconspicuous, male or female; sepals and petals absent; stamen 1, in clusters of 5–18; pistil 1. Fruit: Capsule 3-angled, 1.5–2 mm long.

Notes: The species name *serpyllifolia* means "thyme-leaved."

Cypress Spurge
Euphorbia cyparissias

Habitat: Roadsides and waste areas; introduced ornamental from central Europe. General: Stems 20–80 cm tall, leafy; exuding milky juice when broken; rhizome with numerous pink buds, creeping, forming large patches. Leaves: Alternate, club-shaped, 0.5–3.2 cm long, 1–3 mm wide, dark green; branch leaves scale-like. Flower Cluster: Cyathium umbrella-shaped; rays 5–22, 0.9–4 cm long; bracts in whorls of 10–18, 0.5–2.1 cm long, 1–5 mm wide, yellowish green turning bronze or purple with age; clusters of male flowers surrounding 1 female flower. Flowers: Greenish yellow, inconspicuous; male flowers with 1 stamen and 1 bract; female flowers with 1 pistil. Fruit: Capsule 3-lobed, 2–4 mm long; seeds 1–3.

Notes: The genus was reportedly named in 12 BCE by King Juba II of Mauretania to honour his Greek physician, Euphorbus. The species name *cyparissias* means "cypress."

Toothed Spurge
Euphorbia dentata

Habitat: Waste areas, railways and roadsides; native but weedy. **General:** Stems 20–50 cm tall, branched, hairy; exuding milky juice when broken. **Leaves:** Opposite, linear to oval, 2–8 cm long, 0.5–3 cm wide, dull green above, bluish green below; margins toothed; stalks 0.5–1.5 cm long, hairy. **Flower Cluster:** Cyathium compact; bracts 2–4 mm long, fringed; 2-lipped gland present; clusters of male flowers surrounding 1 female flower. **Flowers:** Whitish green, inconspicuous, 2–4 mm long; male flowers with 1 stamen and 1 small bract; female flowers with 1 pistil. **Fruit:** Capsule 3-lobed, 4–6 mm across; seeds 3, oval, 2.5–3 mm long.

Notes: The species name *dentata* means "toothed," a reference to the leaf margins.

Leafy Spurge
Euphorbia esula

Habitat: Roadsides and railway grades; introduced from the Caucasus region of western Asia in 1889. **General:** Stems 40–100 cm tall, bluish green, hairless; exuding milky juice when broken; rhizome with numerous pink buds, creeping. **Leaves:** Alternate, opposite and whorled, linear to oblong or lance-shaped, 2–7 cm long, 3–5 mm wide; margins smooth; stalkless. **Flower Cluster:** Cyathium umbrella-shaped with 7–12 rays, greenish yellow; bracts opposite, 1.2 cm long, 1 cm wide; 4 crescent-shaped glands in each cluster. **Flowers:** Greenish yellow, inconspicuous; male flowers with 1 stamen, in groups of 5; female flowers with 1 pistil, in groups of 3. **Fruit:** Capsule nodding, 3–5 mm long, exploding at maturity and scattering seeds up to 5 m away; seeds 3.

Notes: Leafy spurge is designated a noxious weed in Ontario. It is poisonous to most livestock except sheep, who feed on it without any ill effect. The milky juice may cause severe skin rashes in humans.

Waterweed
Euphorbia helioscopia

Habitat: Waste areas, roadsides and railway grades;
introduced from Europe.
General: Stems 10–60 cm
tall, erect; exuding milky
juice when broken Leaves:
Alternate and whorled, oval
to spatula-shaped, 2–5 cm
long, 1–1.5 cm wide; margins
finely toothed; leaves below
flower cluster broadly elliptic
to oblong or oval, in a whorl
of 5. Flower Cluster:
Cyathium umbrella-shaped; rays 5, branched; bracts 1–3 mm long. Flowers: Green,
inconspicuous; male flowers with 1 stamen, in groups of 5; female flowers with
1 pistil, in groups of 3. Fruit: Capsule 2.5–4 mm long; seeds 3.

Notes: The species name *helioscopia* means "turning toward the sun."

STAFF-TREE FAMILY · Celastraceae
Family description on p. 59.

Running Strawberry-bush
Euonymus obovatus

Habitat: Rocky woods and ravines.
General: Herb-like shrub; stems 10–30 cm
tall, trailing, angular; bark greyish green to
brown. Leaves: Opposite, 2–5 pairs per
stem, elliptic to oblong or oval, 1–7 cm long,
0.5–3.5 cm wide; margins finely toothed;
stalks grooved, less than 5 mm long. Flower
Cluster: Solitary or 2–3-flowered cyme,
in leaf axils. Flowers: Greenish purple,
inconspicuous, 4–7 mm across; petals 5,
overlapping; sepals 5; stamens 5, bright
orange-yellow; pistil 1. Fruit: Capsule
3-lobed, orangey pink to crimson, spiny
to warty; seeds 3, with bright orange to
scarlet aril.

Notes: The species name means "obovate,"
a reference to the shape of the leaves.

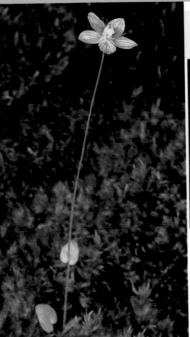

Northern Grass-of-Parnassus
Parnassia palustris

Habitat: Moist meadows and streambanks.
General: Flowering stems 10–30 cm tall, hairless; rhizome short. **Leaves:** Basal, 2–4 cm long, heart-shaped; stem leaf 1, clasping at or below midstem. **Flower Cluster:** Solitary. **Flowers:** White, showy, 1.5–2.5 cm wide; petals 5; sepals 5; stamens 5, alternating with 5 greenish yellow staminodes, each with 9–15 gland-tipped branches; pistil 1. **Fruit:** Capsule egg-shaped, 0.8–1.2 cm long.

Notes: The genus *Parnassia* is named after Mt. Parnassus in Greece.

IMPATIENS FAMILY
Balsaminaceae

The impatiens family includes shrubs and herbs with simple, alternate, opposite or whorled leaves. These succulent plants have watery juice. The flowers, produced singly or in racemes or panicles, are irregular and have 3 petals, 3 sepals (1 spurred sac and 2 small appendages), 5 stamens and 1 pistil. The fruit is an explosive capsule, providing the common name "touch-me-not" for some species.

Garden impatiens (*I. walleriana*) is a common ornamental.

Touch-me-not
Impatiens capensis

Habitat: Marshes, shorelines and moist, wooded areas. General: Stems 30–150 cm tall, succulent, translucent, often red-tinged; juice watery. Leaves: Alternate, oval, 2–10 cm long, 1.5–4 cm wide, often purplish green; margins coarsely toothed; stalked. Flower Cluster: Solitary, on drooping stalks from leaf axils. Flowers: Bright orange, reddish or pale yellow with red or purple spots, irregular, 2–2.5 cm long; petals 3 (2 fused, 1 free); sepals 3 (1 large, spurred sac and 2 small, green appendages), spur 7–8 mm long, bent parallel to sac; stamens 5, partly united; pistil 1. Fruit: Capsule 1.5–2.5 cm long, explosive, ejecting seeds with force.

Notes: The common name refers to the mature capsules, which burst open when touched. • The juice of this plant is a well-known treatment for the itch of poison-ivy (*Toxicodendron radicans*), p. 107, and the rash caused by common stinging nettle (*Urtica dioica*), p. 192.

Himalayan Impatiens
Impatiens glandulifera

Habitat: Edges of ponds, wet meadows and springy, open woods; introduced ornamental from the Himalayan region of Asia. General: Stems 1–3 m tall, succulent, hollow, green tinged red or purple, hairless, angular; adventitious roots at the base. Leaves: Opposite or whorls of 3–4, oblong to egg-shaped, 5–23 cm long, 1–7 cm wide; margins with 20 or more teeth per side; stalks 3–3.5 cm long. Flower Cluster: Cyme 2–14-flowered, on elongated axillary stalks 3–9 cm long; bracts elliptic to oval or lance-shaped, 7–8 mm long. Flowers: Pink, white or purple, irregular, 2–3.5 cm long; petals 5, 2 fused; sepals 3, 2 fused, lower sepal with spur 5–6 mm long; stamens 5; pistil 1. Fruit: Capsule club-shaped, 1.4–3 cm long, 4–8 mm wide, nodding, 4–16-seeded, exploding at maturity, ejecting seeds up to 7 m from parent.

Notes: The genus name *Impatiens* is Latin for "impatient," a reference to the explosive seed capsules. The species name *glandulifera* means "gland-bearing."

Pale Touch-me-not

Impatiens pallida

Habitat: Wet woods and meadows.
General: Stems 30–150 cm tall, succulent, translucent; juice watery. **Leaves:** Alternate, oval, 2–10 cm long, 1.5–4 cm wide; margins coarsely toothed; stalked. **Flower Cluster:** Solitary, on drooping stalks from leaf axils. **Flowers:** Pale yellow, irregular, 2.5–4 cm long; petals 3 (2 fused, 1 free); sepals 3 (1 large, spurred sac and 2 small, green appendages), spur 5–8 mm long, bent parallel to sac; stamens 5, partly united; pistil 1. **Fruit:** Capsule 2–2.5 cm long, explosive, ejecting seeds with force, hence the name touch-me-not.

Notes: The species name *pallida* means "pale green in colour." • The juice of this plant is a well-known treatment for the itch of poison-ivy (*Toxicodendron radicans*), p. 107, and the rash caused by common stinging nettle (*Urtica dioica*), p. 192.

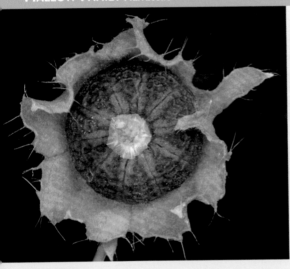

MALLOW FAMILY

Malvaceae

The mallow family includes trees, shrubs and herbaceous plants with simple, alternate leaves. Stipules are present. The flowers have 5 petals, 5 sepals, numerous stamens and 1 pistil with 4 to many carpels. A characteristic feature of this family is the fusion of the stamen stalks into a tube surrounding the pistil. The petals are also fused to the base of the staminal tube. The fruit is a schizocarp that breaks into segments called mericarps. Each mericarp contains 1 or more seeds.

The mallow family is very important economically. Cotton (*Gossypium* spp.) and okra (*Abelmoschus esculentus*) are important crop species, whereas hollyhock (*Alcea* spp.) and hibiscus (*Hibiscus* spp.) are grown for their ornamental value. The family also contains several weedy species.

Velvetleaf, Indian Mallow

Abutilon theophrasti

Habitat: Roadsides and waste areas; introduced from India as a potential fibre crop around 1750. **General:** Stems 1–2.3 m tall, soft-haired throughout; taproot white. **Leaves:** Alternate, heart-shaped, 5–15 cm long, 5–20 cm wide, covered in soft, velvety hairs; margins smooth to slightly toothed; stalks long; leaves drooping at night. **Flower Cluster:** Solitary, in leaf axils. **Flowers:** Orangey yellow, 1.5–2.5 cm across; petals 5; sepals 5; stamens numerous, fused into a column and surrounding the style; pistil 1. **Fruit:** Schizocarp head-like, horned, 2–2.5 cm across, greenish black, breaking into 10–15 mericarps at maturity; mericarps 5–15-seeded.

Notes: The species name commemorates Theophrastus (371–287 BCE), a Greek philosopher and a student of Aristotle and Plato.

Flower-of-an-hour

Hibiscus trionum

Habitat: Roadsides and waste areas; introduced ornamental from southern Europe. **General:** Stems 30–50 cm tall, erect, branched at the base, bristly haired, the hairs star-shaped to forked. **Leaves:** Alternate, upper leaves deeply 3-lobed, the lobes 1–5 cm long, 2.5–3 cm wide; margins coarsely toothed; stalks 1–3 cm long; stipules 3–7 mm long, 1 mm wide.

Flower Cluster: Solitary, in leaf axils; stalk 2–3 cm long, densely haired; bracts 10, narrow, papery, 0.7–1 cm long. **Flowers:** Pale yellow with purplish brown centre, 3.5–4.5 cm wide, opening for only a few hours; petals 5; sepals 5, papery, prominently purple-veined, surface hairy; stamens numerous, anthers orange; pistil 1. **Fruit:** Capsule globe-shaped, 1.2–1.5 cm long, bristly haired, opening at the top; calyx papery, purple-veined, inflated and surrounding the capsule.

Notes: The genus name *Hibiscus* is Greek for "mallow." The species name *trionum* means "three-coloured." • The common name "flower-of-an-hour" refers to the very short bloom time of this species.

Round-leaved Mallow
Malva rotundifolia

Habitat: Waste areas and roadsides; introduced from Europe and Asia before 1880. General: Stems 20–130 cm long, trailing to erect, branched, often forming large mats. Leaves: Alternate, round to kidney-shaped, 2.5–7.5 cm long, 2–5 cm across, shallowly 5-lobed, smooth to sparsely hairy; margins with rounded teeth; stalks 1–1.5 cm long. Flower Cluster: Raceme 2–5-flowered, clustered in leaf axils; bracts 3, small. Flowers: White to pale blue or lilac, 0.8–1.4 cm across; petals 5; sepals 5; stamens numerous, fused and forming a column around the style; pistil 1. Fruit: Schizocarp breaking into 8–15 mericarps at maturity; mericarps 1-seeded; calyx becoming enlarged and veiny.

Notes: *Malva* is the ancient Latin name for "mallow." The species name *rotundifolia* means "round-leaved."

MANGOSTEEN FAMILY
Clusiaceae

The mangosteen family is a large family with over 1200 species worldwide and includes the former families of Hypericaceae and Guttiferae. Members of this family are trees, shrubs and herbs with simple, opposite or whorled leaves. Flowers are composed of 4–6 petals, 4–5 sepals, 9 to many stamens and 1 pistil. The stamens often appear in bundles opposite the petals. The fruit is a berry, drupe or capsule.

A well-known species, mangosteen (*Garcinia mangostana*), is a tropical evergreen cultivated for its edible fruit.

Large Canada St. John's-wort

Hypericum majus

Habitat: Shorelines, marshes and wet meadows. **General:** Stems 10–50 cm tall, erect; rhizome short, leafy, ridged, creeping. **Leaves:** Opposite, oblong to lance-shaped, 1–4 cm long, 0.2–1.1 cm wide, prominently 5–7-nerved, covered with pale brown, glandular dots; margins smooth; stalkless. **Flower Cluster:** Cyme few-flowered; bracts small or absent. **Flowers:** Yellow, 0.5–1 cm wide; petals 5, soon withering; sepals 5; stamens 15–35; pistil 1, styles 3. **Fruit:** Capsule 5–8 mm long, often purplish; seeds yellow.

Notes: The genus name comes from the Greek words *hyper*, "above," and *eikon*, "picture," a reference to people hanging the plant above pictures or paintings to ward off evil spirits. The species name *majus* means "larger."

St. John's-wort

Hypericum perforatum

Habitat: Waste areas, dry, open fields and roadsides; introduced from Europe, western Asia and North Africa. **General:** Stems 30–75 cm tall, leafy, branched; base woody; short, horizontal branches spreading from crown, sterile, 2–10 cm long, often reddish with black glands. **Leaves:** Opposite, numerous, linear-oblong to elliptic, 1–5 cm long, distinctly 3–5-veined, covered in small, translucent dots when held to the light; margins smooth; stalkless. **Flower Cluster:** Cyme flat-topped, 25–100-flowered. **Flowers:** Bright yellow, 2–2.5 cm across, showy; petals 5, black-dotted on margins; sepals 5, green; stamens numerous, in 3–5 bundles; pistil 1, styles 3. **Fruit:** Capsule reddish brown, 0.5–1 cm long, sticky-haired.

Notes: The common name refers to the belief that these plants bloom on St. John's Day (June 24). • This introduced species contains hypericin, a chemical compound that, in conjunction with exposure to the sun, can cause sunburn in people and animals with sensitive skin.

Fraser's Marsh St. John's-wort
Triadenum fraseri

Habitat: Marshes and wet shorelines. **General:** Stems 30–60 cm tall; rhizome creeping. **Leaves:** Opposite, elliptic to oval, 3–6 cm long, 1–3 cm wide, base heart-shaped, underside dotted with translucent glands; margins smooth; stalkless; turning purplish green in late summer. **Flower Cluster:** Cyme few-flowered, axillary or terminal. **Flowers:** Purple to pink or pinkish green, 1–2 cm wide, opening only on bright, sunny days; petals 5; sepals 5; stamens 9, in 3 groups of 3, with 3 small, orange glands alternating with stamen groups; pistil 1, styles 3. **Fruit:** Capsule purplish, 0.7–1.2 cm long; seeds brown.

Notes: The genus name comes from the Greek words *treis*, "three," and *aden*, "gland," a reference to the glands, which alternate with the stamens. The species name commemorates Scottish botanist John Fraser (1750–1811).

ROCKROSE FAMILY · Cistaceae

The rockrose family is a small group of herbs and shrubs with simple, alternate or opposite leaves. Flowers are borne singly or in axillary cymes. The regular flowers are composed of 0–5 petals, 5 sepals, many stamens and 1 pistil with 1–5 styles. The stamens are spirally arranged from the outside inward. The fruit is a capsule.

Long-branched Frostweed, Frost-wort · *Helianthemum canadense*

Habitat: Dry, open meadows and woods.
General: Stems 15–45 cm tall, branches from upper leaf axils overtopping main stem.
Leaves: Alternate, linear to oblong or lance-shaped, 2–3 cm long (leaves on branches often shorter); stalkless. **Flower Cluster:** Solitary or in axillary cymes of 2–4 flowers. **Flowers:** Yellow, showy, 2–4 cm across, lasting a single day, opening only in bright sunlight; petals 5, wedge-shaped; sepals 5 (outer 2 short, inner 3 long); stamens numerous; pistil 1; small, petal-less flowers appearing in axillary clusters after showy flowers fade. **Fruit:** Capsule 4–6 mm long, produced by showy flowers; petal-less flowers producing capsules 3–3.5 mm long.

Notes: The genus name *Helianthemum* comes from the Greek words *helios*, "sun," and *anthemom*, "flower." • The common name "frostweed" refers to the sap crystals that appear near the base of the stem, giving the plant a frosted appearance.

VIOLET FAMILY
Violaceae

The violet family is composed of herbaceous plants with simple, alternate or basal leaves. Flowers may be solitary or appear in racemes, panicles or cymes. The irregular flowers may be of 2 types, either showy or small and inconspicuous. The showy flowers have 5 petals, 5 sepals, 5 stamens and 1 pistil with 3 fused carpels. The small, inconspicuous flowers set seed even though they do not open. A spur may be present at the base of some petals. The fruit is a capsule, berry or nut.

A well-known member of this family is the pansy violet (*Viola tricolor*).

Early Blue Violet
Viola adunca

Habitat: Open meadows and woods.
General: Stems 4–25 cm long, spreading, tufted, leafy; root slender. Leaves: Alternate, oval to round, 1–3 cm long, 1–2 cm wide; margins wavy-toothed; stalks long; stipules 3–10 mm long, linear to lance-shaped with smooth to spiny-toothed margins. Flower Cluster: Solitary, in leaf axils. Flowers: Blue to violet, 0.8–2 cm long; petals 5, lower petal spurred, the spur straight or hooked, 4–6 mm long; sepals 5; stamens 5; pistil 1. Fruit: Capsule 4–5 mm long; seeds dark brown.

Notes: The species name *adunca* means "hooked," a reference to the bent tip of the spur.

Field Violet · *Viola arvensis*

Habitat: Waste areas, gardens and roadsides; introduced ornamental from Europe.
General: Stems 10–35 cm tall, erect to ascending, branched, hairy on the angles, partially shade tolerant; wintergreen-scented when crushed. Leaves: Basal leaves rounded to oblong, 1–1.5 cm long, 0.7–1.2 cm wide, veins on underside sparsely haired, margins toothed, stalks 1–2 cm long, stipules small; stem leaves alternate, oblong to lance-shaped, 2–8 cm long, 1–1.5 cm wide, margins with rounded teeth, stipules leaf-like, divided into 5–9 narrow segments. Flower Cluster: Solitary, on stalks 2–4 cm long from leaf axils. Flowers: Pale yellow to white with purple or mauve markings, 1–1.5 cm long, 0.9–1.2 cm wide; petals 5, lateral petals bearded, lower petal spurred, the spur 2–4 mm long; sepals 5, united, about the same length or longer than petals; stamens 5; pistil 1. Fruit: Capsule globe-shaped, 0.5–1 cm long, light brown; seeds about 75, ejected with force at maturity up to 2.1 m from parent.

Notes: The species name *arvensis* means "of cultivated fields," a reference to this plant's preferred habitat.

Canada Violet

Viola canadensis

Habitat: Moist woods. **General:** Stems 20–60 cm tall, leafy, often forming large colonies by underground stolons. **Leaves:** Alternate, heart-shaped, 3–10 cm long, 4–8 cm wide, reduced in size upward; margins toothed; stalks long; stipules 1–2 cm long, papery with smooth margins. **Flower Cluster:** Solitary, in axils of upper leaves. **Flowers:** White to pale violet with purplish veins and yellow throat, showy, fragrant, 1–2.5 cm across; petals 5, spurred, the spur short; sepals 5; stamens 5; pistil 1. **Fruit:** Capsule 0.8–1.2 cm long.

Notes: The species name *canadensis* refers to this plant's range and its region of botanical discovery.

Sweet White Violet

Viola macloskeyi ssp. *pallens*

Habitat: Wet woods and springy streambanks. **General:** Flowering stems 7–13 cm tall, green; rhizome slender, creeping. **Leaves:** Basal, pale green to yellowish green, heart-shaped, 2–6.3 cm wide; margins shallowly toothed. **Flower Cluster:** Solitary, on leafless stems; bracts 2. **Flowers:** White, not fragrant, 1–1.5 cm wide; petals 5, lower 3 with brownish purple veins; sepals 5; stamens 5; pistil 1; petal-less flowers on erect stems. **Fruit:** Capsule green, 4–5 mm long.

Notes: The species name commemorates George Macloskie (1834–1920), an Irish botanist and professor of biology at Princeton University.

Downy Yellow Violet
Viola pubescens

Habitat: Dry woods.
General: Stems 15–40 cm tall, covered in soft, woolly hairs; rhizome brown, woody; roots coarse, fibrous.
Leaves: Basal leaf 1, long-stalked, heart to kidney-shaped, 4–10 cm long, 4–13 cm wide, covered in soft hairs, margins with rounded teeth; stem leaves 2–4, alternate; stipules broadly oval. **Flower Cluster:** Solitary, on hairy stalks from leaf axils. **Flowers:** Yellow, 1.5–2.5 cm wide; petals 5, lower 3 with brownish purple veins, 2 lateral petals bearded, lowest petal spurred, the spur 3–5 mm long; sepals 5; stamens 5; pistil 1; petal-less flowers producing fruit. **Fruit:** Capsule 1–1.2 cm long, woolly haired.

Notes: The species name *pubescens* means "downy-haired."

Long-spurred Violet
Viola rostrata

Habitat: Rich woods.
General: Stems erect to sprawling, 5–25 cm tall, hairless; rhizome branched, woody.
Leaves: Alternate, stalked, oval to heart-shaped, 2–4 cm long, margins finely toothed; stipules lance-shaped, margins fringed or toothed. **Flower Cluster:** Solitary, on stalks from leaf axils. **Flowers:** Pale violet with dark veins radiating from the centre, 1–1.5 cm across; petals 5, lowest petal spurred, the spur 1–1.6 cm long; sepals 5. **Fruit:** Capsule 5–6 mm long; seeds yellowish brown.

Notes: The species name *rostrata* means "beaked," a reference to the long spur.

Common Blue Violet
Viola sororia

Habitat: Moist meadows and rich woods; native species with a weedy nature, often found in lawns and waste areas. **General:** Flowering stems 7.5–20 cm tall, woolly haired; rhizome creeping, stout. **Leaves:** Basal, heart to kidney-shaped, 1–10 cm wide, underside with woolly hairs; margins toothed; stalks 5–15 cm long, hairy. **Flower Cluster:** Solitary, on stalks to 10 cm long; bracts 2, 3–4 mm long. **Flowers:** Deep blue to lavender or white, 1.8–2.5 cm wide; petals 5, lateral petals bearded, spur usually absent; sepals 5; stamens 5; pistil 1; petal-less flowers short-stalked, producing fruit. **Fruit:** Capsule green mottled with brown and purple, 0.6–1.2 cm long.

Notes: The species name *sororia* means "sisterly," a reference to this plant's resemblance to other violet species. • Common blue violet was designated the state flower of New Jersey in 1913.

Pansy Violet, Johnny-jump-up
Viola tricolor

Habitat: Waste areas and parkland; introduced ornamental from Europe. **General:** Stems 10–40 cm tall, erect, branched. **Leaves:** Basal leaves oval to lance-shaped, 2–6 cm long, long-stalked; stem leaves alternate, oval to oblong or lance-shaped, 2–4 cm long, margins with rounded teeth; stipules 1–4 cm long, leaf-like with pinnately divided margins. **Flower Cluster:** Solitary, on stalks from leaf axils; bracts 2, triangular, small. **Flowers:** Variable with purple, yellow and white markings, 1.5–2 cm across; petals 5, spur 5–8 mm long; sepals 5; stamens 5; pistil 1. **Fruit:** Capsule elliptic to oblong, 0.6–1.2 cm long.

Notes: The species name means "tri-coloured," a reference to the flower.

CACTUS FAMILY
Cactaceae

Members of the cactus family reach their greatest diversity in subtropical and tropical North and South America. Cacti are stem-succulent plants with axillary spines. The highly reduced, alternate leaves are either simple and deciduous or absent. The axillary buds, called areolae, often bear 1 to many spines. The solitary flowers have numerous petals, sepals, stamens and fused pistils. The fruit is a berry or fleshy capsule.

Members of this family vary widely in size and include ball cactus (*Escobaria vivipara*), which is less than 10 cm tall, as well as the large, tree-like saguaro cactus (*Carnegiea gigantea*), which grows to 14 m tall.

Fragile Prickly Pear
Opuntia fragilis

Habitat: Dry, open hillsides; native to western Canada. **General:** Stems spreading, often forming dense mats; fleshy segments oval, nearly circular in cross-section, 2–5 cm long, 1–3 cm wide; areolae with 3–7 spines, the spines 1–2.5 cm long. **Leaves:** Absent; scale-like and reddish when present, withering soon after emerging. **Flower Cluster:** Solitary, borne between the areolae. **Flowers:** Pale yellow, 4–6 cm across; petals, sepals and stamens numerous; pistil 1. **Fruit:** Berry 1.5–2.5 cm long, fleshy, reddish green, covered with spines; edible and sweet at maturity.

Notes: *Opuntia* is the Greek name for a spiny plant that grew near Opuntis in ancient Greece. • As the species and common names imply, this cactus is fragile, with stem segments that are easily separated.

LOOSESTRIFE FAMILY
Lythraceae

The loosestrife family has over 500 species worldwide. Members of this family include herbs, shrubs and trees with simple, opposite or whorled leaves. Stipules may be present or absent. The perfect, regular flowers are composed of 3–16 petals, 3–16 sepals, 6–32 stamens and a single pistil. The fruit is a capsule.

One species, purple loosestrife (*Lythrum salicaria*), is a serious weed of wetlands and marshes. Once widely planted for its ornamental value, many jurisdictions have designated it as a noxious weed.

Swamp Loosestrife
Decodon verticillatus

Habitat: Swamps and marshes.
General: Semi-aquatic, shrub-like; stems 1–2.5 m tall, 4–6-sided, hairy; lower part of stem may be thick and spongy, allowing plants to form a floating bog mat in times of flooding. **Leaves:** Opposite or in whorls of 3–4, lance-shaped, 5–15 cm long, 0.8–3 cm wide; stalks short. **Flower Cluster:** Cyme axillary, densely flowered. **Flowers:** Pinkish purple, bell-shaped, 2–2.5 cm across; petals 5; sepals 5–7, triangular; stamens 10, of 2 different lengths; pistil 1. **Fruit:** Capsule round, 4–6 mm across.

Notes: *Decodon* comes from the Greek words *deka*, "ten," and *odous*, "tooth," a reference to each sepal having 10 teeth. The species name *verticillatus* means "whorled," referring to the leaf arrangement.

Purple Loosestrife
Lythrum salicaria

Habitat: Marshes, wetlands and roadside ditches; introduced ornamental from Europe.
General: Semi-aquatic, shrub-like; stems 50–150 cm tall, square, branched, covered in downy hairs, lower part of stem often spongy and cork-like; root system extensive. **Leaves:** Opposite or in whorls of 3 (upper leaves may be alternate), lance-shaped, 3–10 cm long, 2.5–20 mm wide, slightly hairy; margins smooth; stalkless. **Flower Cluster:** Spike 10–100 cm long, interrupted, leafy; flowers in whorls. **Flowers:** Pinkish purple, 1.5–2 cm across; petals 4–8; sepals 4–8, prominently veined; stamens 8–16; pistil 1. **Fruit:** Capsule 4–7 mm long.

Notes: *Lythrum* comes from the Greek word *lithron*, "blood," a reference to the flower colour. The species name means "like *Salix* (willows)," referring to the leaves. • Once believed to be incapable of producing seeds, purple loosestrife was widely planted in gardens as an ornamental. However, it is highly invasive and is now designated a noxious weed throughout most of North America.

MELASTOME OR MEADOW-BEAUTY FAMILY
Melastomataceae

Members of this tropical family are trees, shrubs and herbs, and only a single species is found in Ontario. These plants have opposite leaves and no stipules. The flowers, borne singly in leaf axils or clustered in racemes or panicles have 4–5 petals, 4–5 sepals, 8 or 10 stamens and 1 pistil. The ovary may be superior or inferior. The fruit is a capsule or berry.

Common Meadow-beauty, Deergrass

Rhexia virginica

Habitat: Wet, sandy or mucky shorelines. **General:** Stems square, winged, 20–60 cm tall; roots tuber-like. **Leaves:** Opposite, oval to lance-shaped, 2–6 cm long, 8–30 cm wide; margins toothed; stalkless. **Flower Cluster:** Cyme 2–8-flowered. **Flowers:** Pink to purple, showy, 2.5–3.8 cm wide; petals 4–5; sepals 4–5; stamens 8, bright yellow; pistil 1. **Fruit:** Capsule urn-shaped; seeds resembling small snail shells.

Notes: The species name means "of Virginia."

EVENING PRIMROSE FAMILY · Onagraceae

The evening primrose family has over 675 species worldwide. Members of this family are herbs or shrubs with simple, alternate or opposite leaves. Stipules may be present or absent. Flower parts (petals, sepals and stamens) occur in multiples of 2 or 4. Plants in this family produce capsules, berries, nutlets or achenes. The name "evening primrose" refers to some species flowering in late afternoon or early evening.

The evening primrose family has limited economic importance. Several species, including evening primrose (*Oenothera* spp.), clarkia (*Clarkia* spp.) and fuchsia (*Fuchsia* spp.), are grown as ornamentals.

Fireweed, Willowherb
Chamerion angustifolium

Habitat: Open woods and riverbanks. **General:** Stems 1.3–3 m tall, unbranched; rhizome creeping, forming colonies. **Leaves:** Alternate, lance-shaped, 1.5–20 cm long, 0.5–35 mm wide, prominently veined; margins smooth to wavy; stalks short. **Flower Cluster:** Raceme 10–60 cm long, 8–80-flowered; nodding in bud. **Flowers:** Pink to light purple (occasionally white), 1.5–3 cm across; petals 4; sepals 4; stamens 8; pistil 1, style nodding, stigma 4-lobed. **Fruit:** Capsule 2.5–10 cm long, pinkish green, often 4-sided; seeds numerous, covered with white hairs.

Notes: The species name *angustifolium* means "narrow-leaved." • The leaves resemble those of willows, hence the common name "willowherb."

Enchanter's-nightshade
Circaea alpina

Habitat: Moist woods. **General:** Stems 10–50 cm tall, simple to branched; rootstock tuber-like. **Leaves:** Opposite, oval to heart-shaped, 2–5 cm long, 1.5–4.5 cm wide; margins smooth to coarsely toothed; stalked. **Flower Cluster:** Raceme 8–12-flowered. **Flowers:** White, 2–5 mm across; petals 2; sepals 2, reflexed; stamens 2; pistil 1. **Fruit:** Nut club-shaped, 2–3 mm long, covered with soft, hooked bristles; fruiting stalks 2–5 mm long.

Notes: The scientific and common names refer to the mythological enchantress Circe. She was believed to use poisonous members of this genus in her sorcery. The species name *alpina* means "alpine."

Broadleaf Enchanter's-nightshade

Circaea lutetiana ssp. *canadensis*

Habitat: Ravines and rich woods.
General: Stems 20–100 cm tall; roots fibrous. **Leaves:** Opposite, oblong to oval, 6–12 cm long, tip pointed; margins toothed; stalks rounded or angular on the lower side. **Flower Cluster:** Raceme 5–20 cm long; flower stalks 0.5–1.2 cm long. **Flowers:** White, 3–6 mm wide; petals 2, deeply lobed; sepals 2, reflexed; stamens 2; pistil 1. **Fruit:** Nut 3.5–5 mm long, surface covered in hooked bristles.

Notes: The species name means "of or from Lutetia," an ancient name for Paris, France.

Hairy Willowherb

Epilobium ciliatum

Habitat: Wet meadows, stream-banks and ditches. **General:** Stems 15–120 cm tall, hairy at leaf nodes and flower cluster; leafy rosettes and fleshy bulblets produced in autumn. **Leaves:** Opposite below, alternate above, oval to lance-shaped, 2.5–9 cm long, 0.8–4.5 cm wide; margins with 15–30 teeth per side; stalks 0–8 mm long. **Flower Cluster:** Raceme erect; flower stalks 0.3–1 cm long. **Flowers:** Pink to whitish, 0.3–1 cm long; petals 4; sepals 4, purplish; stamens 8; pistil 1. **Fruit:** Capsule 3–10 cm long, hairy; seeds with white, silky hairs.

Notes: *Epilobium* comes from the Greek words *epi*, "upon," and *lobos*, "a pod," a reference to the inferior ovary. The species name *ciliatum* means "soft-haired."

Hairy Willowherb, Codlins-and-cream

Epilobium hirsutum

Habitat: Wet, open woods, marshes and roadside ditches; introduced from Europe as a garden plant in the 1850s. **General:** Semi-aquatic, often with cork-like tissue at the base; stems erect, 20–250 cm tall, branched, covered in long, soft hairs; rhizome creeping; stolons rope-like, thick, fleshy. **Leaves:** Opposite, oblong to lance-shaped, 2–12 cm long, 0.5–4.5 cm wide, prominently veined, surfaces with woolly hairs and scattered glandular hairs; margins sharply toothed; stalkless and clasping the stem. **Flower Cluster:** Raceme erect; stalks 0.5–1.8 cm long **Flowers:** Rose purple, erect; petals 4, notched at tip; sepals 4, densely haired; stamens 8; pistil 1. **Fruit:** Capsule 3–10 cm long, stalks 0.5–2 cm long, woolly haired; seeds covered with dull white hairs.

Notes: The species name *hirsutum* means "hairy."

Yellow Evening Primrose

Oenothera biennis

Habitat: Dry, open areas and roadsides; introduced ornamental from Europe. **General:** Flowering stem 50–150 cm tall, reddish green, leafy, often branched; taproot large. **Leaves:** Basal leaves 10–20 cm long, midrib pinkish; stem leaves alternate, oblong to lance-shaped, 2.5–12 cm long, 0.5–5 cm wide, hairy to hairless; margins smooth to toothed; short-stalked to stalkless. **Flower Cluster:** Spike leafy, 10–40 cm long; flower buds erect; bracts large. **Flowers:** Bright yellow, 2–5 cm across, opening in the evening and closing before noon the next day; petals 4; sepals 4; stamens 8; pistil 1. **Fruit:** Capsule erect, 1–3.5 cm long, 4–7 mm wide, hairy.

Notes: The species name *biennis*, "biennial," refers to this plant's life cycle. In the first growing season, a basal rosette of leaves is produced, followed by a flowering stem in the second growing season.

GINSENG FAMILY
Araliaceae

The ginseng family has over 700 species worldwide. Members of this family include herbs, shrubs, woody vines
and trees with simple to compound,
alternate or basal leaves. Flowers are
borne in umbrella-shaped clusters
called umbels. The regular, perfect
flowers are composed of 5 petals,
5 minute sepals, 5 stamens and
1 pistil with 2–5 styles. The fruit
is a berry-like drupe.

The best-known member of this family is ginseng (*Panax ginseng*), a medici-
nal herb grown throughout the world. Ornamental species include English ivy
(*Hedera helix*) and umbrella-plant (*Schefflera arbicola*).

Wild Sarsaparilla
Aralia nudicaulis

Habitat: Moist to dry woods. **General:** Flowering stem 20–40 cm tall; rhizome
thick, woody, aromatic. **Leaves:** Basal, 1, 30–60 cm tall, compound with 3 divisions,
each with 3–5 finely toothed leaflets; leaflets oval to oblong or elliptic, 10–15 cm
long, 4–9 cm wide, margins toothed. **Flower Cluster:** Umbel globe-shaped,
2–5 cm across, 2–7 (commonly 3) per flowering stem. **Flowers:** Greenish white,
2–4 mm across; petals 5; sepals 5, very small; stamens 5; pistil 1. **Fruit:** Drupe
berry-like, green turning purplish black at maturity, 4–6 mm across; inedible.

Notes: The species name *nudicaulis* means "naked stem." • Wild sarsaparilla,
an ingredient in traditional root beer, has medicinal properties. The aromatic
root makes an excellent tea.

False Spikenard
Aralia racemosa

Habitat: Rich, deciduous forests. **General:** Stems 1–3 m tall, leafy, nodes purplish, dying back to ground level each autumn; rhizome aromatic. **Leaves:** Alternate, 30–80 cm long, ternately pinnate with 9–21 leaflets; leaflets oval to heart-shaped, 5–15 cm long, margins toothed. **Flower Cluster:** Raceme of umbels, 20–35 cm long; umbels about 10-flowered; bracts 2–4 mm long. **Flowers:** Greenish white, 1–3 mm across; petals 5; sepals 5; stamens 5; pistil 1. **Fruit:** Drupe dark purple, 5–7 mm across.

Notes: The species name *racemosa* means "having a raceme," a reference to the flower cluster.

CARROT FAMILY
Apiaceae

Formerly called Umbelliferae, the carrot family is large and has the greatest species diversity in northern temperate zones. Members of this family are herbaceous plants with alternate, basal or opposite, simple to compound leaves. The sheathing leaf stalks have oil tubes that give the plants their strong, distinctive scent. Stipules are usually absent. Flowers are borne in umbrella-shaped clusters called umbels or in heads; the umbels may be simple or compound. The regular or irregular flowers have 5 petals, 5 minute sepals (often absent), 5 stamens and 1 pistil with 2 carpels. The base of the style forms a cap on the top of the ovary. This structure is called a stylopodium, which means "style foot." The fruit is a schizocarp that breaks into 2 mericarps.

The carrot family is very important economically. It includes several food crops such as carrot (*Daucus carota*), dill (*Anethum graveolens*), parsley (*Petroselinum crispum*), cumin (*Cuminum cyminum*), fennel (*Foeniculum vulgare*), celery (*Apium graveolens*), parsnip (*Pastinaca sativa*) and caraway (*Carum carvi*). There are also several poisonous plants in this family.

Great Angelica, Purple-stemmed Angelica

Angelica atropurpurea

Habitat: Moist to wet areas along streams and rivers. **General:** Stems 1–3.5 m tall, dark purple to purple-splotched, hollow, strongly scented; roots aromatic, possible poisonous. **Leaves:** Alternate, bipinnately compound with numerous leaflets; leaflets broadly oval, 4–10 cm long, margins toothed; lower leaves to 60 cm wide, upper leaves to 20 cm wide. **Flower Cluster:** Umbel 10–25 cm across; rays 20–45; bracts absent. **Flowers:** Greenish white, 2–4 mm across; petals 5; sepals absent; stamens 5; pistil 1. **Fruit:** Schizocarp 4–6.5 mm long, breaking into 2 mericarps at maturity.

Notes: *Angelica* is Greek for "angelic medicinal properties." The species name *atropurpurea* means "dark purple."

Wild Chervil · *Anthriscus sylvestris*

Habitat: Waste woods, open meadows and forest edges; introduced from Europe. **General:** Stems 60–120 cm tall, erect, hollow, furrowed, hairless above, hairy below; roots thick, tuber-like. **Leaves:** Alternate, fern-like, 10–30 cm long, 2–3 times pinnately compound; leaflets 1–5 cm long, oval, hairy below, coarsely toothed. **Flower Cluster:** Umbel 2–6 cm across; rays 6–15, each 1.5–3 cm long; bracts oval to lance-shaped, 3–6 mm long. **Flowers:** White, 3–4 mm across; petals 5; sepals 5, small; stamens 5; pistil 1; often male flowers only in the centre of the umbel. **Fruit:** Schizocarp 6–9 mm long, oblong, smooth, short-beaked, breaking into 2 mericarps at maturity

Notes: *Anthriscus* is the Latin name for chervil. The species name *sylvestris* means "woods" or "forest," a reference to this plant's habitat.

Caraway
Carum carvi

Habitat: Roadsides and waste areas; introduced from Europe as a garden plant. General: Stems 30–80 cm tall, hairless; taproot spindle-shaped, 5–25 cm long. Leaves: Alternate, bases sheathing, compound, divided 3–4 times; leaflets linear, 0.5–1.5 cm long. Flower Cluster: Umbel 2.5–6 cm across, compound with 7–15 rays, each 2–4 cm long; primary umbel on stalk 5–13 cm long; umbellets 4–15-flowered; bracts thread-like, 1–5 or absent; flower stalks 1–12 mm long. Flowers: White or pinkish, 2–4 mm across; petals 5; sepals absent; stamens 5; pistil 1. Fruit: Schizocarp prominently ribbed, 3–4 mm long, breaking into 2 single-seeded segments at maturity.

Notes: *Carum* comes from *karan*, the Greek word for caraway. The species name *carvi* is Latin for caraway.

Bulb-bearing Water-hemlock
Cicuta bulbifera

Habitat: Swamps, marshes and lakeshores. General: Stems 30–80 cm tall, slender; small bulbs (bulblets) form in axils of upper leaves; root tuber-like; poisonous. Leaves: Alternate, 2–3 times pinnately compound (upper leaves simple); leaflets narrow, linear, 2–6 cm long, margins sparsely toothed. Flower Cluster: Umbel compound, 2–5 cm across. Flowers: Greenish white, 1–3 mm across; petals 5; sepals 5; stamens 5; pistil 1. Fruit: Schizocarp 1.5–2 mm long, breaking into 2 single-seeded segments at maturity.

Notes: *Cicuta* is an ancient Latin name for a poisonous, umbel-bearing plant. The species name *bulbifera* means "bulb-bearing," a reference to the bulblets that are produced in the upper leaf axils. These bulblets, a type of vegetative reproduction, are shed and produce new plants.

Spotted Water-hemlock
Cicuta maculata

Habitat: Marshes, swamps and road-side ditches. General: Stems 60–220 cm tall, hollow, often with purplish spots and streaks; root 3–10 cm long, tuber-like with numerous horizontal chambers. Leaves: Alternate, pinnately compound with several leaflets; leaflets 3–20 cm long, 0.5–3.5 cm wide, divided twice into narrow segments, margins coarsely toothed. Flower Cluster: Umbel 3–10 cm across, compound with 18–28 rays; smaller umbellets 12–25-flowered; bracts few, narrow. Flowers: Greenish white, 2–3 mm across; petals 5; sepals 5, very small; stamens 5; pistil 1. Fruit: Schizocarp 2–4 mm long, breaking into 2 single-seeded segments at maturity.

Notes: The species name *maculata* means "spotted," a reference to the stems. • Spotted water-hemlock is one of the most toxic native plants in Canada. The roots and stems contain cicutoxin and are extremely poisonous to humans and livestock.

Poison Hemlock
Conium maculatum

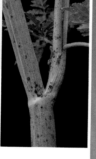

Habitat: Marshes and wet ditches; introduced from Europe and Asia. General: Stems 1–3 m tall, hollow, hairless, pale green, often splotched with purple, ill-scented when crushed; taproot 20–25 cm long, white.

Leaves: Alternate, 20–40 cm long, 8–12 cm wide, 3–4 times pinnately compound with numerous leaflets, fern-like; leaflets oblong to oval, margins toothed; stalks splotched with purple; lower leaves sheathing. Flower Cluster: Umbel 4–8 cm wide, compound with 10–15 rays; bracts oval to lance-shaped, 2–5 cm long. Flowers: White, 2–3 mm wide; petals 5; sepals absent; stamens 5; pistil 1. Fruit: Schizocarp 2–4 mm long, breaking into 2 single-seeded segments at maturity.

Notes: The species name *maculatum* means "spotted," a reference to the stems. • Poison hemlock, as the common name implies, is extremely poisonous.

Canada Honewort
Cryptotaenia canadensis

Habitat: Moist, shady, rich woods. **General:** Stems 30–100 cm tall, erect; roots thickened. **Leaves:** Alternate, compound with 3 leaflets; leaflets oval to elliptic, 4–15 cm long, 1–8 cm wide, margins coarsely toothed; lower leaf stalks 2–10 cm long and sheathing, upper leaves stalkless. **Flower Cluster:** Umbel 5–8 cm across, compound with 2–8 rays of unequal lengths 1–5 cm long; smaller umbellets 3–10-flowered, rays unequal and to 7 mm long; bracts 1–3, lance-shaped, 2–4 mm long. **Flowers:** White, 2–3 mm wide; petals 5; sepals 5, small; stamens 5, anthers yellow; pistil 1. **Fruit:** Schizocarp 5–8 mm long, breaking into 2 single-seeded segments at maturity.

Notes: The genus name comes from the Greek words *kruptos*, "hidden," and *tainia*, "band" or "stripe," a reference to the hidden oil tubes that are visible on the fruit of most members of the carrot family.

Wild Carrot, Queen Anne's Lace
Daucus carota

Habitat: Waste areas and roadsides; introduced from Europe and Asia. **General:** Stems 20–160 cm tall, hollow, bristly haired, carrot-like odour when crushed; taproot woody, white. **Leaves:** Alternate, 5–40 cm long, pinnately compound, divided 2–4 times into narrow segments; lower leaves long-stalked, upper leaves stalkless with papery white basal sheaths. **Flower Cluster:** Umbel 6–15 cm across, compound with numerous small umbellets; bracts finely dissected, appearing in a whorl below the main umbel. **Flowers:** White, 2–3 mm across; petals 5; sepals 5; stamens 5; pistil 1; single dark red or purplish black flower in the centre of the main umbel. **Fruit:** Schizocarp 2–4 mm long, breaking into 2 mericarps; fruiting cluster compact at maturity, resembling a bird's nest.

Notes: *Daucus* is the Latin name for carrot. The species name comes from the Greek word *karoton*, which also means "carrot." • Cultivated carrot (cultivars of *D. c.* ssp. *sativus*) closely resembles wild carrot but is easily distinguished by its edible orange taproot.

Giant Hogweed

Heracleum mantegazzianum

Habitat: Waste areas and roadsides; introduced ornamental from
the Caucasus Mountains of southwestern Asia. General: Stems
4–6 m tall, 3–10 cm across, hollow, often splotched with red or
purple patches or bumps; bristles
near the base containing toxic sap.
Leaves: Basal leaves 30–100 cm wide, compound with 3 leaflets, the leaflets deeply
cut and toothed; stem leaves alternate, lower leaves compound with 3 leaflets,
upper leaves deeply 3-lobed, margins toothed; stalks long, often covered in sharply
pointed bumps. Flower Cluster: Umbel 30–80 cm across, compound with
50–150 rays, each ray 15–40 cm long; smaller umbellets many-flowered; bracts linear to oval; flower stalks 1–2 cm long. Flowers: White, 2–15 mm across; petals 5;
sepals 5; stamens 5; pistil 1. Fruit: Schizocarp 6–18 mm long, 4–10 mm wide;
surface with 3–5 swollen, brown resin canals.

Notes: The species name commemorates Paolo Mantegazza (1831–1910), an Italian
neurologist noted for the isolation of cocaine from coca leaves. • Giant hogweed
produces light-activated chemicals (furanocoumarins) that can cause minor to
severe skin rashes when exposed to sunlight. Blisters appear within 48 hours and
become purplish black, producing scars that can take up to 6 years to heal. This
affliction often leads to long-term skin sensitivity to sunlight.

Cow Parsnip

Heracleum sphondylium
ssp. *montanum*

Habitat: Moist, open
forests and meadows.
General: Stems
1–2.5 m tall, hollow,
hairy, pungent odour
when crushed; taproot thick, woody. Leaves: Basal and alternate, compound with 3 prominently veined leaflets; leaflets
10–40 cm long, 10–30 cm wide, margins lobed to coarsely
toothed; leaf bases sheathing, enlarged. Flower Cluster:
Umbel 15–30 cm across, compound with 15–30 rays; smaller
umbellets several-flowered; bracts lance-shaped, deciduous.
Flowers: White, 3–15 mm across, largest flowers on outer edge of umbel; petals 5;
sepals minute or absent; stamens 5; pistil 1. Fruit: Schizocarp 8–12 mm long, breaking into 2 single-seeded segments at maturity.

Notes: The genus name comes from the Greek *Herakles*, meaning "Hercules."
The species name *sphondylium* means "vertebrae," alluding to the segmented stem.

Woolly Sweet-cicely
Osmorhiza claytoni

Habitat: Moist, wooded areas.
General: Stems 15–90 cm tall,
hairy; taproot thick, anise-
scented. **Leaves:** Alternate,
pinnately compound with 3–11
leaflets; leaflets lance-shaped,
1.5–9 cm long, hairy on both
surfaces, margins coarsely
toothed; lower leaves stalked,
upper leaves stalkless; stipules with hairy margins. **Flower Cluster:** Umbel compound with 3–6 umbellets; umbellets 2–10-flowered. **Flowers:** Greenish white,
1–2 mm wide; petals 5; sepals absent; stamens 5; pistil 1. **Fruit:** Schizocarp black,
bristly, 1–2.5 cm long, breaking into 2 mericarps at maturity.

Notes: The genus name *Osmorhiza* comes from the Greek *osme*, "odour," and
rhiza, "root," a reference to the aromatic root. The species name commemorates
John Clayton (1694–1773), an early plant collector in Virginia. • The common name
also refers to the pleasant odour of the roots.

Wild Parsnip
Pastinaca sativa

Habitat: Waste
areas and roadsides;
introduced from
Europe and Asia.
General: Stems
30–150 cm tall, hol-
low, grooved; tap-
root resembling
a parsnip. **Leaves:**
Alternate, compound with 5–15 leaflets; leaflets
oblong to oval, 5–10 cm long, margins sharply
toothed to lobed; lower leaves long-stalked, upper
leaves sheathing. **Flower Cluster:** Umbel
10–20 cm across, compound with 15–25 rays;
smaller umbellets several-flowered; bracts absent.
Flowers: Yellow, 2–5 mm across; petals 5; sepals
small or absent; stamens 5; pistil 1. **Fruit:** Schizocarp 5–7 mm long, elliptic to oval, breaking into
2 mericarps at maturity.

Notes: *Pastinaca* comes from the Latin word *pastus*, meaning "food." The species
name *sativa* means "cultivated." • There are some reports that the wild form of this
garden vegetable is poisonous. Caution is advised when using this plant.

Black Snakeroot

Sanicula marilandica

Habitat: Moist, wooded areas. **General:** Stems 30–100 cm tall, hairless; rootstock stout; roots fibrous. **Leaves:** Basal leaves 4–20 cm long, long-stalked, palmately compound with 5–7 leaflets; leaflets oblong to lance-shaped, 5–20 cm long, 1–2 cm wide, margins coarsely toothed; stem leaves alternate, palmately compound, stalkless. **Flower Cluster:** Umbel head-like, 0.6–1.5 cm across; compound with 3–9 umbellets, each 12–25-flowered; umbellets with 3 stalkless, perfect flowers and several stalked, male flowers. **Flowers:** Greenish white, 1–2 mm across; petals 5; sepals 5, small; stamens 5; pistil 1. **Fruit:** Schizocarp 5–7 mm long, covered with hooked bristles, breaking into 2 single-seeded segments at maturity.

Notes: The genus name comes from the Greek *sanare*, "to heal," a reference to the plant's medicinal properties. The species name means "of Maryland."

Water Parsnip

Sium suave

Habitat: Marsh edges, wet meadows and ditches. **General:** Stems 50–200 cm tall, hollow, strongly ribbed, branched; roots clustered, spindle-shaped. **Leaves:** Alternate, pinnately compound with 7–17 leaflets; leaflets linear to lance-shaped, 5–10 cm long, 3–15 cm wide, margins toothed; submersed leaves, when present, divided several times into thread-like segments. **Flower Cluster:** Umbel 5–12 cm across, compound with 6 or more rays; bracts lance-shaped, reflexed. **Flowers:** White, 1–4 mm across; petals 5; sepals absent; stamens 5; pistil 1. **Fruit:** Schizocarp oval, 2–3 mm long, breaking into 2 single-seeded segments at maturity.

Notes: The species name *suave* means "sweet" or "agreeable," a reference to the edible root. • Extreme caution should be exercised when collecting water parsnip because of its resemblance to spotted water-hemlock (*Cicuta maculata*), p. 301, an extremely poisonous plant that grows in similar habitats. The roots of water parsnip can be roasted, fried or eaten raw, but consuming even a small portion of spotted water-hemlock can be fatal.

Yellow Pimpernel
Taenidia integerrima

Habitat: Dry, rocky woods. **General:** Stems 30–100 cm tall, erect, hairless, dull green to reddish brown, fragrant; taproot thickened, tuber-like. **Leaves:** Alternate, 10–30 cm long, 5–15 cm wide, stalks 0–13 cm long, 2–3 times compound; leaflets oblong to elliptic or lance-shaped, 1.5–3 cm long, 0.5–1.5 cm wide. **Flower Cluster:** Umbel compound, 10–18 cm across, with 7–12 rays; rays 1–4 cm long, hairless; flower stalks 2–5 mm long. **Flowers:** Yellow, 1–4 mm wide, often male or female; petals 5; sepals 5, small; stamens 5, long-stalked; pistil 1. **Fruit:** Schizocarp 4–5 mm long, prominently ribbed, breaking into 2 mericarps at maturity.

Notes: The genus name comes from the Greek word *tainidion*, "little band," a reference to the ribbed fruit. The species name *integerrima* means "most entire," alluding to the smooth leaf margins.

Golden Alexanders, Golden Meadow-parsnip
Zizia aurea

Habitat: Moist meadows and open thickets. **General:** Stems 20–60 cm tall, hairless; taproot thick. **Leaves:** Basal and alternate, pinnately compound, divided 2–3 times; leaflets 3–9, oval to lance-shaped, 2.5–5 cm long, 1–2.5 cm wide, margins toothed. **Flower Cluster:** Umbel compound with 10–18 umbellets, each 6–20-flowered; bracts absent; innermost flower stalkless. **Flowers:** Yellow, 1–3 mm across; petals 5; sepals 5; stamens 5; pistil 1. **Fruit:** Schizocarp 3–4 mm long, oval, breaking into 2 mericarps at maturity.

Notes: The genus name commemorates German botanist Johann Baptist Ziz (1779–1829). The species name *aurea* means "golden."

DOGWOOD FAMILY · Cornaceae

Family description on p. 70.

Bunchberry
Cornus canadensis

Habitat: Cool, moist woods, often beneath evergreens. **General:** Stems 8–18 cm tall, erect; rhizome creeping, often forming large colonies. **Leaves:** Whorl of 4–6, elliptic to oblong or diamond-shaped, 2–8 cm long, 1.5–2.5 cm wide; margins smooth; stalks short; 2 small, opposite stem leaves may appear below the whorl. **Flower Cluster:** Cyme open, 5–15-flowered; bracts 4, white, 1–2.5 cm long, petal-like. **Flowers:** White, 1–2 mm across; petals 4; sepals 4; stamens 4; pistil 1. **Fruit:** Drupe berry-like, red, 6–8 mm across; edible but tasteless.

Notes: The species name means "of Canada."

HEATHER FAMILY · Ericaceae

The heather family is a diverse group of shrubs and herbaceous plants with simple, alternate, opposite, basal or whorled leaves. The flowers, borne singly or in racemes, corymbs, umbels or panicles, may be regular or irregular, unisexual or perfect. Flowers have 2–5 petals, 3–5 sepals, 2 to many stamens and 1 pistil with 2–10 carpels. The ovary may be superior or inferior. The fruit is a berry, drupe or capsule.

The wintergreen (Pyrolaceae) and Indian-pipe (Monotropaceae) families are now included in the heather family. The main Ontario subfamilies are as follows:

Ericoideae/Vaccinioideae/Arbutoideae	Pyroloideae	Monotropoideae
Shrubs Fruit a capsule (Ericoideae), berry (Vaccinioideae), or drupe (Arbutoideae)	Evergreen herbs Fruit a capsule	Herbs without chlorophyll Fruit a capsule

Well-known shrubs of the heather family include blueberry (*Vaccinium myrtilloides*) and cranberry (*Vaccinium oxycoccus*). Several other species, including rhododendrons (*Rhododendron* spp.) and azaleas (*Azalea* spp.), are grown for their ornamental value.

Pipsissewa, Prince's-pine
Chimaphila umbellata

Habitat: Dry, open forests, often on sandy soil. **General:** Shrub-like herb; flowering stems 10–30 cm tall, erect, leafy; rhizome creeping. **Leaves:** Opposite or whorled, evergreen, thick, leathery, elliptic to oblong or lance-shaped, 3–7 cm long, 0.5–2 cm wide, shiny and dark green above; margins toothed; stalks short. **Flower Cluster:** Umbel 2–10-flowered, long-stalked. **Flowers:** White to pinkish, bell-shaped, 0.8–1.4 cm across; petals 5; sepals 5; stamens 10, swollen at the base; pistil 1. **Fruit:** Capsule 4–8 mm across, erect at maturity.

Notes: The genus name comes from the Greek words *cheima*, "winter weather," and *phelein*, "to love," a reference to the evergreen habit of this species. The species name *umbellata* refers to the flower cluster.

Mayflower, Trailing Arbutus · *Epigaea repens*

Habitat: Sandy to rocky woods. **General:** Shrub-like herb; stems 2–7 cm tall, branched, covered with bristly, brown hairs; rhizome short, creeping. **Leaves:** Alternate, evergreen, oval to oblong, base rounded to heart-shaped, 1–7 cm long, 1–4 cm wide; margins smooth, with brownish hairs; stalks 0.4–5 cm long, short-haired. **Flower Cluster:** Spike 2–5 cm long, 3–5-flowered; bracts 2, oval, as long as sepals. **Flowers:** White to pink, spicy-fragrant, tubular, 0.8–1.2 cm long, throat hairy; male or female; petals 5; sepals 5; stamens 10; pistil 1. **Fruit:** Capsule berry-like, yellowish orange, globe-shaped, 4–6 mm long, fleshy inside.

Notes: The genus name comes from the Greek words *epi*, "upon," and *ge*, "the earth," a reference to the growth habit. The species name *repens* means "creeping."

One-flowered Wintergreen
Moneses uniflora

Habitat: Cool, moist, coniferous forests.
General: Stems 5–10 cm tall; rhizome creeping. **Leaves:** Opposite or in whorls of 3, crowded near the base, ever-green, leathery, round, 1–2 cm long. **Flower Cluster:** Solitary, on a leafless stalk. **Flowers:** White, nodding, fragrant, 1–2 cm wide; petals 5, waxy; sepals 5; stamens 8–10; pistil 1. **Fruit:** Capsule 5-lobed, 6–8 mm across.

Notes: The genus name comes from the Greek words *monos*, "one," and *hesis*, "delight," a reference to the single, showy flower. The species name *uniflora* also means "one-flowered."

Pinesap
Monotropa hypopithys

Habitat: Moist to dry pine forests. **General:** Stems brownish yellow to pinkish red, 10–40 cm tall, covered with soft hairs. **Leaves:** Alternate, scale-like, 1–1.5 cm long; stalkless. **Flower Cluster:** Raceme 5–32 cm long, nodding, becoming erect with age. **Flowers:** Yellowish orange to reddish, 1–1.2 cm long; petals 4–5; sepals 4–5 (occasionally absent); stamens 8–10; pistil 1. **Fruit:** Capsule 4–5-valved, 0.6–1 cm long.

Notes: The species name *hypopithys* means "growing under pines."

Indian-pipe
Monotropa uniflora

Habitat: Shady woods with deep leaf litter. **General:** Stems white, waxy, 10–30 cm tall, somewhat succulent, single or in clusters, turning black at maturity; saprophytic (living on decaying matter); roots forming a ball-like mass. **Leaves:** Alternate, white, scale-like, 0.5–1 cm long; stalkless. **Flower Cluster:** Solitary; bracts white. **Flowers:** White, nodding, bell-shaped, 1.3–2.5 cm long; petals 5; sepals 2–4, bract-like; stamens 10; pistil 1. **Fruit:** Capsule erect, 5–7 mm across, opening along the sides.

Notes: The genus name comes from the Greek words *monos*, "one," and *tropos*, "turn," a reference to the solitary, nodding flower. The species name *uniflora* means "one-flowered."

One-sided Wintergreen
Orthilia secunda

Habitat: Dry to moist, mossy forests. **General:** Stems 10–20 cm tall; rhizome creeping. **Leaves:** Alternate but appearing basal, evergreen, round to oval, 1.5–6 cm long; margins smooth to wavy; stalks long. **Flower Cluster:** Raceme nodding, 1-sided, 6–20-flowered. **Flowers:** Greenish white, bell-shaped, 5–7 mm wide; petals 5, waxy; sepals 5; stamens 10; pistil 1. **Fruit:** Capsule globe-shaped, 4–8 mm across.

Notes: The species name *secunda* means "one-sided," a reference to the flower cluster.

Pink Wintergreen
Pyrola asarifolia

Habitat: Moist woods.
General: Flowering stems
10–30 cm tall; rhizome creeping. **Leaves:** Alternate but
appearing basal, evergreen,
leathery, round to oval,
3–6 cm long; margins smooth
to wavy-toothed; stalks long;
may have 1–3 scale-like leaves
on flowering stem. **Flower
Cluster:** Raceme 3–20 cm
long, 4–22-flowered. **Flowers:** Pink, showy, 0.8–3 cm
across; petals 5; sepals 5; stamens 10; pistil 1, style
curved, 0.5–1 cm long. **Fruit:** Capsule globe-
shaped with persistent style.

Notes: The species name *asarifolia* means "leaves
like *Asarum*," which is wild ginger.

Greenish-flowered
Wintergreen · *Pyrola chlorantha*

Habitat: Dry
woods. **General:**
Stems 10–25 cm
tall; rhizome
creeping. **Leaves:**
Alternate but
appearing basal,
evergreen, leath-
ery, 1–3 cm long;
margins smooth
to wavy. **Flower
Cluster:** Raceme
5–10 cm long, 3–10-flowered. **Flowers:**
Greenish white, 8–15 cm wide; petals 5; sepals 5;
stamens 10; pistil 1, style curved upward. **Fruit:**
Capsule globe-shaped.

Notes: The species name *chlorantha* means
"green-flowered."

Small Bog Cranberry, Swamp Cranberry

Vaccinium oxycoccos

Habitat: Mossy, wet bogs.
General: Evergreen, herb-like shrub; stems 10–50 cm long, trailing, copper-coloured.
Leaves: Alternate, evergreen, leathery, oval to lance-shaped, 2–10 mm long, 1.5–2 mm wide;

margins rolled under. **Flower Cluster:** Raceme 1–3-flowered, at branch tips; bracts 2, red, scale-like; flower stalks 1–4 cm long. **Flowers:** Pink, nodding; petals 4, reflexed, 5–6 mm long; sepals 4, reflexed; stamens 8; pistil 1. **Fruit:** Berry reddish, speckled, 0.5–1 cm across; edible.

Notes: The species name *oxycoccos* means "with acidic berries," a reference to the tart flavour of the fruit.

PRIMROSE FAMILY

Primulaceae

The primrose family has over 1000 species worldwide, with the majority of the species occurring in northern temperate regions. Members of this family are herbaceous plants with simple, alternate, opposite, whorled or basal leaves. Stipules are absent. The perfect, regular flowers are composed of 4–9 petals, 4–5 united sepals, 4–9 stamens and 1 pistil. The fruit is a capsule that opens by 2–6 valves.

Several species are grown for ornamental value and include primroses (*Primula* spp.), shooting stars (*Dodecatheon* spp.) and cyclamens (*Cyclamen* spp.)

Scarlet Pimpernel
Anagallis arvensis

Habitat: Sandy, open areas and gardens; introduced from Europe and Asia as a garden plant. **General:** Stem square, trailing to erect with several branches, 5–30 cm long; often rooting at nodes. **Leaves:** Opposite (occasionally in whorls of 3), elliptic to oval, 0.5–2.5 cm long, 0.4–1.8 cm wide, underside often black- or purple-dotted; margins smooth; stalkless. **Flower Cluster:** Solitary, on stalks 1–5 cm long originating in leaf axils. **Flowers:** Orange to scarlet, white or blue, 0.5–1.4 cm wide, nodding; petals 5, fringed; sepals 5, fused; stamens 5, the filaments with soft, pale blue hairs; pistil 1. **Fruit:** Capsule round, 3–6 mm across.

Notes: The genus name *Anagallis* is the Greek word for pimpernel. The species name *arvensis* means "of cultivated fields."
• Scarlet pimpernel is also known as "poorman's weatherglass" because the flowers only open on sunny days.

Fringed Loosestrife
Lysimachia ciliata

Habitat: Shorelines, marshes and moist woods. **General:** Stems 40–100 cm tall, hairless; rhizome slender. **Leaves:** Opposite, light green, oval to lance-shaped, 3–10 cm long, 1–4 cm wide; margins smooth to wavy; stalks 0.6–2 cm long, fringed, with a row of hairs on 1 side of the stalk. **Flower Cluster:** Raceme 2–3-flowered, in leaf axils. **Flowers:** Bright yellow with reddish centre, saucer-shaped, 1.5–2.5 cm across; petals 5–6, fringed; sepals 5–6; stamens 5 (5 small, sterile stamens alternating with 5 fertile stamens); pistil 1. **Fruit:** Capsule oval, 4–6 mm across.

Notes: The genus name *Lysimachia* comes from the Greek words *lysis,* "to release," and *mache,* "strife," hence the name loosestrife. The species name *ciliata* means "fringed," a reference to the row of hairs on the leaf stalks.

Creeping Jenny, Moneywort
Lysimachia nummularia

Habitat: Moist, open areas and rich woods; introduced ornamental from Europe and Asia. **General:** Evergreen to semi-evergreen; stems trailing, 10–40 cm long, branched from the base, winged. **Leaves:** Opposite, round to oval, 1.3–2.8 cm long and wide, dark green above, light green below, hairless; margins smooth; stalks 4–6 mm long. **Flower Cluster:** Solitary, in leaf axils; stalks 1.7–2.3 cm long. **Flowers:** Yellow dotted with dark red, 2–3 cm across; petals 5–6; sepals 5–6, green; stamens 5, filaments and anthers yellow; pistil 1. **Fruit:** Capsule globe-shaped, 7–9 mm across.

Notes: The species name *nummularia* means "money" or "coins," a reference to the shape and size of the leaves.

Swamp Candles
Lysimachia terrestris

Habitat: Swamps, streambanks and lakeshores. **General:** Stems 25–100 cm tall; rhizome thick. **Leaves:** Opposite, elliptic to lance-shaped, 3–10 cm long, 0.5–2 cm wide; margins smooth; stalks less than 1 cm long. **Flower Cluster:** Raceme 10–30 cm long; flower stalks 0.5–2 cm long. **Flowers:** Yellow with red centre, 0.8–1.3 cm wide; petals 5–6; sepals 5–6; stamens 5; pistil 1. **Fruit:** Capsule 3–3.5 mm long, dark-dotted.

Notes: The species name *terrestris* means "on land."
• In late fall, bulblets are produced in the upper leaf axils.

Tufted Loosestrife
Lysimachia thyrsiflora

Habitat: Marshes, swamps and moist, open areas. **General:** Stems 20–60 cm tall, erect, hairless; rhizome creeping. **Leaves:** Opposite, elliptic to oblong or lance-shaped, 3–12 cm long, 0.6–3.5 cm wide, dotted with dark glands; stalkless. **Flower Cluster:** Spike globe-shaped, many-flowered, 1–3 cm long, in leaf axils. **Flowers:** Yellow, 5–7 mm long; petals 5–6, often purple- or black-dotted; sepals 5–6; stamens 5; pistil 1. **Fruit:** Capsule 2–3 mm across, covered with glandular hairs.

Notes: The species name *thyrsiflora* means "flowers in a thyrse," which is a compact, multi-branched flower cluster.

Garden Yellow Loosestrife
Lysimachia vulgaris

Habitat: Marshes, swamps and roadside ditches; introduced ornamental from Europe and Asia. **General:** Stems erect, 60–150 cm tall, densely glandular-haired; rhizome creeping. **Leaves:** Opposite or in whorls of 3–4, elliptic to lance-shaped, 3–12 cm long, 1–5 cm wide, dotted with black or orange glands; margins smooth; stalks 0.2–1 cm long. **Flower Cluster:** Panicle, on stalk 1.5–4.5 cm long; bracts 2–8 mm long; flower stalks 0.3–1.2 cm long. **Flowers:** Yellow, 1.2–2 cm wide; petals 5; sepals 5, green with reddish brown margins; stamens 5; pistil 1. **Fruit:** Capsule globe-shaped, 3–4 mm long.

Notes: The species name *vulgaris* means "common."

Dwarf Lake Primrose
Primula mistassinica

Habitat: Streambanks, lakeshores and calcareous meadows. **General:** Stems 5–15 cm tall, often covered in white powder. **Leaves:** Basal, elliptic to spatula-shaped, 5–7 cm long, 0.2–1.6 cm wide, prominently veined; margins smooth to toothed; stalks winged. **Flower Cluster:** Raceme umbel-like, 1–10-flowered; flower stalks 0.5–2 cm long; bracts reflexed. **Flowers:** Lavender, trumpet-shaped, 5–8 mm long; petals 5, notched at tip; sepals 5, green; stamens 5; pistil 1. **Fruit:** Capsule globe-shaped.

Notes: The species name refers to the location where this species was discovered, Lake Mistassini in north-central Québec.

Northern Starflower
Trientalis borealis

Habitat: Moist woods. **General:** Stems 6–18 cm tall, leafy; rootstock creeping; tubers 1–2 cm long, 2–6 mm thick. **Leaves:** Whorl of 5–10, lance-shaped, 3–10 cm long, 0.7–1.2 cm wide; margins smooth to finely toothed; stalkless; may have a few alternate, scale-like leaves below the whorl. **Flower Cluster:** Solitary, on long stalks from the axil of whorled leaves; 1–3 flowers per plant.

Flowers: White, 0.8–1.5 cm across, star-shaped; petals 7; sepals 7; stamens 7; pistil 1. **Fruit:** Capsule globe-shaped, 6–8 mm across.

Notes: The genus name comes from the Latin word *triens*, meaning "one-third," an obscure reference to the length of the flower stalk in comparison to the height of the plant. The species name *borealis* means "northern." • The common name "starflower" refers to the shape of the flowers.

GENTIAN FAMILY
Gentianaceae

The gentian family has over 500 species worldwide. Members of this family are primarily herbs, with a few shrubs. The simple, entire leaves are alternate, opposite or whorled. Our species are annual, biennial and perennial herbs with opposite leaves. Stipules are absent. The perfect, regular flowers are composed of a 4–5-lobed corolla, a 4–5-lobed calyx, 4–5 stamens and a single pistil. The corolla is tubular to wheel-shaped. The fruit, a capsule, has 2 compartments.

Closed Gentian, Bottle Gentian
Gentiana andrewsii

Habitat: Moist woods and open meadows. **General:** Stems 20–80 cm tall, unbranched, hairless, pale green to purplish. **Leaves:** Opposite and in whorls of 3–7 below the flower cluster, oval to lance-shaped, 2–15 cm long, 0.5–4 cm wide, prominently 3–7-nerved; margins smooth; stalkless. **Flower Cluster:** Cyme terminal, 3–12-flowered. **Flowers:** Dark blue to purple or white, bottle-shaped, 3–4 cm long, closed at the tip; petals 5, with plaited and irregularly toothed tips; sepals 5; stamens 5; pistil 1. **Fruit:** Capsule many-seeded; seeds winged.

Notes: The genus name commemorates King Gentius of Illyria, who discovered the

medicinal qualities of yellow gentian (*G. lutea*). The species name commemorates Henry Charles Andrews (1794–1830), an early English botanical artist.

Fringed Gentian
Gentianopsis crinita

Habitat: Moist meadows. **General:** Stems 5–80 cm tall, erect, rarely branched. **Leaves:** Basal leaves oblong to lance-shaped, 0.5–3.5 cm long, 0.2–1 cm wide; stem leaves opposite, linear to lance-shaped, 2–5 cm long, 0.3–2 cm wide, tip pointed, stalkless. **Flower Cluster:** Solitary, on long, erect stalks. **Flowers:** Blue to white, tubular, 4–6 cm long; petals 4, margins fringed; sepals 4; stamens 4; pistil 1; opening only in sunlight. **Fruit:** Capsule urn-shaped, 3–4 cm long.

Notes: The genus name *Gentianopsis* means "gentian-like." Species in this genus were once included in the genus *Gentian*. The species name *crinita* means "with hairs," a reference to the fringed petals.

Spurred Gentian
Halenia deflexa

Habitat: Damp woods. **General:** Stems 20–50 cm tall. **Leaves:** Basal leaves spatula-shaped; stem leaves opposite, oblong to lance-shaped, 2–5 cm long, prominently 3-nerved; margins smooth; stalkless. **Flower Cluster:** Cyme 5–9-flowered. **Flowers:** Purplish green to bronze, 1–1.2 cm long; petals 4, spurred, each spur 1–5 mm long; sepals 4; stamens 4; pistil 1. **Fruit:** Capsule bottle-shaped, 0.6–1.5 cm long.

Notes: The genus name commemorates Jonas Halenius (1727–1810), a Swedish botanist and a student of Linnaeus. The species name *deflexa* means "bent backward," a reference to the spurs.

DOGBANE FAMILY

Apocynaceae

The dogbane family has over 4500 species worldwide. Members of this family are herbs, shrubs, woody vines and trees, most with milky sap. The simple, entire leaves are opposite or whorled. The flowers are composed of 5 fused petals, 5 sepals, 5 stamens and a single pistil. The fruit is a pod-like follicle that contains numerous seeds with long, silky hairs.

Recent taxonomic research includes the milkweed family (Asclepiadaceae) in the dogbane family. In this subfamily, Asclepiadoideae, the stamens are fused to the style and stigma, forming a structure called a gynostegium. Horn-like appendages on the gynostegium are referred to as the corona.

The 2 main subfamilies are distinguished below:

Apocynoideae	Asclepiadoideae
Stamen filaments free Pollen grains free Nectaries usually absent	Stamen filaments fused to style and stigma Pollen in masses (pollinia) Nectaries present

Common periwinkle (*Vinca minor*), a well-known ornamental groundcover, belongs to the dogbane family. Another important member is common milkweed (*Asclepias syriaca*), the food source for the monarch butterfly.

Spreading Dogbane

Apocynum androsaemifolium

Habitat: Dry, sandy, open forests and meadows.
General: Stems 20–100 cm tall, reddish green, branched above; exuding milky juice when broken; rhizome creeping. **Leaves:** Opposite, bright green, oval to oblong, 2.5–12 cm long, 0.5–6 cm wide, often drooping; margins smooth to wavy; stalks short; turning yellow to red in autumn. **Flower Cluster:** Cyme 2–5-flowered, at branch ends; bracts oblong to lance-shaped. **Flowers:** Pink or with pinkish stripes, fragrant, cup-shaped, 6–9 mm long; petals 5; sepals 5; stamens 5; pistil 1. **Fruit:** Follicle pod-like, reddish green, 2–20 cm long; in pairs; poisonous.

Notes: The genus name *Apocynum* is Greek for "away from dog," a reference to this species' ancient use as a dog poison and the origin of the common name "dogbane." • The milky juice of this plant yields a type of latex, and several attempts have been made to grow spreading dogbane commercially for the production of rubber.

Indian Hemp · *Apocynum cannabinum*

Habitat: Moist, open woods and shorelines.
General: Stems 30–80 cm tall, erect, branched
above; exuding milky juice when broken. **Leaves:**
Opposite, oval to lance-shaped, 5–12 cm long,
0.7–4.5 cm wide; margins smooth to wavy; stalked
or stalkless. **Flower Cluster:** Cyme, at branch
ends. **Flowers:** Greenish white, tube- or urn-
shaped, 2–5 mm long; petals 5, united, the lobes
not reflexed; sepals 5; stamens 5; pistil 1. **Fruit:**
Follicle pod-like, 6–12 cm long; in pairs; poisonous.

Notes: The species name means "like *Cannabis* or hemp," a reference to the
fibrous stems. • As the common name implies, the stems of Indian hemp were
a natural source of twine and rope.

Swamp Milkweed · *Asclepias incarnata*

Habitat: Wet meadows and swamps. **General:**
Stems 40–150 cm tall, branched; exuding milky sap
when broken; rhizome creeping. **Leaves:** Opposite,
oblong to lance-shaped, 7.5–15 cm long, 1–4 cm
wide, prominently veined; margins smooth; stalks
short. **Flower Cluster:** Umbel flat-topped,
7–10 cm across. **Flowers:** Pinkish purple, 5–7 mm
across; petals and sepals bent backward, exposing
the corona; corona composed of 5 "hoods," each
2–4 mm long; horns incurved, longer than hoods. **Fruit:** Follicle pod-like, 5–10 cm
long, surface smooth; seeds with silky, white hairs.

Notes: The genus name commemorates Asklepios, the Greek god of medicine. The
species name *incarnata* means "flesh-coloured," a reference to the flower colour.

Common Milkweed
Asclepias syriaca

Habitat: Waste areas, roadsides and open fields. **General:** Stems 50–200 cm tall, hairy, unbranched; exuding milky juice when broken; rhizome creeping. **Leaves:** Opposite, elliptic to oval or oblong, 7–26 cm long, 5–18 cm wide, underside covered in soft hairs; margins smooth; stalks 0.5–1.5 cm long. **Flower Cluster:** Umbel 20–130-flowered, globe-shaped, terminal or axillary. **Flowers:** Greenish to pinkish purple, 0.8–1 cm long, 1.1–1.4 cm wide, fragrant; petals and sepals bent backward, exposing the corona; corona with 5 pale purple "hoods," each 6–8 mm long; horns short, incurved. **Fruit:** Follicle pod-like, spindle-shaped, 8–12 cm long, greyish green, surface with white, woolly hairs and warty bumps; seeds with silky hairs.

Notes: The species name *syriaca* means "of Syria." • This plant's milky sap contains alkaloids and several other poisonous substances. Monarch butterfly larvae feed on the sap without harm. However, the toxins accumulate in the insect's body and provide a natural defence against predators.

Butterflyweed, Orange Milkweed · *Asclepias tuberosa*

Habitat: Dry, open meadows. **General:** Stems 30–75 cm tall, hairy, branched; exuding watery (not milky) juice when broken; roots tuberous. **Leaves:** Alternate or opposite, linear to lance-shaped, 5–10 cm long, 1–3 cm wide, base wedge- to heart-shaped; margins smooth; stalkless. **Flower Cluster:** Umbel 7–9 cm across, 20–40-flowered; bracts 0.8–1.2 cm long; flower stalks 1.5–2.5 cm long. **Flowers:** Yellowish to orangey red, 0.7–1 cm long; petals and sepals bent backward, exposing the corona; corona with 5 yellowish orange "hoods," each 5–7 mm long; horns short, straight. **Fruit:** Follicle pod-like, spindle-shaped, 8–12 cm long, surface with soft hairs; seeds with silky hairs.

Notes: The species name means "tuberous," a reference to the large root. • A tea made from the root was used to treat lung inflammations such as pleurisy, asthma and bronchitis, hence another common name, "pleurisy-root."

Common Periwinkle

Vinca minor

Habitat: Waste areas, road-sides and forests; introduced from southern Europe as a groundcover. **General:** Evergreen; stems trailing, 30–100 cm long, often forming mats. **Leaves:** Opposite, elliptic to lance-shaped, 1–5 cm long, 0.5–2.5 cm wide, leathery, dark green; margins smooth; stalks short. **Flower Cluster:** Solitary, in leaf axils; flower stalk 1–1.5 cm long. **Flowers:** Lilac to blue, 2–3 cm across, 0.8–1.2 cm long; petals 5; sepals 5; stamens 5; pistil 1. **Fruit:** Follicle pod-like, narrow; in pairs; seeds hairless.

Notes: The genus name *Vinca* comes from the Latin word *Vincapervinca*, an ancient name for this species.

Dog-strangling Vine, Pale Swallow-wort

Vincetoxicum rossicum

Habitat: Dry, open areas, hill-sides and forest edges; introduced ornamental from Ukraine and Russia; rapidly spreading throughout southern Ontario. **General:** Stems 60–250 cm long, erect to twining or scrambling; rhizome short; crown woody. **Leaves:** Opposite, oval to elliptic, 6–12 cm long, 2.5–7 cm wide, hairy on underside margins and veins; margins smooth; stalks 0.5–2 cm long. **Flower Cluster:** Cyme umbel-like, 5–20-flowered; flower stalks 1.5–5 cm long, hairy. **Flowers:** Pink, red or maroon to brown, 5–7 mm across; petals 5, twisted in bud; sepals 5; corona fleshy, lobed, pink to orange or yellow; gynostegium pale yellow to greenish yellow; stamens 5; pistils 2. **Fruit:** Follicle pod-like, 2.8–7 cm long, often producing 2 fruits per flower; seeds light to dark brown, 4–6.5 mm long, winged.

Notes: The species name *rossicum* means "of Russia," this plant's native range.

MORNING-GLORY FAMILY · Convolvulaceae

The morning-glory family has over 1500 species worldwide. Members of this family are twining herbs, shrubs and trees with milky sap. The alternate leaves are simple to compound. The regular flowers are composed of 5 united petals, 5 sepals, 5 stamens and 1 pistil. The corolla is tubular or funnel-shaped. The fruit is a capsule.

The dodder family (Cuscutaceae) is now included in the morning-glory family. The subfamilies are distinguished below:

Convolvuloideae	Cuscutoideae
Plants green Leaves well-developed	Plants usually yellow, orange or brown; parasitic Leaves scale-like

Several morning-glory species are grown for their horticultural and ornamental value, including sweet potato (*Ipomoea batatas*) and garden morning-glory (*I. purpurea*). Several species from both subfamilies are troublesome weeds on agricultural land.

Hedge Bindweed, Wild Morning-glory

Calystegia sepium

Habitat: Roadsides, forest edges and waste ground. **General:** Climbing vine; stems 1–3 m long, twining, extensively branched; rhizome creeping. **Leaves:** Alternate, triangular to arrowhead-shaped, 4–15 cm long, 1.2–12 cm wide, tips pointed; margins smooth; stalks long. **Flower Cluster:** Raceme of 1–4 flowers per leaf axil, on 4-angled stalks; bracts 2, 1–5 cm long. **Flowers:** White to pinkish, tubular to funnel-shaped, 4–5 cm long, 3–6 cm across; petals 5; sepals 5; stamens 5; pistil 1. **Fruit:** Capsule 0.8–1 cm across; seeds black.

Notes: *Calystegia* is Greek for "concealing the calyx." The species name *sepium* is Latin for "of fences and hedges," a reference to the plant's habitat.

Field Bindweed
Convolvulus arvensis

Habitat: Waste areas, roadsides and landscaped areas; introduced from Europe. **General:** Climbing vine; stems 1–7 m long, twining or trailing; rhizome deep, cord-like. **Leaves:** Alternate, triangular to arrowhead-shaped, 2–6 cm long, 1–4 cm wide, tips rounded; margins smooth; stalks long.

Flower Cluster: Solitary or occasionally a raceme of 1–4 flowers, in leaf axils; bracts 2, small. **Flowers:** White to pinkish, funnel-shaped, 2–3 cm across, 1.5–2 cm long; petals 5; sepals 5; stamens 5; pistil 1. **Fruit:** Capsule egg-shaped, 3–9 mm across.

Notes: The genus name *Convolvulus* comes from the Latin word *convolvere*, "to roll together," a reference to the twining stems. The species name *arvensis* means "of cultivated fields," alluding to this plant's weedy nature.

Swamp Dodder, Gronovius' Dodder
Cuscuta gronovii

Habitat: Moist, open areas and wet meadows. **General:** Parasitic on wetland plants; stems orangey yellow, 1–10 m long, thread-like; forming large masses of stems and covering the host plant. **Leaves:** Alternate, scale-like, 1–2 mm long. **Flower Cluster:** Cyme compact, many-flowered. **Flowers:** Creamy to yellowish white, bell- to globe-shaped, 2.5–4 mm long; petals 4–5; sepals 4–5; stamens 5; pistil 1. **Fruit:** Capsule round, 2–4 mm across; seeds 4–8.

Notes: *Cuscuta* is derived from the Arabic name for dodder. The species name commemorates Dutch botanist Jan Frederik Gronovius (1686–1762).

Ivy-leaved Morning-glory
Ipomoea hederacea

Habitat: Roadsides and waste areas; introduced ornamental from tropical North America. **General:** Climbing vine; stems 50–200 cm long, twining or trailing, hairy. **Leaves:** Alternate, 5–12 cm long and wide, deeply 3-lobed (occasionally 5-lobed), hairy, base heart-shaped; margins smooth; stalks long. **Flower Cluster:** Raceme 1–3-flowered, in leaf axils; flower stalks 2–4 cm long. **Flowers:** Pale blue turning pinkish purple upon opening, funnel-shaped, 3–5 cm long; petals 5, united; sepals 5, lance-shaped, 1.5–2.5 cm long, densely haired to bristly; stamens 5; pistil 1. **Fruit:** Capsule egg-shaped, partly surrounded by the calyx; seeds 4–6, dark brown to black.

Notes: The genus name comes from the Greek words *ips*, "worm," and *homoios*, "resembling," a possible reference to the stems. The species name means "like *Hedera* (ivy)," a reference to the genus *Hedera*.

PHLOX FAMILY
Polemoniaceae

The phlox family has over 300 species worldwide. Members of this family are herbs and shrubs with alternate or opposite, simple or compound leaves. The showy flowers are composed of 5 fused petals, 5 sepals, 5 stamens and 1 pistil with a 3-lobed style. The stamens are inserted on the tubular corolla. The fruit is a capsule.

Ornamental members of this family include Jacob's-ladder (*Polemonium* spp.), cup-and-saucer vine (*Cobaea scandens*) and scarlet gilia (*Ipomopsis aggregata*).

Wild Blue Phlox, Blue Phlox

Phlox divaricata

Habitat: Moist to dry, rocky woods. **General:** Stems erect, 15–50 cm tall, trailing and leafy near the base; rhizome creeping. **Leaves:** Opposite (usually 4 pairs per stem), oval to lance-shaped, 2.5–5 cm long, 0.5–2.5 cm wide, hairy; margins smooth; stalkless to clasping. **Flower Cluster:** Cyme loosely branched, often covered in sticky hairs. **Flowers:** Pale blue to purple, tubular with 5 spreading lobes, 2–3.8 cm across, 1–2 cm long; petals 5; sepals 5; stamens 5; pistil 1. **Fruit:** Capsule with 3 compartments, many-seeded.

Notes: The genus name comes from the Greek word *phlox*, meaning "a flame." The species name *divaricata* means "spreading."

WATERLEAF FAMILY

Hydrophyllaceae

The waterleaf family is a small family of herbs, shrubs and trees with simple or compound, alternate or opposite leaves. Stipules are absent. The flowers are composed of 5 united petals, 5 united sepals, 5 stamens and 1 style. The fused petals form a cup or funnel. The stamens are attached to and alternate with the petals. The fruit is a capsule.

Ornamental members of this family include baby blue-eyes (*Nemophila menziesii*) and scorpion-weeds (*Phacelia* spp.).

Virginia Waterleaf
Hydrophyllum virginianum

Habitat: Moist, wooded areas. **General:** Stems 20–80 cm tall, leafy, covered in dense, stiff hairs; rhizome fleshy. **Leaves:** Alternate, stalked, 10–20 cm long, mottled green, pinnately divided into 3–7 broadly oval to triangular lobes; margins coarsely toothed. **Flower Cluster:** Cyme scorpioid or head-like, densely flowered; flower stalks 3–5 mm long. **Flowers:** White to dark violet, bell-shaped, 0.7–1 cm long; petals 5; sepals 5, hairy; stamens 5, long-stalked, hairy; pistil 1. **Fruit:** Capsule globe-shaped, many-seeded.

Notes: The genus name comes from the Greek words *hydor*, "water," and *phyllon*, "leaf." The species name means "of Virginia." • The mottled leaves look as though they are water-stained, hence the common name "waterleaf."

BORAGE FAMILY Boraginaceae

BORAGE FAMILY
Boraginaceae

Members of this family are herbs and shrubs with simple, alternate leaves. Stipules are present. A distinguishing feature of the borages is the hairy texture of the leaves and stem, and also the flowers, which usually appear in one-sided clusters. The clusters often resemble a scorpion's tail and are technically referred to as a "scorpiod cyme." The flowers are composed of 5 fused petals, 5 sepals, 5 stamens inserted on the corolla and a single pistil. The style is often inserted at the base of the ovary. The fruit is a nutlet.

Well-known species include borage (*Borago officinalis*), forget-me-nots (*Myosotis* spp.), lungworts (*Pulmonaria* spp.) and viper's bugloss (*Echium vulgare*).

Hound's-tongue
Cynoglossum officinale

Habitat: Roadsides, waste areas and open fields; introduced from Europe. **General:** Stems 40–120 cm tall, erect, leafy, covered with soft hairs; taproot black. **Leaves:** Basal leaves oblong to lance-shaped, 10–30 cm long, 2–5 cm wide, stalked; stem leaves alternate, 2–5 cm long, stalkless, clasping the stem; covered with soft, velvety hairs. **Flower Cluster:** Cyme scorpioid, raceme-like, 5–35-flowered, 5–20 cm long, leafy. **Flowers:** Reddish purple, funnel-shaped, 0.5–1 cm across; petals 5; sepals 5; stamens 5; pistil 1. **Fruit:** Nutlet brown, covered with hooked prickles; 4 nutlets per flower; fruiting cluster 1–1.5 cm across.

Notes: The genus name *Cynoglossum* comes from the Greek words *kuon* and *glossa*, meaning "dog's tongue." The species name *officinale* means "medicinal."

Viper's Bugloss, Blueweed
Echium vulgare

Habitat: Roadsides, waste areas and railway grades; introduced from North Africa as an antidote for snakebites. **General:** Stems 20–80 cm tall, often reddish green, covered with bristly hairs, the hairs with swollen red or black bases; taproot stout, black. **Leaves:** Alternate, linear to lance-shaped, 1–15 cm long, 0.5–3 cm wide, covered with long, stiff hairs, the hairs with enlarged red or black bases; basal rosette of leaves produced in first season. **Flower Cluster:** Cyme scorpioid, 1-sided; bracts lance-shaped, bristly haired. **Flowers:** Blue, funnel-shaped, 1.5–2 cm long; petals 5; sepals 5; stamens 5, long stalked; pistil 1. **Fruit:** Nutlet angular; 4 nutlets per flower.

Notes: *Echium* comes from *echion*, the Greek name for this plant. The species name *vulgare* means "common." • The dried plant has been used as a remedy for snakebites, hence the common name "viper's bugloss."

Blue-bur
Lappula squarrosa

Habitat: Roadsides, parking lots and waste areas; introduced from Europe and Asia. **General:** Stems 10–50 cm tall, branched, covered with stiff, white hairs, mousy-scented. **Leaves:** Alternate, linear to oblong or lance-shaped, greyish green, 1–7 cm long, covered with stiff hairs; margins smooth; stalkless. **Flower Cluster:** Cyme scorpioid, leafy. **Flowers:** Pale blue with a yellow throat, 3–4 mm across; petals 5; sepals 5; stamens 5; pistil 1. **Fruit:** Nutlet 2–5 mm long, with 2 rows of hooked prickles; 4 nutlets per flower.

Notes: The genus name comes from the Latin word *lappa*, meaning "bur," referring to the bur-like fruit. The species name *squarrosa* means "rough" and "spreading in all directions," a reference to the surface of the nutlets.

Gromwell, European Stoneseed
Lithospermum officinale

Habitat: Waste areas and roadsides; introduced from southern Europe. **General:** Stems 20–100 cm tall, branched near the top, densely hairy throughout; taproot thick. **Leaves:** Alternate, oval to lance-shaped, 3–8 cm long, 0.5–1.5 cm wide, prominently veined; margins smooth; stalks short. **Flower Cluster:** Cyme leafy, to 15 cm long at fruiting. **Flowers:** White to pale yellow, tubular, 4–8 mm across; petals 5; sepals 5, 5–7 mm long at fruiting; stamens 5; pistil 1. **Fruit:** Nutlet oval, white or yellowish brown, 3–3.5 mm long, smooth, shiny.

Notes: The genus name comes from the Greek words *lithos*, "stone," and *sperma*, "seed," a reference to the stone-like seeds. The species name *officinale* means "medicinal."

Tall Lungwort · *Mertensia paniculata*

Habitat: Moist, open areas and woods. **General:** Stems 20–60 cm tall, hairless to slightly hairy. **Leaves:** Alternate, oval to lance-shaped, 5–14 cm long, dark green, covered in soft hairs; margins smooth; stalks winged. **Flower Cluster:** Cyme scorpioid; stalks drooping. **Flowers:** Blue to purple or white, bell-shaped, 1–1.5 cm long; petals 5, united; sepals 5; stamens 5, attached to corolla; pistil 1. **Fruit:** Nutlet; 4 per flower.

Notes: The genus name commemorates German botanist Franz Karl Mertens (1764–1831). The species name *paniculata* refers to the panicle-like flower cluster.

Virginia Bluebells
Mertensia virginica

Habitat: Moist woods. **General:** Stems 20–60 cm tall, hairless, pale green. **Leaves:** Alternate, oval, 2–20 cm long, 1–11 cm wide, dull green above, silvery below; margins smooth; stalks 0–8 cm long, winged. **Flower Cluster:** Cyme compact, elongating in fruit; flower stalks 2–3 mm long, to 12 mm long in fruit. **Flowers:** Blue, trumpet-shaped, 1–3 cm long, 0.5–2 cm wide; petals 5; sepals 5; stamens 5, anthers bluish white; pistil 1. **Fruit:** Nutlets 2–4 mm long, wrinkled.

Notes: The species name means "of Virginia."

Forget-me-not
Myosotis sylvatica

Habitat: Wet woods and streambanks; introduced ornamental from Europe. General: Stems 15–30 cm tall, erect to sprawling, hairy throughout, often rooting from lower nodes. Leaves: Alternate, oblong to lance-shaped, 2.5–8 cm long, 0.3–2 cm wide, surface hairy; margins smooth; stalkless.

Flower Cluster: Raceme scorpioid, 5–20 cm long; bracts absent. Flowers: Blue with yellow throat, 5–9 mm wide; petals 5; sepals 5, 3–5 mm long at fruiting, bristly haired; stamens 5; pistil 1. Fruit: Nutlet smooth, shiny, 1–2 mm long.

Notes: The genus name comes from the Greek words *mus*, "mouse," and *ous*, "ear," a reference to the ear-shaped leaves of some species in this genus.

Prickly Comfrey
Symphytum asperum

Habitat: Roadsides and waste areas; introduced from Europe and Asia. General: Stems erect, 40–100 cm tall, angular, hollow, covered in bristly hairs; taproot thick, black. Leaves: Alternate, oval to lance-shaped, 5–30 cm long, reduced in size upward, upper leaves stalkless, surface with short, bristly hairs, base not extending down the stem; margins smooth. Flower Cluster: Cyme terminal; bracts absent. Flowers: Blue to purple (pink in bud), 1.5–2 cm long, tubular; petals 5; sepals 5; stamens 5; pistil 1. Fruit: Nutlet brown, shiny, 3–5 mm long.

Notes: *Symphytum* comes from the Greek words symphysis, "the growing together of bones," and *phyton*, "plant," a reference to the plant's use in the treatment of broken bones. The species name *asperum* means "rough-leaved."

LOPSEED FAMILY
Phrymaceae

The lopseed family is a small family of herbs and shrubs with about 190 species worldwide. The stems, often 4-sided or winged, have simple, opposite leaves. The regular or irregular flowers are borne in terminal spikes or axillary cymes. The tubular, 2-lipped flowers have 5 fused petals, 5 fused sepals, 2 or 4 stamens and 1 pistil. The stigmas are 2-lobed and close together when a pollinator is present. The fruit is a capsule, achene or berry. *Mimulus,* the largest genus, contains about 120 species. Several are ornamentals sold under the common name "monkeyflower."

Square-stemmed Monkeyflower
Mimulus ringens

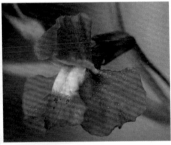

Habitat: Wet, low-lying areas and shorelines. **General:** Stems 30–90 cm tall, square; rhizome creeping. **Leaves:** Opposite, oblong to lance-shaped, 5–10 cm long, 1–2 cm wide, reduced in size upward, base round to heart-shaped; margins finely toothed; stalkless and clasping the stem. **Flower Cluster:** Solitary, on stalks 2–4 cm long, from leaf axils. **Flowers:** Bluish purple, 2–4 cm long; petals 5, upper lip 2-lobed and bent backward, lower lip 3-lobed with 2 yellow spots; sepals 5; stamens 4; pistil 1. **Fruit:** Capsule cylindric, 0.9–1.2 cm long.

Notes: The genus name comes from the Greek word *mimos,* "mimic," a reference to the flowers' resemblance to a monkey's face. The species name *ringens* means "stiff" or "rigid," referring to the stem.

Lopseed, Slender-spiked Lopseed · *Phryma leptostachya*

Habitat: Rich, deciduous forests. **General:** Stems 40–100 cm tall, occasionally branched, bristly haired. **Leaves:** Opposite, oval to lance-shaped, 6–15 cm long, 1–6 cm wide; margins coarsely toothed; lower leaves stalked, upper stalkless. **Flower Cluster:** Spike terminal, 5–15 cm long; bracts 3, awl-shaped. **Flowers:** Pale purple to white, 6–8 mm long, opposite on the stem and at right angles to those above and below; petals 5, united, upper lip short and 2-lobed, lower lip long and 3-lobed; sepals 5; stamens 4; pistil 1. **Fruit:** Achene 2–4 mm long, drooping, enclosed by persistent calyx.

Notes: The species name *leptostachya* means "thin-spiked," a reference to the long, narrow flower clusters. • The common name "lopseed" refers to the hanging fruits.

VERVAIN FAMILY
Verbenaceae

The vervain family has over 2600 species worldwide. Members of this family are primarily tropical herbs, shrubs, woody vines and trees. The opposite to whorled leaves are simple to compound and lack stipules. The perfect flowers are composed of 4 to many fused petals, 4–8 fused sepals, 2–5 stamens and 1 pistil. The fruit is a drupe or nutlet.

A well-known species, lantana (*Lantana camara*), is widely planted in the Toronto area as an ornamental, though it is an invasive weed throughout much of the tropics. Another important member of this family is teak (*Tectona grandis*).

Blue Vervain, False Vervain
Verbena hastata

Habitat: Moist, open areas and woods. **General:** Stems 50–150 cm tall, square, branched near the top, covered in bristly hairs. **Leaves:** Opposite, 4–18 cm long, 1.5–5 cm wide, oblong to lance-shaped, often with 3 lower lobes, both surfaces smooth to hairy; margins coarsely toothed to 3-lobed; stalks short. **Flower Cluster:** Panicle of densely flowered spikes, each 5–10 cm long. **Flowers:** Blue to violet, funnel-shaped, 2–5 mm across; petals 5; sepals 5; stamens 4; pistil 1. **Fruit:** Nutlet narrow, 1–2 mm long; in clusters of 4.

Notes: The species name *hastata* means "arrowhead- or spearhead-shaped," a reference to the shape of the lobed leaves. • The common name "vervain" comes from the Celtic words *fer*, "to take away," and *faen*, "stone," a medicinal reference to the plant's use in the treatment of gallstones.

Hoary Vervain
Verbena stricta

Habitat: Disturbed sites, roadsides and open fields. **General:** Stems 20–150 cm tall, 4-angled, covered in soft, velvety to bristly hairs. **Leaves:** Opposite, oval, 1–10 cm long, 0.5–5 cm wide, bristly haired; margins coarsely toothed; stalks 0–5 mm long. **Flower Cluster:** Panicle of densely flowered spikes, each 5–40 cm long; bracts 2–5 mm long. **Flowers:** Blue to purple or pinkish, trumpet-shaped, 0.5–1 cm across; petals 5; sepals 5; stamens 4; pistil 1. **Fruit:** Nutlet 2–3 mm long; surface with network of veins.

Notes: The species name *stricta* means "stiff" or "erect." • The common name "hoary vervain" refers to the whitish appearance of the plant, which results from the presence of the hairs.

White Vervain
Verbena urticifolia

Habitat: Open woods and moist meadows.
General: Stems 40–200 cm tall, 4-angled, branched from the base, almost hairless. **Leaves:** Opposite, oblong to lance-shaped, 8–20 cm long, 1–8 cm wide, winged; margins coarsely toothed; stalks 0.5–3 cm long.
Flower Cluster: Panicle of spikes 5–25 cm long, sparsely flowered; bracts 1–1.5 mm long. **Flowers:** White, 2–3 mm across; petals 5; sepals 5; stamens 4; pistil 1. **Fruit:** Nutlet 1–3 mm long, ribbed.

Notes: The species name means "leaves like *Urtica* (stinging nettle)."

MINT FAMILY · Lamiaceae

This family was previously called Labiatae for the structure of the 2-lipped flowers. Members of the mint family are shrubs and herbaceous plants with simple, opposite leaves and square stems. The opposite leaves, arranged at 90° to those above and below, are referred to as decussate. The regular to irregular flowers are borne singly, in cymes in the axils of the upper leaves or in terminal racemes. The flowers have 4–5 fused petals, 4–5 fused sepals, 2 or 4 stamens and 1 pistil with 2 carpels. The corolla is often 2-lipped, with 2 upper and 3 lower petals. The fruit is a cluster of 4 nutlets.

Economically important members of this family include basil (*Ocimum basilicum*), mint (*Mentha* spp.), oregano (*Origanum vulgare*), rosemary (*Rosmarinus officinalis*), sage (*Salvia officinalis*) and thyme (*Thymus* spp.). Species grown for their ornamental value include coleus (*Coleus* spp.), lavender (*Lavendula* spp.) and catnip (*Nepeta cataria*). Several species in this family are noxious weeds.

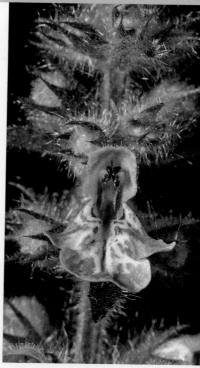

Giant Hyssop
Agastache foeniculum

Habitat: Open grasslands and woods.
General: Stems 50–100 cm tall, square, hairless, crushed stems and leaves with anise-like odour; rootstock creeping.
Leaves: Opposite, oval to triangular, 2–7 cm long, underside whitish; margins coarsely toothed; stalks short. **Flower Cluster:** Spike terminal, 2–10 cm long, 1.2–1.8 cm wide, interrupted, crowded, hairy. **Flowers:** Blue or violet, irregular, 2-lipped, 0.6–1.2 cm long; petals 5, united; sepals 5, prominently 15-nerved; stamens 4, long-stalked; pistil 1. **Fruit:** Nutlet brown; 4 nutlets per flower.

Notes: The genus name comes from the Greek words *agan*, "very much," and *stachys*, "an ear of corn or wheat," referring to the numerous flowering spikes. The species name is a diminutive form of the Latin word *foenum*, meaning "hay," a reference to the plant's scent.

Wild Basil
Clinopodium vulgare

Habitat: Woods and roadsides.
General: Stems 10–40 cm tall, occasionally branched. **Leaves:** Opposite, oval to oblong or lance-shaped, 2–4 cm long, 0.7–1.5 cm wide; margins smooth or with a few small teeth; stalks 0.3–1 cm long.
Flower Cluster: Cyme head-like, densely flowered; bracts stiff-haired. **Flowers:** Pink to pale purple, 1.2–1.5 cm long, 2-lipped; petals 5, united; sepals 5, fused into a tube, 0.9–1 cm long; stamens 4; pistil 1. **Fruit:** Nutlets smooth.

Notes: *Clinopodium* is Greek for "bed-foot," a possible reference to the flowerheads resembling bed castors. The species name *vulgare* means "common."

Hemp Nettle
Galeopsis tetrahit

Habitat: Waste
areas and roadsides;
introduced from
Europe and Asia.
General: Stems
30–80 cm tall, square,
bristly haired, often
swollen below nodes.
Leaves: Opposite,
oval to lance-shaped, 3–12 cm long, 1–5 cm wide,
both surfaces bristly haired; margins with 5–10
coarse teeth per side; stalked. **Flower Cluster:**
Spike terminal, axillary; flowers in whorls. **Flowers:** White, pink or variegated, with 2 yellow
spots, 2-lipped, 1.5–2.2 cm long; petals 5, united;
sepals 5, spine-tipped, prominently 5-ribbed;
stamens 4, hairy; pistil 1. **Fruit:** Nutlet 3–4 mm
long, greyish brown, egg-shaped; 4 nutlets per
flower.

Notes: *Galeopsis* is the ancient Latin name given to plants with a 2-lipped corolla.
The species name *tetrahit* means "4-angled," a reference to the square stem.

Ground-ivy, Creeping Charlie, Gill-over-the-ground
Glechoma hederacea

Habitat: Lawns, waste areas and
roadsides; introduced ornamental from
Europe. **General:** Stems 20–50 cm
long, square, creeping, rooting at nodes
when in contact with soil, rancid, mint-
like odour when crushed. **Leaves:**
Opposite, round to kidney-shaped,
often purple-tinged, 1–3 cm long,
1–4 cm wide; margins wavy-toothed;
stalks 1–10 cm long. **Flower Cluster:** Spike 2–7-flowered, in leaf axils.
Flowers: Purplish blue, 2-lipped,
1–2 cm long; petals 5, united; sepals 5,
prominently 15-nerved; stamens 4; pistil 1. **Fruit:** Nutlet smooth, about 1 mm
long; 4 nutlets per flower.

Notes: *Glechoma* is the Greek name for a type of mint. The species name *hederacea*
and the common name "ground-ivy" both refer to this plant's resemblance to ivy
(*Hedera* spp.).

Henbit

Lamium amplexicaule

Habitat: Fields, roadsides and waste areas; introduced from Europe. General: Stems 10–40 cm long, square, creeping, tips curved upward, hairless. Leaves: Opposite, round to oval, 0.5–3 cm long, 1–5 cm wide, upper surface hairy; margins with 2–4 rounded teeth per side; lower leaves long-stalked, upper leaves stalkless and clasping. Flower Cluster: Spike crowded, in whorls in upper leaf axils. Flowers: Pink to purple, helmet-shaped, 2-lipped, 1–1.6 cm long; petals 5, united (upper lip 3–5 mm long and hairy, lower lip spotted); sepals 5, prominently 5-nerved; stamens 4; pistil 1. Fruit: Nutlet 1–2 mm long, greyish brown, speckled; 4 nutlets per flower.

Notes: *Lamium* is the Latin name for "dead-nettle." The species name *amplexicaule* means "clasping the stem," a reference to the leaves.

Motherwort · *Leonurus cardiaca*

Habitat: Waste areas and roadsides; introduced from Europe and Asia. General: Stems stiff, 40–150 cm tall, square, hollow, aromatic when crushed; rhizome creeping. Leaves: Opposite, 2–12 cm long, 3–15 cm wide, palmately 3–7-lobed, dull green, underside bristly haired; margins toothed or lobed; stalks 3–14 cm long, 4-angled; upper leaves less deeply lobed, short-stalked. Flower Cluster: Verticel (dense whorl) of 6–15 flowers per node; bracts of upper flowers 3-toothed. Flowers: Pink to pale purple, 0.8–1.5 cm long, 2-lipped; petals 5, united; sepals 5, prickly; stamens 4; pistil 1. Fruit: Nutlet bristly haired; 4 nutlets per flower, partially surrounded by persistent, prickly calyx.

Notes: The genus name comes from the Greek words *leon*, "lion," and *oura*, "tail." The species name *cardiaca* means "treating heart ailments," a reference to this plant's medicinal usage. • This plant was traditionally used to treat menstrual and childbearing afflictions, hence the common name "motherwort."

Water Horehound
Lycopus asper

Habitat: Moist, open areas.
General: Stems 20–70 cm
tall, square, hairy, nodes often
purplish; roots producing
tubers. **Leaves:** Opposite,
oblong to lance-shaped,
2–8 cm long; margins with
6–12 teeth per side; stalkless.
Flower Cluster: Cyme terminal, axillary, densely flowered. **Flowers:** White,
cup-shaped to tubular, 2–4 mm
long; petals 4, united; sepals 4–5; stamens 4, 2 fertile and 2 sterile (often absent); pistil 1. **Fruit:** Nutlet about 2 mm long; 4 nutlets per flower.

Notes: The genus name comes from the Greek words *lukos*, "wolf," and *pous*,
"foot." The species name *asper* means "rough."

Wild Mint · *Mentha arvensis*

Habitat: Marshes, wet meadows and
ditches. **General:** Stems 10–60 cm tall,
greenish purple, square, hairy on the angles,
distinctive mint-like scent; rhizome creeping.
Leaves: Opposite, oblong to oval or lance-
shaped, 1–8 cm long, 1–1.5 cm wide, both surfaces gland-dotted; margins toothed;
stalks short. **Flower Cluster:** Verticel (dense whorl) crowded, flowers in whorls
in upper leaf axils. **Flowers:** Purple to pink or white, tubular, 3–7 mm long; petals 4,
united; sepals 4, prominently 10–13-nerved; stamens 4, long-stalked; pistil 1. **Fruit:**
Nutlet oval; 4 nutlets per flower.

Notes: *Mentha* is the Latin name for mint. The species name *arvensis* means "of cultivated fields." • While walking in a wetland, the smell of wild mint is often apparent
long before the plant is seen.

Wild Bergamot
Monarda fistulosa

Habitat: Open woods and meadows. General: Stems 30–120 cm tall, square, hairless or smooth, pleasantly scented. Leaves: Opposite, oval to triangular or lance-shaped, 2–10 cm long, 1–3 cm wide, both surfaces with grey hairs; margins coarsely toothed; stalks 1–5 cm long. Flower Cluster: Spike terminal, head-like, 2–6 cm across; bracts lance-shaped, pinkish green. Flowers: Pink to lilac, tubular, 2-lipped, 2–3.5 cm long; petals 5, united, hairy, bearded; sepals 5, prominently 15-nerved; stamens 2, long-stalked; pistil 1. Fruit: Nutlet oblong, smooth; 4 nutlets per flower.

Notes: The genus name commemorates Nicholas Monardes, a 15th-century Spanish physician and botanist. The species name *fistulosa* means "tube- or pipe-shaped," a reference to the flower.

Catnip · *Nepeta cataria*

Habitat: Waste areas, fields and roadsides; introduced from Europe as a garden plant. General: Stems 30–80 cm tall, square, branched above, covered with white hairs, strongly scented. Leaves: Opposite, oval, 2–8 cm long, 1.5–5 cm wide, underside densely haired; margins coarsely toothed; stalked. Flower Cluster: Spike crowded, terminal, 2–5 cm long. Flowers: White with pink or purple spots, tubular, 2-lipped, 0.8–1.2 cm long; petals 5, united; sepals 5, prominently 15-nerved; stamens 4; pistil 1. Fruit: Nutlet smooth; 4 nutlets per flower.

Notes: *Nepeta* is the Latin name for catnip. The species name *cataria* means "connected to or possessed by cats."

Obedient-plant
Physostegia virginiana

Habitat: Damp to moist, open woods and meadows. **General:** Stems 50–100 cm tall, square, hairless. **Leaves:** Opposite, oblong to lance-shaped, 1–12 cm long, 0.5–2 cm wide, reduced in size upward, spine-tipped; margins coarsely toothed; stalks short. **Flower Cluster:** Spike terminal, 10–20 cm long. **Flowers:** Pinkish purple to white, tubular, 2–3 cm long; petals 5, united; sepals 5, united, becoming inflated with age; stamens 4; pistil 1. **Fruit:** Nutlet smooth; 4 nutlets per flower.

Notes: The genus name comes from the Greek words *physa*, "bladder," and stege, "covering," a reference to the inflated calyx.
• The common name "obedient-plant" refers to the flowers, which if bent, remain in that position for a period of time.

Heal-all, Self-heal
Prunella vulgaris

Habitat: Moist, open woods; both native and Eurasian populations exist in the natural environment. **General:** Stems 10–30 cm tall, square, branched, somewhat hairy; rhizomes creeping. **Leaves:** Opposite, oval to oblong or lance-shaped, 2.5–5 cm long, 0.7–4 cm wide; margins smooth to slightly toothed; stalks 0.5–3 cm long. **Flower Cluster:** Spike terminal, crowded, 2–10 cm long,

0.7–4 cm across; bracts toothed, hairy. **Flowers:** Purplish blue, pink or white, tubular, 2-lipped, 0.8–1.6 cm long; petals 5, united; sepals 5, bristle-tipped; stamens 4; pistil 1. **Fruit:** Nutlet smooth; 4 nutlets per flower.

Notes: The genus name *Prunella* is possibly derived from the Latin *prunum*, "purple," a reference to the flower colour, or from the German *bruen*, "quinsy," a throat disorder that was treated using this plant. The species name *vulgaris* means "common." • The common names "heal-all" and "self-heal" both pertain to the medicinal value of this plant.

Virginia Mountain-mint
Pycnanthemum virginianum

Habitat: Moist to dry, open, sandy meadows. **General:** Stems 30–100 cm tall, branched, bushy, often reddish green, square, scattered hairs on the angles; rhizome short, creeping. **Leaves:** Opposite, numerous, linear to lance-shaped, 3–6.5 cm long, 0.3–1.3 cm wide; margins smooth; stalkless; strong mint odour when crushed. **Flower Cluster:** Cyme head-like, 1–1.8 cm across, 5–50-flowered. **Flowers:** White with purple dots, 0.5–1 cm long, 2-lipped; petals 5, united; sepals 5; stamens 4; pistil 1. **Fruit:** Nutlet dull black; 4 nutlets per flower.

Notes: The genus name comes from the Greek words *puknos*, "dense," and *anthemon*, "flower," a reference to the densely flowered heads.

Skullcap • *Scutellaria galericulata*

Habitat: Wet meadows and marshes. **General:** Stems 10–80 cm tall, square, somewhat hairy on the angles; rhizome creeping. **Leaves:** Opposite, oblong to lance-shaped, 2–6 cm long, 0.6–2.5 cm wide, base heart-shaped; margins wavy; stalks 1–4 mm long. **Flower Cluster:** Solitary, on stalks 2–3 mm long from upper leaf axils. **Flowers:** Blue, tubular, 2-lipped, 1.2–2.5 cm long; petals 5, united, upper lip helmet-shaped; sepals 5; stamens 4, lower pair longer and hairy; pistil 1. **Fruit:** Nutlet wrinkled; 4 nutlets per flower.

Notes: The genus name is from the Latin word *scutella*, "a small dish," a reference to the shape of the calyx. The species name *galericulata* means "helmet-like." • The common name "skullcap" refers to the resemblance of the corolla to the top of a skull.

Hedge Nettle
Stachys palustris

Habitat: Moist meadows, ditches and streambanks. **General:** Stems 30–80 cm tall, square, unbranched, pale green, covered with soft, bristly hairs; rhizome creeping, producing succulent tubers. **Leaves:** Opposite, oval to oblong, 2–15 cm long, 1–5 cm wide; margins with rounded teeth; short-stalked to stalkless. **Flower Cluster:** Spike terminal, crowded, 2–25 cm long. **Flowers:** Pink to purple, mottled with pale and dark spots, irregular, 2-lipped, 1–1.5 cm long; petals 5, united; sepals 5; stamens 4; pistil 1. **Fruit:** Nutlet dark brown, 1–3 mm long; 4 nutlets per flower.

Notes: *Stachys* is Greek for "spike," a reference to the flower cluster. The species name *palustris* means "marsh-loving."

NIGHTSHADE FAMILY
Solanaceae

With over 2800 species worldwide, the nightshade family, also called the potato family, reaches its greatest diversity in the tropics. Species in this family have simple to compound, alternate to subopposite leaves without stipules. The flowers are composed of 4–7 petals, 4–7 sepals, 5 stamens and a single pistil. The petals may be partly or completely fused and form a funnel-shaped corolla. The fruit is a berry or capsule.

The nightshades are an economically important family of plants. One genus, *Solanum*, provides potato (*S. tuberosum*), tomato (*S. lycopersicum*) and eggplant (*S. melongena*). Other food crops include peppers (*Capsicum* spp.) and Cape gooseberry (*Physalis peruviana*). Ornamental species include petunias (*Petunia* x *hybrida*), Chinese lanterns (*Physalis alkekengi*) and angel's trumpets (*Brugmansia* spp.). Tobacco (*Nicotiana tabacum*) is also a member of the nightshades. Poisonous members of this family include henbane (*Hyoscyamus niger*), jimsonweed (*Datura stramonium*) and deadly nightshade (*Atropa belladonna*).

Garden Datura
Datura inoxia

Habitat: Waste areas, roadsides and gardens; introduced ornamental from India. General: Stems 60–200 cm tall, covered in sticky hairs. Leaves: Alternate, broadly oval, 10–18 cm long, 4–15 cm wide, covered in felt-like hairs; margins wavy to irregularly toothed; stalks 3–8 cm long. Flower Cluster: Solitary, on stalks 1–8 cm long. Flowers: White, 9–18 cm long, 7–10 cm across, funnel-shaped; petals 5; sepals 5; stamens 5; pistil 1. Fruit: Capsule globe-shaped, 3–4 cm across, covered in prickles; seeds brown.

Notes: The genus name comes the Arabic word for jimsonweed (below), another species in the same genus. The species name *inoxia* means "not spiny."

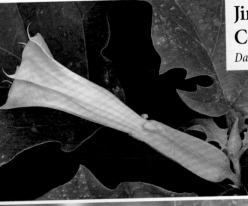

Jimsonweed, Common Thorn-apple
Datura stramonium

Habitat: Waste areas and roadsides; introduced from India. General: Stems 0.5–2 m tall, purplish green, branched near the top, ill-scented when crushed. Leaves: Alternate, oval to triangular, 7–20 cm long, 3–10 cm wide; margins irregularly toothed; stalks 3–5.5 cm long. Flower Cluster: Solitary, on stalks 0.5–1.2 cm long, from leaf axils. Flowers: White to pale purple, funnel-shaped, 5–12.5 cm long, 2–5 cm wide; petals 5; sepals 5; stamens 5; pistil 1. Fruit: Capsule egg-shaped, 3–6 cm wide, spiny; seeds black.

Notes: The species name *stramonium* means "spiny-fruited."
• All parts of this plant, especially the seeds and fruit, contain poisonous compounds that may cause death if ingested.

Clammy Ground-cherry
Physalis heterophylla

Habitat: Dry, sandy, open areas. **General:** Stems 20–90 cm tall, branched, covered in sticky hairs. **Leaves:** Alternate, oval to lance-shaped, 3–8 cm long, surfaces with scattered hairs; margins smooth; stalks long. **Flower Cluster:** Solitary, in leaf axils. **Flowers:** Greenish yellow with a purplish brown centre, 1.7–2.2 cm across, bell-shaped, nodding; petals 5; sepals 5; stamens 5, anthers yellow; pistil 1. **Fruit:** Berry 1–1.5 cm across, yellow, enclosed by inflated, bladder-like calyx; fruiting calyx 3–4 cm long.

Notes: The genus name *Physalis* comes from the Greek word *physa*, meaning "a bladder," a reference to the fruit. The species name *heterophylla* means "bearing leaves of different forms." • The mature yellow fruit is edible and can be made into pies and jellies.

Horse-nettle
Solanum carolinense

Habitat: Waste areas and roadsides; introduced from the southern U.S. and Mexico. **General:** Stems 20–120 cm tall, erect, branched, covered in star-shaped hairs (4–8-rayed) and yellow spines 0.6–1.2 cm long; rhizome deep, fleshy. **Leaves:** Alternate, oval to egg-shaped, 7–15 cm long, 3–6 cm wide, dark green, tips pointed;

margins sharply lobed (2–5 lobes per side) or toothed; stalks, veins and midribs covered in rough, star-shaped hairs and prickly spines; spines yellow, flattened, 2–5 mm long. **Flower Cluster:** Cyme raceme-like, 1-sided, 5–20-flowered; stalks prickly. **Flowers:** White to violet, 1.5–2.5 cm across; lower flowers perfect, upper flowers male; petals 5, united; sepals 5, united; stamens 5; pistil 1. **Fruit:** Berry globe-shaped, 0.9–1.5 cm across, green turning yellow at maturity; calyx persistent, spiny, partially surrounding the fruit.

Notes: The species name means "of Carolina." • The common name "horse-nettle" refers to the poisonous toxins in this plant, which have been responsible for numerous horse and cattle deaths.

Black Nightshade · *Solanum nigrum*

Habitat: Waste areas, roadsides and gardens; native species with a weedy nature. **General:** Stems erect, 10–100 cm tall, widely branched, somewhat hairy, ill-scented. **Leaves:** Alternate, oval to triangular, 1–8 cm long, 0.5–5 cm wide, dark green; margins wavy or irregularly toothed; stalks long. **Flower Cluster:** Raceme 5–10-flowered, on short stalk connected directly to stem. **Flowers:** White with yellow centre, 3–8 mm across; petals 5; sepals 5; stamens 5; pistil 1. **Fruit:** Berry globe-shaped, 0.9–1.5 cm across, green turning bluish black at maturity; poisonous.

Notes: The species name *nigrum* means "black," a reference to the colour of the fruit.

Buffalo-bur

Solanum rostratum

Habitat: Roadsides and waste areas; introduced from the southwestern U.S. **General:** Stems 30–100 cm tall, branched, covered in spiny, star-like hairs; spines 1–15 mm long. **Leaves:** Alternate, 6–20 cm long, 2–7 cm wide, 5–7-lobed to deeply divided, surfaces with yellow star-like hairs, spiny on veins and both surfaces; margins smooth; stalked. **Flower Cluster:** Raceme 1–5-flowered, elongating with maturity; flower stalks prickly. **Flowers:** Yellow, 2–3.5 cm wide, wheel-shaped; petals 5; sepals 5, spiny, becoming bur-like in fruit; stamens 5; pistil 1. **Fruit:** Berry 0.8–1.2 cm across, dark brown to black, enclosed by spiny calyx; seeds 50–120.

Notes: The species name *rostratum* means "beaked or curved at the end."

FIGWORT FAMILY
Scrophulariaceae

Members of the figwort family are found throughout the world in a variety of habitats. Several species are partly parasitic on the roots of other plants. These shrubs and herbaceous plants have alternate, opposite or whorled leaves. The leaves may be simple, deeply divided or compound. The perfect flowers, borne in spikes or racemes, may be regular or irregular. Flowers have 4–5 fused petals, 4–5 fused sepals, 4 (sometimes 2 or 5) stamens and 1 pistil. The corolla is often 2-lipped. The fruit is a capsule.

One species, butterfly-bush (*Buddleja davidii*), is often grown as an ornamental. A few species in this family are noxious weeds.

Moth Mullein
Verbascum blattaria

Habitat: Dry, open areas and roadsides; introduced from Russia. **General:** Stems 30–150 cm tall, hairless, unbranched. **Leaves:** Alternate, 1–10 cm long, 0.5–4 cm wide, oblong to lance-shaped; margins toothed to pinnately lobed; stalks short. **Flower Cluster:** Raceme 10–50 cm long, many-flowered; flower stalks 0.5–1.5 cm long. **Flowers:** Yellow, 2–3 cm wide; petals 5; sepals 5; stamens 5, reddish purple; pistil 1. **Fruit:** Capsule oval, 7–8 mm long; calyx persistent.

Notes: *Verbascum* is the ancient Roman name for mullein. The species name comes from the Latin word *blatta*, meaning "moth."

Common Mullein
Verbascum thapsus

Habitat: Dry, open areas; introduced from Greece. **General:** Flowering stems 1–2.5 m tall, covered with felt-like hairs; taproot deep. **Leaves:** Basal leaves elliptic to oblong or lance-shaped, 15–45 cm long, 1–10 cm wide, covered with dense, woolly or felt-like hairs; stem leaves alternate, 10–40 cm long, 1–4 cm wide, reduced in size upward, leaf bases running down the stem, appearing stalkless; margins smooth. **Flower Cluster:** Raceme spike-like, terminal, 10–50 cm long, 2.5–3.5 cm across. **Flowers:** Yellow, numerous, 1.5–2.5 cm across; petals 5; sepals 5; stamens 5, the filaments covered with yellow hairs; pistil 1. **Fruit:** Capsule globe-shaped, 0.3–1 cm long, woolly haired.

Notes: The species name refers to Thapsus, the former name of Magnisi, a town on the east coast of Sicily. • The ground leaves of common mullein were used as a fish-control agent in the mid-1700s.

BROOMRAPE FAMILY Orobanchaceae

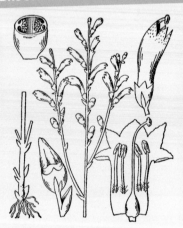

BROOMRAPE FAMILY
Orobanchaceae

The broomrape family has 2060 species worldwide and now includes several genera from the figwort family (Scrophulariaceae). A family of parasitic and semi-parasitic plants, many species attach themselves to the roots of other plants. The stems often lack chlorophyll and are usually yellow, brown or purplish. The alternate leaves are reduced to small scales. The irregular, perfect flowers are composed of 5 united petals, 4–5 sepals, 4 stamens and 1 style. The flowers produce numerous microscopic seeds.

Branched broomrape (*Orobanche ramosa*) is a serious parasitic weed in tomato, canola and sunflower fields. It is also found in tobacco, pepper, cabbage, cauliflower, celery and onion crops.

Slenderleaf False Foxglove
Agalinis tenuifolia

Habitat: Moist meadows, sandy shorelines and low, wooded areas. **General:** Stems 15–60 cm tall, branched, green to reddish purple; partially parasitic on roots of other plants. **Leaves:** Opposite, linear, 1–7.5 cm long, 2–6 mm wide, green to reddish or purplish green; margins smooth; stalkless. **Flower Cluster:** Solitary, from upper leaf axils. **Flowers:** Pink, 1.2–1.8 cm across, 1–1.5 cm long, irregular, weakly 2-lipped; petals 5; sepals 5; stamens 4, hairy; pistil 1. **Fruit:** Capsule globe-shaped, 3–5 mm long.

Notes: The genus name *Agalinus* means "remarkable flax," a possible reference to the flax-like leaves. The species name *tenuifolia* means "slender-leaved."

Yellow False Foxglove
Aureolaria flava

Habitat: Rich, deciduous woodlands. **General:** Stems green, leafy, 50–200 cm tall, hairless; semi-parasitic on roots of oaks (*Quercus* spp.), pp. 35–38. **Leaves:** Opposite and alternate, 3–10 cm long, lower leaves pinnately lobed, upper leaves lance-shaped with smooth margins; stalked. **Flower Cluster:** Solitary, from leaf axils, on stalks 0.4–1 cm long. **Flowers:** Yellow, showy, funnel-shaped, irregular, 2-lipped, 3.5–5 cm long; petals 5; sepals 5; stamens 4; pistil 1; flowers turning black when picked. **Fruit:** Capsule oval, 1.2–2 cm long.

Notes: *Aureolaria* means "golden," and *flava* means "yellow," both referencing the flower colour.

Scarlet Painted-cup, Indian Paintbrush · *Castilleja coccinea*

Habitat: Moist sand and peaty meadows, to dry tall-grass prairie. **General:** Stems 20–60 cm tall, leafy, covered in soft hairs; semi-parasitic on roots of other plants. **Leaves:** Alternate; lower leaves elliptic with smooth margins; upper leaves deeply 3–5-lobed, 2.5–7.5 cm long, stalkless. **Flower Cluster:** Spike densely flowered, 4–6 cm long, elongating to 20 cm in fruit; bracts showy, scarlet red-tipped. **Flowers:** Greenish yellow, 2–3 cm long, tubular, irregular, 2-lipped; petals 5 (upper lip 2-lobed, lower lip 3-lobed and shorter); sepals 4, fused; stamens 4; pistil 1. **Fruit:** Capsule 0.5–1.5 cm long.

Notes: The genus name *Castilleja* commemorates Spanish botanist, Domingo Castillejo (1744–93). • The common name "paintbrush" comes from the plant's resemblance to a paintbrush that has been dipped in paint.

Cancer-root

Conopholis americana

Habitat: Dry, deciduous forests; parasitic on roots of oaks (*Quercus* spp.), pp. 35–38, and beech (*Fagus grandifolia*), p. 35. **General:** Stems 10–20 cm tall, unbranched, yellow to creamy white, turning brown with age. **Leaves:** Alternate, reduced to scales, less than 1.3 cm long, somewhat fleshy. **Flower Cluster:** Spike densely flowered; bracts 0.8–1.3 cm long, turning brown with age. **Flowers:** Creamy white, tubular, 1.2–2.5 cm long, irregular, 2-lipped; petals 5 (upper lip 2-lobed, lower lip 3-lobed); sepals 5; stamens 4, anthers grey; pistil 1. **Fruit:** Capsule globe-shaped, creamy white; bracts brown, papery.

Notes: The genus name comes from the Greek words *conos*, "cone," and *pholis*, "a scale," a reference to this plant's resemblance to a pinecone.

Beechdrops, Virginia Beechdrops

Epifagus virginiana

Habitat: Rich, deciduous forests; range coinciding with beech (*Fagus grandifolia*), p. 35. **General:** Stems brownish purple, branched, 15–45 cm tall; parasitic on roots of beech. **Leaves:** Alternate, triangular, scale-like, 2–4 mm long. **Flower Cluster:** Raceme 10–35 cm long; bracts small. **Flowers:** White with brownish purple stripes, tubular, 8–1.2 cm long, showy; petals 5; sepals 5; stamens 4; pistil 1; lower flowers brownish purple, 3–6 mm long, inconspicuous, not opening. **Fruit:** Capsule 4–5 mm long, produced only by lower inconspicuous flowers.

Notes: The genus name comes from the Greek words *epi*, "on," and *phagos*, "beech," referring to the plant's parasitic nature on the roots of beech trees.

Canada Wood Betony, Head Betony · *Pedicularis canadensis*

Habitat: Sandy, open woodlands. **General:** Stems 10–40 cm tall, covered in woolly hairs; semi-parasitic on roots of other plants. **Leaves:** Basal and alternate, oblong to lance-shaped, 7–13 cm long, reduced in size upward, deeply divided; margins toothed or lobed; lower leaves stalked, upper stalkless. **Flower Cluster:** Spike 3–5 cm long, compact. **Flowers:** Red, yellow or purplish yellow, tubular, 1.5–2.5 cm long, irregular, 2-lipped; sepals 5; petals 5 (upper lip 2-lobed, lower lip 3-lobed); stamens 4, 2 long-stalked and 2 short-stalked; pistil 1. **Fruit:** Capsule 1.2–1.6 cm long; fruiting cluster to 20 cm long.

Notes: The genus name comes from the Greek word *pediculus*, meaning "louse," a reference to a superstition that livestock contracted lice by eating this plant. The species name means "of Canada."

BLADDERWORT FAMILY
Lentibulariaceae

The bladderwort family has over 200 species worldwide. Bladderworts are carnivorous herbs that grow in water and on moist ground. Small insects and aquatic life are trapped by the sticky leaves or in the hollow, bladderlike traps. The simple to dissected leaves may be alternate, basal or whorled. The flowers are highly modified and are composed of 5 united petals, 2–5 united sepals, 2 stamens and 1 pistil. The fruit is a 2-valved capsule.

Common Butterwort
Pinguicula vulgaris

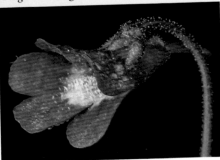

Habitat: Bogs, marshes and mossy streambanks. **General:** Flowering stems to 12 cm tall; insectivorous. **Leaves:** Basal, yellowish green, elliptic, 2–5 cm long, 1–2 cm wide, upper surface sticky; margins inrolled. **Flower Cluster:** Solitary, terminal. **Flowers:** Purple, resembling a violet, 1.2–1.8 cm long, irregular, 2-lipped; petals 5, spurred; sepals 5; stamens 2; pistil 1. **Fruit:** Capsule 2-valved.

Notes: The sticky upper surface of the leaves traps insects, which are used to supplement the plant's nutritional requirements.

PLANTAIN FAMILY
Plantaginaceae

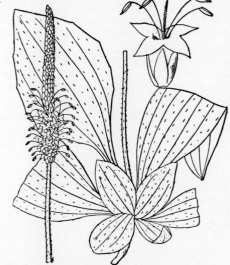

The plantain family has 1900 species worldwide. Members of this family are herbs (occasionally aquatic), with leaves that are simple to dissected and alternate (appearing basal), opposite or whorled. The inflorescence is a spike, raceme or head. The perfect, regular, often 2-lipped flowers are composed of 4–5 united petals, 4–5 united sepals, 4 stamens (a staminode may be present) and I pistil. The fruit is a capsule or nutlet.

The plantain family now includes members of the mare's-tail (Hippuridaceae) and water starwort (Callitrichaceae) families, as well as several genera that were previously in the figwort family (Scrophulariaceae).

A few members of the plantain family are grown for their ornamental value. One species, purple foxglove (*Digitalis purpurea*), was the original source of the cardiac medication digoxin. Several species in this family are noxious weeds.

Dwarf Snapdragon
Chaenorrhinum minus

Habitat: Gravelly or sandy soils, often on railway grades and roadsides; introduced from Europe and Asia. **General:** Stems 5–40 cm tall, profusely branched, surface with short, sticky hairs. **Leaves:** Opposite and alternate, oblong to linear or lance-shaped, 1–2.5 cm long, 1–5 mm wide; margins smooth; stalkless. **Flower Cluster:** Solitary, in axils of upper leaves, on stalks 1–1.5 cm long. **Flowers:** Light blue to purple with yellow throat, 6–8 mm long, funnel-shaped, 2-lipped; petals 5, spurred, the spur 1.5–2 mm long; sepals 5; stamens 4; pistil 1. **Fruit:** Capsule round, 4–6 mm across.

Notes: The genus name comes from the Greek words *chaino*, "to gape," and *rhis*, "snout," a reference to the open-mouthed corolla. The species name *minus* means "small."

353

Turtlehead, Smooth Balmony · *Chelone glabra*

Habitat: Moist, open woods and marshes. **General:** Stems 20–80 cm tall, hairless, branched. **Leaves:** Opposite, linear to lance-shaped, 8–20 cm long, 1–4 cm wide; margins sharply toothed; stalks short. **Flower Cluster:** Spike 2–8 cm long, compact; bracts 2–3, leaf-like. **Flowers:** White tinged with greenish yellow, pink or purple, tubular, 2.5–3.8 cm long, 2-lipped but appearing closed; petals 5 (upper lip 2-lobed, arched over 3-lobed lower lip); sepals 5; stamens 5, 4 densely haired and 1 hairless (staminode); pistil 1. **Fruit:** Capsule 1–1.5 cm long; seeds flat with winged margins.

Notes: The genus name *Chelone* is the Greek word for "turtle," a reference to the flower's resemblance to a turtle's head. The species name *glabra* means "smooth" or "hairless."

Grecian Foxglove
Digitalis lanata

Habitat: Dry, open hillsides; introduced ornamental from southern Europe. **General:** Stems 50–100 cm tall, erect, rarely branched. **Leaves:** Alternate, oblong to lance-shaped, 5–25 cm long; margins smooth; stalkless. **Flower Cluster:** Spike densely haired, many-flowered. **Flowers:** White to pale yellow with purplish brown veins, 2.5–3 cm long, 2-lipped, strongly inflated; petals 5; sepals 5, densely haired; stamens 4; pistil 1. **Fruit:** Capsule globe-shaped.

Notes: *Digitalis* is Latin for "finger," a reference to the shape of the flower. The species name *lanata* means "woolly."

Toadflax
Linaria vulgaris

Habitat: Roadsides, railway grades, ditches and
waste areas; introduced
ornamental from Europe.
General: Stems
20–130 cm tall, leafy,
hairless; rootstock
extensive, creeping, often
forming patches. **Leaves:**
Alternate (may appear
opposite or whorled),
crowded, linear to lance-
shaped, 2–10 cm long,
1–5 mm wide, pale

green; margins smooth; stalkless. **Flower Cluster:** Raceme terminal, 10–30 cm
long, many-flowered. **Flowers:** Yellow, showy, resembling snapdragons, 2-lipped,
2–3.5 cm long; petals 5, spurred; sepals 5; stamens 4; pistil 1; spur 2–3 cm long.
Fruit: Capsule oval to egg-shaped, 0.8–1.2 cm long.

Notes: The genus name *Linaria* comes from the Latin word *linum*, "flax," referring
to the numerous, narrow leaves, which resemble those of true flax (*Linum* spp.).
The species name *vulgaris* means "common."

Hairy Beard-tongue · *Penstemon hirsutus*

Habitat: Dry, rocky, open areas and hillsides. **General:**
Stems 10–90 cm tall, erect to ascending, lower part hairless, upper part covered in woolly hairs; rhizome slender.
Leaves: Opposite, oblong to lance-shaped, 5–12 cm long,
light green; margins smooth to finely toothed; stalkless.
Flower Cluster: Raceme, in leaf axils. **Flowers:** Pale
violet with white lobes, trumpet-shaped, 2–3 cm long,
2-lipped; sepals 5; petals 5
(upper lip 2-lobed and erect,
lower lip 3-lobed); stamens 5,
4 fertile and 1 sterile and
densely haired (staminode);
pistil 1. **Fruit:** Capsule
7–10 mm long.

Notes: The genus name
comes from the Greek words
pente, "five," and *stemon*,
"stamen," a reference to
the flower's fifth stamen or
staminode. The species name
hirsutus means "hairy."

Ribgrass
Plantago lanceolata

Habitat: Lawns, roadsides and waste areas; introduced from Europe. **General:** Flowering stems 10–80 cm tall, leafless; rootstalk short. **Leaves:** Basal, lance-shaped, grass-like, 8–30 cm long, 1–4 cm wide, prominently 3–7-veined, tufts of brown hairs at the base; margins smooth. **Flower Cluster:** Spike cylindric to globe-shaped, 2–12 cm long; bracts papery, brownish. **Flowers:** Brownish greenish, 4–6 mm across; petals 4, united; sepals 4, united; stamens 4, yellow, longer than petals; pistil 1. **Fruit:** Capsule elliptic, 3–4 mm long.

Notes: The genus name *Plantago* comes from the Latin word *planta*, meaning "foot-print." The species name *lanceolata* means "lance-shaped," a reference to the shape of the leaves.

Common Plantain
Plantago major

Habitat: Lawns, roadsides and waste areas; introduced from Europe. **General:** Flowering stems 10–60 cm tall, leafless, hairless to finely haired; rootstock short. **Leaves:** Basal, oval, 5–30 cm long, 2–12 cm wide, underside prominently 3–7-ribbed; margins smooth to wavy; stalked. **Flower Cluster:** Spike 7–30 cm long; bracts oval. **Flowers:** Greenish white, 2–3 mm across; petals 4; sepals 4; stamens 4; pistil 1. **Fruit:** Capsule egg-shaped, 2–4 mm long.

Notes: The species name *major* means "larger." • First Nations peoples called this plant "whiteman's-foot" because it seemed to appear in the footsteps of European explorers and settlers.

Flaxseed Plantain

Plantago psyllium

Habitat: Roadsides, waste areas and railway grades; introduced from Europe and Asia. General: Stems erect, 10–50 cm tall, covered with soft hairs. Leaves: Opposite or in whorls of 3, linear, 3–10 cm long, 3–4 mm wide, surfaces with soft hairs; margins smooth; stalkless. Flower Cluster: Spike elliptic to globe-shaped, 1–2 cm long, from leaf axils; stalk 3–8 cm long; bracts brown with green tips, 2–5 mm long. Flowers: Greenish white, 2–5 mm across; petals 4; sepals 4; stamens 4; pistil 1. Fruit: Capsule 2–6 mm long; seeds black, shiny.

Notes: The species name *psyllium* comes from the Greek word for "flea," a reference to the size and shape of the seeds.

American Brooklime

Veronica americana

Habitat: Marshy ground and stream-banks. General: Stems 20–60 cm long, trailing to ascending, hairless, often rooting at lower nodes. Leaves: Opposite (upper leaves occasionally alternate), oval to oblong, 2–7 cm long, 0.6–3 cm wide; margins finely toothed; stalks short. Flower Cluster: Raceme short, in upper leaf axils. Flowers: Blue to pale violet or white, 4–6 mm across; petals 4; sepals 4; stamens 2, long-stalked; pistil 1. Fruit: Capsule flattened, round, 3.5–4.5 mm wide.

Notes: The genus name *Veronica* commemorates St. Veronica, the woman who gave her handkerchief to Jesus so that he could wipe his face while carrying the cross to Calvary. The species name means "of America," the native range of this plant.

Common Speedwell, Corn Speedwell
Veronica arvensis

Habitat: Waste areas and roadsides; introduced from Europe and Asia. **General:** Stems 5–30 cm tall, simple or branched. **Leaves:** Opposite, 3–5 pairs per stem, oval to round, 0.5–1.5 cm long, 0.4–1 cm wide; margins with rounded teeth; stalks 0–3 mm long. **Flower Cluster:** Raceme many-flowered, elongating to 20 cm long at maturity; bracts alternate, leaf-like; flower stalks 1–2 mm long. **Flowers:** Bluish purple, 1–3 mm long; petals 4; sepals 4; stamens 2; pistil 1. **Fruit:** Capsule heart-shaped, 2.5–3.5 mm long.

Notes: The species name *arvensis* means "of cultivated fields," a reference to this plant's weedy nature throughout the world.

Common Gypsyweed
Veronica officinalis

Habitat: Moist woodlands and open meadows. **General:** Stems prostrate to ascending, 10–30 cm long, hairy throughout. **Leaves:** Opposite, elliptic, 2.5–5 cm long, 1–2.5 cm wide; margins toothed; stalks short. **Flower Cluster:** Raceme spike-like, 3–6 cm long; bracts linear to lance-shaped. **Flowers:** Pale blue to violet with dark veins, 5–7 mm wide; petals 4; sepals 4; stamens 2; pistil 1. **Fruit:** Capsule heart-shaped, 4.5–5 mm wide.

Notes: The species name *officinalis* means "medicinal," a reference to this plant's use to treat a number of ailments.

Bird's-eye Speedwell, Persian Speedwell
Veronica persica

Habitat: Moist, open areas; introduced from southwestern Asia. General: Stems 10–20 cm tall, often densely haired. Leaves: Opposite, usually 3–4 pairs per stem, oval to lance-shaped or rounded, 1–2 cm long, 0.8–1.5 cm wide; margins with rounded teeth, 3–6 per side; stalks 1–8 mm long. Flower Cluster: Raceme terminal; bracts alternate, leaf-like; fruiting stalks 1.5–3 cm long. Flowers: Blue, wheel-shaped, 0.8–1.1 cm across; petals 4; sepals 4, 5–8 mm long in fruit; stamens 2; pistil 1. Fruit: Capsule heart-shaped, 4–6 mm long, 6–9 mm wide, conspicuously net-veined.

Notes: The species name means "of or from Persia," a reference to this plant's native range.

Culver's-root
Veronicastrum virginicum

Habitat: Moist, open areas and woodlands. General: Stems 80–200 cm tall, branched above. Leaves: Whorl of 4–7, oblong to linear or lance-shaped, 5–12 cm long, 0.5–2 cm wide, hairless above, densely haired below; margins toothed; stalks 3–10 mm long. Flower Cluster: Raceme spike-like, 5–20 cm long, densely flowered; bracts about 1 mm long; flower stalks 0.5 mm long. Flowers: White (occasionally pinkish), 4–7 mm long, 2-lipped; petals 4, upper lip formed by 1 petal, lower lip 3-lobed; sepals 4–5; stamens 2, anthers orange; pistil 1. Fruit: Capsule oval, 4–5 mm long.

Notes: The genus name comes from *Veronica* and the suffix -*astrum*, alluding to this plant's resemblance to several species of *Veronica* (pp. 357–59). The species name means "of Virginia."

BEDSTRAW OR MADDER FAMILY
Rubiaceae

The bedstraw family, often called the madder family, has over 6500 species worldwide. Most members of this family are tropical trees and shrubs with opposite or whorled leaves. Species found in temperate regions are usually herbaceous plants with opposite or whorled leaves. The leaves have stipules that are similar in shape and size to the leaves. The flowers are composed of 4–5 petals fused into a funnel-shaped corolla, 4–5 small sepals, 4–5 stamens and 1 pistil. The fruit is a berry, drupe, capsule or schizocarp.

The bedstraws are economically important, and members of this family include coffee (*Coffea arabica*), gardenias (*Gardenia* spp.) and quinine (*Cinchona officinalis* and *C. pubescens*).

Cleavers · *Galium aparine*

Habitat: Waste areas; introduced from Europe.
General: Stems 20–150 cm long, weak, trailing, square; backward-pointing hairs at leaf nodes enable the plant to cling onto surrounding vegetation for support. **Leaves:** Whorls of 6–8, oblong to lance-shaped, 1–8 cm long, 2–3 mm wide; margins and midrib with bristly hairs; stalks short. **Flower Cluster:** Cyme 3–9-flowered, in leaf axils. **Flowers:** Greenish white, 1–3 mm across; petals 4; sepals 4; stamens 4; pistil 1. **Fruit:** Nutlet 3–4 mm across, surface with hooked bristles; in pairs.

Notes: The genus name *Galium* comes from the Greek word *gala*, meaning "milk," a reference to the use of yellow spring bedstraw (*G. verum*) to curdle milk to make cheese. The species name *aparine* means "scratching," alluding to the bristly hairs on the stem.

Northern Bedstraw
Galium boreale

Habitat: Moist, open woods, meadows and forest edges. **General:** Stems 30–80 cm tall, square, leafy; rootstock slender, brown. **Leaves:** Whorls of 4, linear to lance-shaped, 2–5 cm long, 2–8 mm wide, prominently 3-nerved; margins smooth; stalks short. **Flower Cluster:** Panicle of numerous cymes, crowded. **Flowers:** White, faintly fragrant, 2–4 mm across; petals 4; sepals 4; stamens 4; pistil 1. **Fruit:** Nutlet 1–2 mm across, surface densely white-haired; in pairs.

Notes: The species name means "boreal" or "northern." • The common name "bedstraw" refers to this plant's use as mattress stuffing by early settlers.

Small Bedstraw
Galium trifidum

Habitat: Marshy ground. **General:** Stems 10–30 cm long, weak, trailing, square; branches numerous, forming large mats, covered with stiff, bristly hairs. **Leaves:** Whorls of 4, linear to spatula-shaped, 6–14 mm long, 1–3.5 mm wide; margins and underside bristly haired; stalks short. **Flower Cluster:** Cyme 2–3-flowered, in leaf axils. **Flowers:** Greenish white, 1–3 mm across; petals 3; sepals 3; stamens 4; pistil 1. **Fruit:** Nutlet 1–2 mm across, hairless; in pairs.

Notes: The species name *trifidum* means "3-parted," a reference to the flowers, which occur in groups of 3, as well as to the 3 sepals and 3 petals.

Sweet-scented Bedstraw

Galium triflorum

Habitat: Moist woodlands. **General:** Stems 30–100 cm long, weak, trailing, square, smooth to bristly haired. **Leaves:** Whorls of 5–6, oblong to lance-shaped, 2–6 cm long, bristle-tipped, sweet-scented; margins rough-haired; stalks short. **Flower Cluster:** Cyme 3-flowered, repeatedly branched, in leaf axils. **Flowers:** White, 2–4 mm across; petals 4; sepals 4; stamens 4; pistil 1. **Fruit:** Nutlet 2–3 mm across, surface with hooked bristles; in pairs.

Notes: The species name *triflorum* means "3-flowered," a reference to the flower cluster.

Creeping Partridge-berry, Twinberry

Mitchella repens

Habitat: Moist to dry forests. **General:** Evergreen; stems 10–30 cm long, trailing, often rooting at nodes. **Leaves:** Opposite, round to oval, 1–2 cm long and wide, shiny, green with white veins; margins smooth; stalks short. **Flower Cluster:** In pairs, in leaf axils. **Flowers:** White, tubular, 1–1.6 cm long, fragrant; petals 4, inner surface fringed with hairs; sepals 4; stamens 4; pistil 1. **Fruit:** Berry 2-lobed, 5–8 mm across; edible but tasteless.

Notes: The genus name commemorates American physician and botanist Dr. John Mitchell (1711–68). The species name *repens* means "creeping," a reference to the plant's growth habit.

Honeysuckle Family · Caprifoliaceae

Family description on p. 121.

Twinflower
Linnaea borealis

Habitat: Moist woodlands. **General:** Herb-like shrub; stems to 2 m long, trailing, 3–10 cm tall; branches leafy, green to reddish brown. **Leaves:** Opposite, evergreen, oblong to oval, 0.8–2 cm long, 0.3–1.5 cm wide; 2 pairs of notches near tip; stalks 2–5 mm long. **Flower Cluster:** Pair of flowers at the top of a Y-shaped stem. **Flowers:** Pinkish white, funnel-shaped, 0.8–1.5 cm long, pendent, fragrant; petals 5; sepals 5; stamens 4; pistil 1. **Fruit:** Capsule 4–5 mm long; 1-seeded.

Notes: The genus name *Linnaea* honours Swedish botanist Carolus Linnaeus (1707–78), who developed the scientific classification system that is used today. The species name means "boreal" or "northern."

TEASEL FAMILY · Dipsacaceae

The teasel family has over 270 species worldwide. Members of this family are herbaceous plants with opposite or whorled leaves. The leaves are simple to pinnately compound. Flowers are borne in heads and resemble those of the aster family (Asteraceae). The flowers are composed of 4–5 petals, 5 sepals, 2–4 stamens and 1 pistil. Teasels are distinguished from members of the aster family by the presence of sepals (asters have none). The sepals may be divided into 5–20 bristles. The fruit is an achene.

Ornamental members include pincushion flower (*Scabiosa columbaria*) and sweet scabious (*S. stellata*). A few species are introduced weeds.

Wild Teasel, Draper's Teasel

Dipsacus fullonum ssp. *sylvestris*

Habitat: Roadsides and waste areas; introduced from Russia in 1877 for use in carding wool. **General:** Stems 50–200 cm tall, prickly. **Leaves:** Opposite, lance-shaped, 10–40 cm long, 3–6 cm wide, underside prickly on midrib; margins toothed, prickly; stalkless and clasping the stem; upper leaves often fused at the base. **Flower Cluster:** Head cylindric, 3–10 cm long, 2–5 cm wide; bracts stiff, spiny. **Flowers:** Pale purple to pink, tubular, 1–1.5 cm long; petals 4; sepals 5; stamens 4, long-stalked; pistil 1; flowering begins in the middle of the head, with flowers blooming upward and downward at the same time. **Fruit:** Achene 2–5 mm long, surrounded by greyish brown calyx.

Notes: The genus name *Dipsacus* comes from the Greek word *dipsa*, "thirst," a reference to water collecting in the leaf bases. The species name *fullonum* refers to fullers, people who cleaned and thickened (felted) freshly woven cloth.

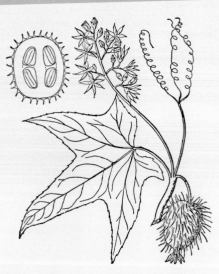

GOURD FAMILY
Cucurbitaceae

The gourd or cucumber family has over 700 species worldwide. Members of this family are mainly tropical and subtropical perennial vines with alternate, palmately veined leaves and tendrils. Flowers are of 2 types, male and female. Male flowers are composed of 4–6 petals, 4–6 sepals and 1–5 stamens. Female flowers have similar petals and sepals and a single pistil. The fruit may be a berry, pepo, capsule or achene.

The gourd family has several economically important members, including squashes and pumpkins (*Cucurbita* spp.), cucumbers (*Cucumis sativus*), watermelon (*Citrullus* spp.) and luffa (*Luffa aegyptiaca*), whose mature fruits are used for bath sponges.

Wild Cucumber, Prickly Cucumber · *Echinocystis lobata*

Habitat: Waste areas, road-sides and thickets; introduced from Europe. **General:** Climbing or trailing stems 4–8 m long, angular, surface with a few hairs. **Leaves:** Alternate, 5–13 cm across, palmately 3–7-lobed; margins toothed; stalks long; tendrils long, 3-branched, opposite the upper leaves. **Flower Cluster:** Panicle of male flowers; female flowers solitary in leaf axils.

Flowers: Male flowers white, 0.8–1.6 cm across, petals 6, sepals 6, stamens 2–3; female flowers greenish yellow, petals 6, sepals 6, pistil 1. **Fruit:** Pepo berry-like, oblong, 2.5–5 cm long, pale green, mottled, surface with weak prickles; interior fibrous, mesh-like; seeds 4.

Notes: The genus name *Echinocystis* comes from the Greek words *echinos* and *kustis*, meaning "hedgehog bladder," a reference to the spiny fruits. The species name *lobata* means "lobed."

Bur-cucumber, One-seeded Bur-cucumber · *Sicyos angulatus*

Habitat: Moist woodlands and waste areas. **General:** Climbing stems 1–6 m long, covered in sticky hairs; native plant with a weedy nature. **Leaves:** Alternate, 4–17 cm long and wide, palmately 3–5-lobed, rough-textured; margins toothed; stalks 4–7 cm long; tendril 3-branched, opposite the leaves.

Flower Cluster: Panicle of male flowers; female flowers in head-like clusters. **Flowers:** Greenish white, 0.8–1 cm across; male flowers with 5 petals, 5 sepals and 3 stamens, on long stalks; female flowers with 5 petals, 5 sepals and 1 pistil, on stalks 5–8 cm long. **Fruit:** Pepo resembling an achene, 1.2–1.7 cm long, covered in stiff bristles and spines; in clusters of 4–10.

Notes: The genus name *Sicyos* comes from the Greek word *sikus*, meaning "cucumber." The species name means "angular," a reference to the leaf margins.

HAREBELL FAMILY
Campanulaceae

The harebell or bellflower family has over 2000 species worldwide and now includes the lobelia family (Lobeliaceae). Members of this family are herbs with alternate, simple leaves. Stipules are absent. Flowers are composed of 5 united petals, 5 united sepals, 5 stamens and 1 pistil. The flowers are regular in *Campanula*, whereas those in *Lobelia* are irregular. The fruit is a capsule that releases seeds through pores on its side.

Well-known ornamental species include balloon-flower (*Platycodon* spp.), harebells (*Campanula* spp.) and lobelias (*Lobelia* spp.)

Creeping Harebell
Campanula rapunculoides

Habitat: Waste areas, yards and roadsides; introduced from Europe as a garden plant. **General:** Stems 20–100 cm, erect, unbranched, often forming dense clumps; rhizome white, tuber-like, creeping. **Leaves:** Alternate; lower leaves oval to heart-shaped, long-stalked; upper leaves lance-shaped, 3–7 cm long, 2–5 cm wide, stalkless; margins coarsely toothed. **Flower Cluster:** Solitary, in leaf axils, often appearing on 1 side of stem. **Flowers:** Blue to light purple, nodding, bell-shaped, 2–3 cm long; petals 5; sepals 5; stamens 5; pistil 1. **Fruit:** Capsule round, 6–9 mm across; seeds numerous.

Notes: The species name *rapunculoides* means "like *Campanula rapunculus*," a European species often used as a vegetable.

Harebell
Campanula rotundifolia

Habitat: Dry, open areas. **General:** Stems 20–45 cm tall, slender, erect, several rising from the crown. **Leaves:** Alternate, numerous, linear to lance-shaped, 1–10 cm long, 0.2–1 cm wide, margins smooth, upper leaves stalkless; basal leaves, when present, round to oval, long-stalked. **Flower Cluster:** Cyme 1–5-flowered; bracts linear; flower stalks slender. **Flowers:** Blue, bell-shaped, nodding, 1.5–2.5 cm long; petals 5; sepals 5; stamens 5; pistil 1. **Fruit:** Capsule papery, 3–4 mm across.

Notes: The genus name comes from the Latin *campana*, "bell," a reference to the flower shape. The species name *rotundifolia* means "round-leaved," referring to the shape of the basal leaves, which are often absent at flowering.

American Bellflower
Campanulastrum americanum

Habitat: Moist, open forests, wet meadows and roadside ditches. **General:** Stems 50–200 cm tall, hollow, slightly winged; exuding milky sap when broken. **Leaves:** Alternate, oblong to lance-shaped, 5–17 cm long, 1–6 cm wide; margins toothed; stalks short or absent. **Flower Cluster:** Solitary or 2–3 flowers per axil; bracts 3 per flower. **Flowers:** Blue to purple with white centre, wheel-shaped, 1.5–2.5 cm across; petals 5; sepals 5; stamens 5; pistil 1. **Fruit:** Capsule 7–10 mm long, 3–5 mm wide, somewhat 5-angled.

Notes: The genus name *Campanulastrum* means "like *Campanula*." The species name means "of America" and refers to the plant's range.

Cardinal-flower, Red Betty
Lobelia cardinalis

Habitat: Moist meadows and swamps. **General:** Stems 50–150 cm tall, unbranched; exuding milky sap when broken. **Leaves:** Alternate, lance-shaped, 7–16 cm long, 1–5 cm wide; margins irregularly toothed; stalks short. **Flower Cluster:** Raceme 10–40 cm long. **Flowers:** Brilliant red, 3–4.5 cm long, 2-lipped; petals 5, upper lip 2-lobed, lower lip 3-lobed; sepals 5; stamens 5, fused into a tube and surrounding the style; pistil 1. **Fruit:** Capsule 0.5–1.2 cm long, many-seeded.

Notes: The genus name commemorates Belgian botanist Matthias de L'Obel (1538–1616). The species and common names both refer to the bright red robes worn by Roman Catholic cardinals.

Indian Tobacco
Lobelia inflata

Habitat: Moist, rich woodlands. **General:** Stems 20 to 75 cm tall, winged, milky sap present; roots fibrous. **Leaves:** Alternate, spatula-shaped below, oblong to oval or lance-shaped above, 2–7 cm long, 0.5–3.5 cm wide; margins toothed; stalked. **Flower Cluster:** Raceme 10–20 cm long; bracts leaf-like; flower stalks 3–6 mm long. **Flowers:** Light blue to purplish or white, 6–8 mm long, 2-lipped; petals 5, upper lip 2-lobed, lower lip 3-lobed; sepals 5; stamens 5, anthers purple; pistil 1. **Fruit:** Capsule inflated, 4–8 mm long, 5–6 mm wide, sepals persistent; poisonous.

Notes: The species name means "inflated," a reference to the fruit.

Kalm's Lobelia
Lobelia kalmii

Habitat: Wet meadows. **General:** Stems 10–50 cm tall, slender, occasionally branched. **Leaves:** Basal leaves spatula-shaped, 1–2.5 cm long; stem leaves alternate, linear to lance-shaped, 1–5 cm long, 1–5 mm wide; margins smooth to obscurely toothed. **Flower Cluster:** Raceme 1-sided, 2–9-flowered. **Flowers:** Blue with white centre, 0.8–1.5 cm long, 2-lipped; petals 5, upper 2 petals small and narrow, lower 3 petals larger; sepals 5; stamens 5; pistil 1. **Fruit:** Capsule globe-shaped, 3–5 mm long.

Notes: The species name commemorates Swedish botanist Pehr Kalm (1715–79), a student of Linneaus.

Great Lobelia, Blue Cardinal-flower · *Lobelia siphilitica*

Habitat: Moist woodlands and swamps. **General:** Stems 50–120 cm tall, somewhat hairy and angular. **Leaves:** Alternate, oval to lance-shaped, 5–15 cm long, 2–6 cm wide; margins smooth to wavy or toothed; stalkless. **Flower Cluster:** Raceme many-flowered; bracts 0.5–5 cm long; flower stalks 0.4–1 cm long. **Flowers:** Blue, 2.5–3.5 cm long, 2-lipped; petals 5, upper lip 2-lobed, lower lip 3-lobed and white-striped; sepals 5; stamens 5, fused and forming a tube around the style; pistil 1. **Fruit:** Capsule 0.5–1 cm long, many-seeded.

Notes: The species name *siphilitica* is a reference to Native peoples using great lobelia to treat syphilis.

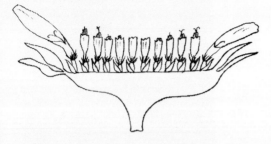

ASTER FAMILY
Asteraceae

Often called the daisy, sunflower or composite family, the aster family is the second largest family of flowering plants in the world. With over 20,000 species, it is a diverse group of plants ranging from herbs to shrubs, and a few species are trees or woody vines. The simple to compound leaves may be alternate, basal, opposite or whorled. Stipules are absent.

The distinguishing feature of this family is the inflorescence, often referred to as a head or capitulum. Resembling a single flower with rows of floral bracts, the head is composed of several small flowers called florets. These florets are of 2 types, ray and disc. The strap-like ray florets resemble individual petals. The disc florets are tubular and often form the centre or button of the "flower." Ray and disc florets form 3 types of flowerheads. Heads composed of ray florets only are called ligulate, whereas those with disc florets only are called discoid. When ray and disc florets are present in the same head, it is called radiate.

Each floret has 5 fused petals, 5 stamens and 1 pistil. The pollen-producing anthers are fused into a tube that surrounds the style. The sepals are highly modified into a structure called the pappus. The pappus may be absent or is represented by scales, hairs or bristles. The fruit, an achene, is single-seeded and does not open at maturity.

Economically important species in this family include common annual sunflower (*Helianthus annuus*), artichoke (*Cynara scolymus*), lettuce (*Lactuca sativa*), echinacea (*Echinacea purpurea*) and marigolds (*Tagetes* spp.). Several species are noxious weeds.

Yarrow
Achillea millefolium

Habitat: Thickets, meadows and roadsides. **General:** Stems 30–80 cm tall, woolly haired, aromatic; rootstock short. **Leaves:** Basal leaves 1–2-pinnately lobed, feather-like, 3.5–35 cm long, 0.5–3.5 cm wide, stalked; stem leaves alternate, 2–10 cm long, stalkless, pinnately divided into numerous segments 1–2 mm wide. **Flower Cluster:**

Heads 10–100, 3–7 mm across, 4–5 mm long, in a flat-topped cluster; ray florets 5–12, white to pinkish white; disc florets 10–30, greyish white; bracts 18–22, in 3–4 overlapping rows, margins pale to dark. **Fruit:** Achene 1–2 mm long, flattened; pappus absent.

Notes: The genus *Achillea* is named after the Greek hero Achilles. The species name *millefolium* means "thousands of leaves," a reference to the numerous leaf segments.

White Snakeroot
Ageratina altissima var. *altissima*

Habitat: Moist, open areas and woodlands. **General:** Stems 30–120 cm tall, often covered in small, crinkled hairs; roots fibrous. **Leaves:** Opposite, triangular to oval or lance-shaped, 4–13 cm long, 2.5–9 cm wide; margins coarsely toothed; stalks 0.5–5 cm long. **Flower Cluster:** Heads 2–4 mm across, in a flat-topped cluster, on hairy stalks 1–5 mm long; heads; ray florets absent; disc florets 10–20, white; bracts 3–5 mm long, in a single row. **Fruit:** Achene black, 2–3 mm long; pappus of capillary bristles 3–4 mm long.

Notes: The genus name *Ageratina* is a diminutive form of *Ageratum*, another Asteraceae genus with similar but larger flowerheads. The species name *altissima* means "very tall" or "tallest."

Common Ragweed
Ambrosia artemisiifolia

Habitat: Roadsides and waste areas; native species with a weedy nature. **General:** Stems 15–150 cm tall, greyish green, bristly haired to hairless, branched. **Leaves:** Alternate (upper) and opposite (lower), 2.5–10 cm long, 2–3 cm wide, divided 1–2 times into narrow segments; margins smooth; stalks 2.5–3.5 cm long. **Flower Cluster:** Heads of 2 types: terminal spikes of 15–150 nodding male heads and axillary clusters of female heads; male heads 2–5 mm wide, 12–20-flowered, ray florets absent, disc florets green, bracts 5–30, green, in 8 rows; female heads 2–5 mm across, ray florets absent, disc floret 1, bracts 5–16, in a single row. **Fruit:** Achene 3–5 mm long, surrounded by woody, bur-like bracts; beak surrounded by a ring of short spines; pappus absent.

Notes: The species name *artemisiifolia* means "with leaves like *Artemisia*," another genus in the aster family. • The male flower heads of common ragweed produce large amounts of pollen, which is the major cause of hayfever.

Great Ragweed
Ambrosia trifida

Habitat: Waste areas and roadsides; introduced from Europe and Asia. **General:** Stems 75–250 cm tall, rough-haired, branched. **Leaves:** Opposite and alternate, prominently 3–5-lobed, 4–25 cm long, 5–25 cm wide, rough and sandpapery; margins coarsely toothed; stalks 1–7 cm long. **Flower Cluster:** Heads of 2 types: terminal spikes of male flowers and axillary clusters of female heads; male heads with ray florets absent, disc florets 3–25, bracts 5–12, black and prominently 3-ribbed; female heads 6–13 mm long, ray florets absent, disc floret 1, enclosed by 5–16 nut-like bracts. **Fruit:** Achene 5.5–6.5 mm long, with prominent beak and a ring of 5 spines; pappus absent.

Notes: The species name *trifida* means "3-parted," a reference to the leaves. • Great ragweed produces large amounts of pollen and is a major cause of hayfever in late summer and early fall.

Pearly Everlasting
Anaphalis margaritacea

Habitat: Dry, open woodlands.
General: Stems 20–90 cm tall, leafy, woolly haired; rhizome woody, creeping; plants often male or female.
Leaves: Alternate, linear to lance-shaped, 3–12 cm long, 0.5–2 cm wide, dark green above, white-woolly below; margins smooth, often rolled under; stalkless. **Flower Cluster:** Heads 0.6–1 cm across, in a dense, terminal cluster to 15 cm across; ray florets absent; disc florets male or female, 50–150, yellow; bracts pearly white, 5–7 mm long, in several overlapping rows. **Fruit:** Achene 0.5–1 mm long; pappus of white capillary bristles.

Notes: The species name *margaritacea* means "pearl-like," a reference to the shiny flower bracts.

Stinking Mayweed
Anthemis cotula

Habitat: Waste areas and roadsides; introduced from Europe. **General:** Stems 10–60 cm tall, profusely branched throughout, hairless, often reddish green; taproot short, thick. **Leaves:** Alternate, 2–6 cm long, 0.5–3 cm wide, pale green to yellowish green, finely divided into narrow segments, ill-scented when crushed; stalkless, margins smooth. **Flower Cluster:** Heads resembling daisies, 1.2–2.6 cm across, solitary at branch tips, on stalks 4–6 cm long; ray florets 10–20, white, 5–11 mm long, 3-toothed, sterile; disc florets 60–300, yellow, 2–2.5 mm long, disc 5–10 mm across; bracts 21–35, in 3–5 rows. **Fruit:** Achene 1–2 mm long, 10-ribbed, glandular-dotted; pappus absent.

Notes: *Anthemis* comes from the Greek word *anthemon*, "flower," a reference to the profuse blooming habit of this species. The species name *cotula* comes from the Greek word *kotule*, "small cup," alluding to the cup-like structure at the base of the leaves.

Common Burdock
Arctium minus

Habitat: Waste ground, railway grades and roadsides; introduced from Europe and Asia. **General:** Stems 80–300 cm tall, erect, hollow, purplish green, branched; taproot large, fleshy. **Leaves:** Basal leaves oblong to oval, 10–60 cm long, 5–40 cm wide, lower surface woolly haired, stalks 15–50 cm long and hollow at the base; stem leaves alternate, reduced in size upward, margins smooth to wavy. **Flower Cluster:** Heads globe-shaped, 1–3 cm across, in a large panicle; ray florets absent; disc florets 20–40, pinkish purple, 7–12 mm long; bracts lance-shaped with hooked tips, in 9–12 overlapping rows. **Fruit:** Achene 5–8 mm long; pappus of short, yellow bristles 1–3.5 mm long, deciduous.

Notes: The genus name *Arctium* comes from the Greek word *arktos,* "bear," a possible reference to the rough-textured bracts. The species name *minus* means "smaller." • In Japan, the roots of this species are harvested and sold as vegetables.

Woolly Burdock
Arctium tomentosum

Habitat: Waste areas and roadsides; introduced from Europe and Asia. **General:** Stems 80–250 cm tall, coarse, branched; taproot stout. **Leaves:** Basal leaves oblong to oval or heart-shaped, 20–50 cm long, 16–40 cm across, stalks hollow, 10–15 cm long; stem leaves alternate, reduced in size upward, margins smooth to wavy. **Flower Cluster:** Heads 1.5–2.5 cm across, in a terminal, flat-topped cluster; ray florets absent; disc florets 25–40, pinkish purple, 9–13 mm long; bracts woolly haired, tips hooked, in 9–17 overlapping rows. **Fruit:** Achene 5–8 mm long; pappus of bristles 1–3 mm long, deciduous.

Notes: The species name *tomentosum* means "woolly," a reference to the woolly haired bracts.

Tuberous Indian Plantain

Arnoglossum plantagineum

Habitat: Wet prairies and meadows. **General:** Stems 50–100 cm tall, angular. **Leaves:** Basal leaves broadly elliptic, 5–17 cm long, palmately nerved, margins smooth to wavy, long-stalked; stem leaves stalked, margins smooth to wavy-toothed. **Flower Cluster:** Heads cylindric, 10–14 mm long, 2.5–5 mm wide, in a corymb-like array; ray florets absent; disc florets 5, white to greenish white; bracts 5, pale green, winged, in 1–2 rows. **Fruit:** Achene spindle- to club-shaped, dark brown, 4–5 mm long, 12–15-ribbed; pappus of 6–9 white bristles.

Notes: The genus name comes from the Greek words *arnos*, "lamb," and *glossum*, "tongue." The species name *plantagineum* refers to the leaves' resemblance to those of plantains (*Plantago* spp.)

Sand Wormwood, Canada Wormwood · *Artemisia campestris*

Habitat: Dry, open, sandy areas. **General:** Stems 30–100 cm tall, reddish green, smooth to hairy; crown woody. **Leaves:** Basal leaves crowded, 3–10 cm long, 0.7–4 cm wide, long-stalked, divided 2–3 times into narrow segments; stem leaves alternate, stalkless, reduced in size upward, divided into narrow segments less than 2 mm across. **Flower Cluster:** Heads 2–4 mm wide, in a narrow, spike-like panicle; ray florets absent; disc florets 5–50, male or female, yellowish green, 1–3 mm long; bracts 2–5 mm long, 2–20 in 4–7 overlapping rows. **Fruit:** Achene 0.8–1 mm long; pappus absent.

Notes: The genus name commemorates the Greek goddess Artemis. The species name *campestris* means "of fields or plains."

Prairie Sagewort, Silver Wormwood
Artemisia ludoviciana

Habitat: Dry, open fields and meadows. General: Stems 15–80 cm tall, greyish green, woolly haired, aromatic when crushed; rhizome creeping. Leaves: Alternate, oblong to lance-shaped, 1–11 cm long, 0.5–4 cm wide, surfaces white-woolly; margins smooth to toothed; stalks absent to very short. Flower Cluster: Heads 3–4 mm high, bell-shaped, in a narrow, spike-like panicle; ray florets absent; disc florets greyish white to yellow, outer flowers 5–12 and female, inner flowers 6–45 and perfect; bracts greyish green, hairy, 2–20 in 4–7 overlapping rows. Fruit: Achene less than 1 mm long; pappus absent.

Notes: The species name *ludoviciana* means "of Louisiana."

Mugwort, Common Wormwood · *Artemisia vulgaris*

Habitat: Waste ground and roadsides; introduced medicinal plant from Europe and Asia. General: Stems 50–190 cm tall, red to brown or purplish, ridged, angular, much-branched and leafy, somewhat woody at the base, clump-forming, ill-scented when crushed; rhizome coarse. Leaves: Alternate, green, 3–10 cm long, 2–8 cm wide, simple to divided and coarsely toothed, hairless above, white-woolly below; lower leaves stalked, upper leaves stalkless; aromatic when crushed. Flower Cluster: Heads 2–3 mm across, 3–4 mm long, 15–30-flowered, erect, in a dense spike 20–30 cm long and 7–15 cm wide; ray florets absent; disc florets 1.5–3 mm long, greenish yellow to reddish brown, outer 7–10 florets female, inner florets 5–20, perfect; bracts 2–20, lance-shaped, margins membranous, in 4–7 rows. Fruit: Achene 1–2 mm long; pappus absent.

Notes: The species name *vulgaris* means "common."

English Daisy

Bellis perennis

Habitat: Lawns, roadsides and waste places; introduced from England. General: Flowering stems 5–20 cm tall, hairy; rhizome short. Leaves: Mostly basal, oblong to spatula-shaped, 1–10 cm long, 0.4–2 cm wide, stalks winged, about as long as blade, margins with rounded teeth; stem leaves, if present, alternate and scale-like. Flower Cluster: Heads solitary, 2–3 cm across; ray florets pinkish to purplish, 35–90, in 3–4 series, each 4–11 mm long; disc florets 60–80, pale yellow, 1–2 mm long; bracts 13–14, in 2 rows. Fruit: Achene flattened, 1–2 mm long; pappus absent.

Notes: The genus name comes from the Latin word *bellus*, meaning "pretty." The species name means "perennial," a reference to this plant's life cycle.

Nodding Beggar-ticks

Bidens cernua

Habitat: Wet or marshy ground. General: Stems 20–100 cm tall, hairless, branched, often rooting at lower nodes. Leaves: Opposite, linear to lance-shaped, 5–20 cm long, 0.5–4.5 cm wide; margins coarsely toothed; stalkless, often fused at the base and forming a cup. Flower Cluster: Heads nodding, 2–3.5 cm across, solitary on a leafy stalk; ray florets 6–8, yellow, 7–15 mm long; disc florets 40–100, yellowish brown, 3–4 mm long; outer bracts 5–8, leaf-like, bent backward; inner bracts yellow with red streaks. Fruit: Achene blackish brown, 5–7 mm long; pappus of 4 barbed awns, each 2–4 mm long.

Notes: The genus name *Bidens* means "2-toothed," a reference to the tooth-like pappus. The species name *cernua* means "nodding."

Plumeless Thistle · *Carduus acanthoides*

Habitat: Roadsides and waste areas; introduced from Europe and North Africa. **General:** Stems 30–400 cm tall, branched, densely haired, winged with spines 2–8 mm long, hollow at the base; taproot large. **Leaves:** Alternate, 10–30 cm long, 2–8 cm wide, lower surface covered in woolly hairs; margins deeply lobed and spiny, the spines 2–5 mm long; lower leaves with winged stalks, upper leaves stalkless. **Flower Cluster:** Heads 1.5–2.5 cm wide, 1.8–2.5 cm high, solitary or in a flat-topped cluster of 2–5, on spiny-winged stalks 4–10 cm long; ray florets absent; disc florets 50–80, pink to purple, 1.3–2 cm long; bracts numerous, linear, 1–2 mm wide, in 7–10 rows; outer bracts spiny; inner bracts soft, rarely spiny. **Fruit:** Achene grey to golden brown, 2.5–3 mm long; pappus of white, barbed bristles, 1.1–1.3 cm long.

Notes: The species name *acanthoides* means "resembling *Acanthus*," a genus of thistle-like plants.

Nodding Thistle
Carduus nutans

Habitat: Waste areas and roadsides; introduced from Europe, Asia and North Africa. **General:** Stems 60–250 cm tall, winged with leaf bases extending down the stem, spines 2–10 mm long; taproot large, fleshy. **Leaves:** Basal leaves produced in first season, 6–8 per rosette, 10–30 cm long, 3–15 cm wide, deeply lobed, spiny; stem leaves alternate, 10–40 cm long, 5–15 cm wide, deeply lobed with 3–5 spiny points, prominently yellow- or white-spined at the tip; covered in fine, woolly hairs. **Flower Cluster:** Heads 1–3, nodding, 1.5–8 cm across, 2–6 cm long, on wingless stalks 2–30 cm long; ray florets absent; disc florets numerous, red to purple, 1.5–1.8 cm long; bracts in 7–10 overlapping rows; outer bracts 2–8 mm wide, spine-tipped, prominently ribbed; inner bracts narrow, often purplish, rarely spiny. **Fruit:** Achene pale yellowish brown, 3–5 mm long; pappus of barbed, white bristles 1.3–2.5 cm long.

Notes: *Carduus* is the ancient Latin name for thistle. The species name *nutans* means "nodding."

Diffuse Knapweed · *Centaurea diffusa*

Habitat: Roadsides and railway tracks; introduced from Europe and Asia. **General:** Stems 10–60 cm tall, profusely branched, covered with fine, cobwebby hairs; mature plants forming tumbleweeds. **Leaves:** Basal leaves stalked, absent at flowering; stem leaves alternate, 10–20 cm long, 1–5 cm wide, deeply divided 1–2 times into narrow segments; upper leaf margins often smooth. **Flower Cluster:** Heads 4–7 mm across, 1–1.5 cm high, in a panicle; ray florets absent; disc florets 25–35, creamy white to yellow or pinkish, 1.2–1.5 cm long; bracts in 6–8 rows, 1–1.3 cm long, margins spiny-haired and prominently spine-tipped, the spines 1.5–4 mm long. **Fruit:** Achene 2–3 mm long; pappus absent.

Notes: The species name means "diffuse," a reference to the profusely branched stems. • Research has shown that the roots of this plant produce a chemical that prevents other species from growing nearby. This trait has enabled diffuse knapweed to infest over 75,000 acres in British Columbia and over 3.3 million acres in Idaho, Oregon and Washington.

Brown Knapweed

Centaurea jacea

Habitat: Roadsides and waste areas; introduced from Europe and Asia. **General:** Stems 30–150 cm tall, branched near the top, covered in woolly hairs. **Leaves:** Alternate, elliptic to oblong or lance-shaped, 5–25 cm long; margins smooth to shallowly toothed; stalks short. **Flower Cluster:** Heads in a flat-topped cluster 1.5–2.5 cm across; ray florets absent; disc florets 40–100, purple, 1.5–1.8 cm long, outer florets often

larger and sterile; bracts in 6 or more rows, margins membranous to irregularly toothed or lobed. **Fruit:** Achene tan, 2–3 mm long; pappus absent.

Notes: The genus name comes from the Greek word *kentaur*, meaning "centaur." These mythical creatures were reputed to have used these plants for their medicinal qualities. The species name *jacea* is the former genus name.

Spotted Knapweed
Centaurea stoebe ssp. *micranthos*

Habitat: Dry, open roadsides and railway tracks; introduced from Europe. **General:** Stems 30–150 cm tall, often purple-streaked; covered with thin, cobwebby hairs, giving the plant a sandpapery texture. **Leaves:** Alternate, 5–25 cm long, divided into narrow segments, hairy, dotted with translucent glands; margins of upper leaves smooth or slightly toothed; stalks short. **Flower Cluster:** Heads 1.5–2.5 cm across, in a panicle; ray florets absent; disc florets purplish pink, outer florets larger and sterile; bracts 1–1.4 cm long, margins comb-like with 5–7 pairs of black hairs 1–2 mm long. **Fruit:** Achene greyish brown, 2.5–3.5 mm long; pappus of numerous comb-like or feathery bristles 1–3 mm long.

Notes: The species name comes from the Greek word *stoibe*, meaning "stuffing." Spotted knapweed resembles plants from the genus *Stoebe*.

Chicory, Wild Succory
Cichorium intybus

Habitat: Roadsides and waste areas; introduced food plant from the Mediterranean region of Europe. **General:** Stems 20–100 cm tall, stiff, branched, hollow, covered in rough hairs; taproot long, thick; all parts exuding milky sap when broken. **Leaves:** Basal leaves oblong to lance-shaped, 7–35 cm long, 1–12 cm wide, margins toothed to deeply lobed; stem leaves alternate, lance-shaped, 3–7 cm long, clasping the stem, margins smooth. **Flower Cluster:** Heads 2–4 cm wide, showy, 1–4 per leaf axil; ray florets 8–25, blue to white; disc florets absent; outer bracts 5, inner bracts 8–10 with spiny margins. **Fruit:** Achene greyish brown, 1–3 mm long; pappus of 1–3 rows of bristle-like scales.

Notes: *Cichorium* is the latinized Arabic name for chicory. The species name *intybus* is derived from the Egyptian word *tybi*, "January," the month in which chicory is traditionally harvested.

Canada Thistle
Cirsium arvense

Habitat: Roadsides, waste areas and gardens; introduced from France in the late 1600s. **General:** Stems 30–150 cm tall, hollow, leafy, nearly hairless; rhizome extensive, creeping; plants male or female. **Leaves:** Alternate, lance-shaped, 5–30 cm long, 1–6 cm wide, lower surface often with woolly hairs; margins deeply lobed and prickly, the spines 1–7 mm long; stalks short. **Flower Cluster:** Heads 1–2.5 cm across, male or female, in a flat-topped cluster, on stalks 2–70 mm long; ray florets absent; disc florets purplish pink, 2–3 mm wide; bracts green, with weak spines, in 6–8 overlapping rows. **Fruit:** Achene light brown, 3–4 mm long; pappus of feathery, white bristles 1.3–3.2 cm long.

Notes: The genus name comes from the Greek word *kirsion*, meaning "a kind of thistle." The species name *arvense* means "of cultivated fields," a reference to this plant's weedy nature on agricultural land.

Swamp Thistle · *Cirsium muticum*

Habitat: Swamps, wet meadows and marshes. **General:** Stems 60–200 cm tall, hollow, not prickly, woolly when young; root turnip-like. **Leaves:** Basal leaves long-stalked, 15–55 cm long, 4–20 cm wide, lobes lance-shaped, weakly spined, underside often covered in cobwebby hairs; stem leaves alternate, reduced in size upward, margins often spinier than basal leaves, short-stalked. **Flower Cluster:** Heads 1.7–3 cm long, 1–3 cm wide, with cobwebby hairs, terminal on stalks 1–15 cm long; ray florets absent; disc florets numerous, lavender to purple or white, 1.6–3.2 cm long; bracts in 8–12 over-lapping rows. **Fruit:** Achene dark brown, 4.5–5.5 mm long; pappus of white bristles 1.2–2 cm long.

Notes: The species name *muticum* means "blunt," a reference to the tips of the floral bracts.

Bull Thistle · *Cirsium vulgare*

Habitat: Roadsides, waste areas and lawns; introduced from Europe, Asia and North Africa. **General:** Stems 50–300 cm tall, winged, branched, leafy; surface with woolly hairs. **Leaves:** Alternate, 7–15 cm long, 0.6–1.2 cm wide, spiny above, finely haired below, bases continuing down stem; margins deeply lobed and prickly, the spines 2–10 mm long. **Flower Cluster:** Heads 3–7.5 cm across, solitary or in groups of 2–3, at branch tips, on stalks 1–6 cm long; ray florets absent; disc florets pinkish purple, 2.5–3.5 cm long; bracts with spines 2–5 mm long, surface with cobwebby hairs, in 10–12 overlapping rows. **Fruit:** Achene greyish brown, 3–4 mm long; pappus of feathery bristles 1.4–3 cm long, deciduous.

Notes: The species name *vulgare* means "common," a reference to this plant's abundance.

Canada Fleabane, Horseweed · *Conyza canadensis*

Habitat: Roadsides and waste areas; native species with a weedy habit. **General:** Stems 30–200 cm, bristly haired, leafy; crushed stem and leaves with a carrot odour. **Leaves:** Alternate, oblong to lance-shaped, 2–10 cm long, 0.4–1.5 cm wide, reduced in size upward; margins smooth; short-stalked to stalkless. **Flower Cluster:** Heads 2–5 mm across, in a profusely branched terminal cluster; ray florets white, often shorter than bracts; disc florets 8–30, yellow; bracts 20–40, 2–4 mm long, in 2–3 overlapping rows. **Fruit:** Achene 1–2 mm long, yellowish brown; pappus of 15–25 greyish white hairs 2–3 mm long.

Notes: The genus name *Conyza* comes from the Greek word *konops*, "flea," a reference to this plant's insecticidal properties. • The leaves contain herpene, a compound that irritates the nostrils of horses, hence the common name "horseweed."

Narrow-leaved Hawksbeard

Crepis tectorum

Habitat: Roadsides and waste areas; introduced from Siberia. **General:** Stems 10–100 cm tall, hairless, leafy; exuding milky sap when broken. **Leaves:** Basal leaves lance-shaped, 10–15 cm long, 1–4 cm wide, stalked, margins deeply lobed to toothed, the teeth backward-pointing; stem leaves alternate, 1–4 cm long, 0.5–1.5 cm wide, margins smooth, stalkless and clasping the stem. **Flower Cluster:** Heads 1–1.5 cm across, 5–20 in a panicle, often in groups of 5; ray florets 30–70, yellow, 1–1.3 cm long; disc florets absent; bracts 5–18, 6–9 mm long, in 1–2 rows, inner bracts 12–15, 5–9 mm long, surface with bristly, black hairs. **Fruit:** Achene dark purplish brown, 3–4 mm long; pappus of white capillary bristles 4–5 mm long.

Notes: The genus name comes from the Greek word *krepis*, meaning "slipper" or "sandal." The species name *tectorum* means "of roofs."

Flat-topped Aster

Doellingeria umbellata

Habitat: Moist, open areas and forest edges. **General:** Stems 50–200 cm tall, sparsely haired; rhizome creeping. **Leaves:** Alternate, elliptic, 6–11 cm long, 1.3–3.5 cm wide, surfaces sparsely haired, but veins hairy; margins smooth; stalkless. **Flower Cluster:** Heads 1.5–2.5 cm across, 20–300 in a flat-topped cluster, on stalks 1–10 mm long; ray florets 2–16, white, 7–10 mm long; disc florets 11–50, yellow becoming brown, 3.5–6 mm long; bracts green, in 3–4 overlapping rows. **Fruit:** Achene brown, 1.5–3 mm long, 4–6-ribbed; pappus of greyish white bristles, outer bristles less than 1 mm long, inner bristles 3.5–6 mm long.

Notes: The genus name honours German botanist Ignatz Doellinger (1770–1841). The species name means "umbel-like," a reference to the flower cluster.

Globe Thistle
Echinops sphaerocephalus

Habitat: Waste areas; introduced ornamental from Europe and Asia. **General:** Stems 80–200 cm tall, covered in sticky or woolly hairs. **Leaves:** Basal and alternate, oblong to elliptic, 5–30 cm long, 2–20 cm wide, covered in greyish white, woolly hairs; margins lobed and spine-tipped, the spines 2–4 mm long; stalks reduced in length upward. **Flower Cluster:** Heads 3–6 cm across, globe-shaped, solitary on long stalks; ray florets absent; disc florets white to pale blue, 1.2–1.4 cm long; bracts in several rows. **Fruit:** Achene 7–10 mm long; pappus of scale-like hairs 1–1.5 mm long.

Notes: The genus name comes from the Greek words *echinos*, "hedgehog," and *opsis*, "resembling," a reference to the spiny flowerheads. The species name *sphaerocephalus* means "with a spherical head."

Philadelphia Fleabane
Erigeron philadelphicus

Habitat: Moist, open areas. **General:** Stems 20–90 cm tall, bristly haired; rootstalk woody. **Leaves:** Basal leaves oblong to lance-shaped, 2–15 cm long, 1–4 cm wide, tapering to a short stalk, margins wavy to lobed; stem leaves alternate, lance-shaped, stalkless and clasping the stem, margins smooth. **Flower Cluster:** Heads 1.3–2.5 cm across, 3–35 in a flat-topped cluster; ray florets 150–400, white to purplish, 0.5–1 cm long, less than 1 mm wide; disc florets numerous, yellow; bracts 4–6 mm long, in 2–3 rows. **Fruit:** Achene 1 mm long; pappus of 15–20 capillary bristles.

Notes: The genus name *Erigeron* comes from the Greek words *eri*, "early," and *geron*, "old man," a reference to some species flowering and setting fruit early in the growing season. The species name means "of Philadelphia."

Perfoliate Thoroughwort,
Boneset · *Eupatorium perfoliatum*

Habitat: Wet meadows, swamps and marshes. **General:** Stems 10–120 cm tall, hairy; rhizome short; roots fibrous. **Leaves:** Opposite, united at the base and surrounding the stem, lance-shaped, 10–20 cm long, 1.5–4.5 cm wide, wrinkled, sparsely haired above, hairy below; margins toothed. **Flower Cluster:** Heads 4–7 mm long, in a flat-topped cluster; ray florets absent; disc florets 7–23, white, 2–3 mm long; bracts 7–10, 4–6 mm long, glandular-dotted, in 1–2 rows. **Fruit:** Achene 3–4.5 mm long; pappus of 20–30 bristles 3–4 mm long.

Notes: The genus name *Eupatorium* commemorates Mithridates Eupator (132–63 BCE), king of Pontus (northeastern Turkey). The species name *perfoliatum* refers to the fused leaf bases.

Bigleaf Aster
Eurybia macrophylla

Habitat: Open woodlands. **General:** Stems reddish green, zigzagged, 15–120 cm tall; rhizome branched, creeping. **Leaves:** Basal leaves 1–4 (absent at flowering), 11–25 cm long, 5.5–15 cm wide, heart-shaped, upper surface rough-hairy, margins toothed, stalks 8–17 cm long; stem leaves alternate, oblong, 2.2–10 cm long, 1–8 cm wide, margins smooth to toothed, stalks 0.3–11 cm long, somewhat clasping. **Flower Cluster:** Heads 2–3 cm across, 8–90 in a flat-topped, terminal cluster, on stalks 0–4 cm long; ray florets 9–20, pale blue to purple, 8–15 mm long, 1–2 mm wide; disc florets 20–40, yellow turning purple, 6–8 mm long; bracts 32–35, purplish green, 7–10 mm long, in 5–6 rows. **Fruit:** Achene brown, 2.5–4.5 mm long; pappus of white to tan bristles 5–7.5 mm long.

Notes: The genus name *Eurybia* is derived from the Greek words *eurys,* "wide," and *baios,* "few," a reference to the small number of ray florets. The species name *macrophylla* means "large-leaved."

Grass-leaved Goldenrod
Euthamia graminifolia

Habitat: Moist, open areas. General: Stems 30–150 cm tall, branched at the top, sparsely haired; rhizome branched, creeping. Leaves: Alternate, numerous, grass-like, 4–13 cm long, 0.2–1.2 cm wide, prominently 3–5-nerved; margins smooth; stalkless. Flower Cluster: Heads 3–5 mm long, in a flat-topped cluster 1.5–28 cm across; ray florets 7–25, yellow, 2–5 mm long; disc florets 5–12, yellow, 2.5–3.5 mm long; bracts yellowish green, in several overlapping rows. Fruit: Achene 0.6–1 mm long; pappus of white capillary bristles to 3.4 mm long.

Notes: The genus name comes from the Greek words *eu*, "well," and *thama*, "crowded." The species name *graminifolia* means "grass-leaved." • The leaves of this plant often have black spots, the result of a fungal infection.

Spotted Joe-Pye Weed
Eutrochium maculatum

Habitat: Open woodlands and marshy ground. General: Stems 60–200 cm tall, purple or green with purple speckles. Leaves: Whorl of 3–6, elliptic to lance-shaped, 6–30 cm long, 2–9 cm wide, glandular-dotted; margins sharply toothed; stalks 0.5–2 cm long. Flower Cluster: Heads 6–9 mm across, in a flat-topped cluster 15–20 cm across; ray florets absent; disc florets 9–22, pink to purple; bracts 10–22, pinkish purple, in 5–6 overlapping rows. Fruit: Achene 3–4.5 mm long, glandular-dotted; pappus of white capillary bristles.

Notes: The genus name comes from the Greek words *eu*, "well," and *trocho*, "wheel-like," alluding to the whorled leaves. The species name *maculatum*, "spotted," refers to the speckled stem.

• The common name "Joe-Pye weed" honours Joe Pye, an early American settler who promoted herbal remedies developed by Native people.

Hairy Galinsoga
Galinsoga quadriradiata

Habitat: Roadsides, waste areas and gardens; introduced from Peru. **General:** Stems 8–60 cm tall, branched, covered with grey, woolly hairs, often rooting at lower nodes. **Leaves:** Opposite, oval to triangular or lance-shaped, 2–6 cm long, 1.5–4.5 cm wide, both surfaces hairy; margins coarsely toothed; stalks 0.2– 1.5 cm long. **Flower Cluster:** Heads 5–8 mm across, 5 in a cyme, on stalks 0.5–2 cm long; ray florets 3–8 (commonly 5), white to pale pink, 3-lobed; disc florets 8–50, yellow; bracts 6–16, oval, surface with sticky hairs, in 2 overlapping rows. **Fruit:** Achene dark brown to black, 1.5–3 mm long; pappus of 15–20 greyish white, awned scales.

Notes: The genus name commemorates Spanish physician and botanist Mariano Martinez de Galinsoga (1766–97). The species name comes from the Latin words *quadri*, "square," and *radiatus*, "with shining rays."

Gumweed
Grindelia squarrosa

Habitat: Wet meadows and roadsides; native to western North America. **General:** Stems 30–100 cm tall, whitish to purplish green, branched, hairless. **Leaves:** Alternate, oblong to lance-shaped, 1–7 cm long, 1–7 mm wide, dotted with small glands; margins smooth to wavy or coarsely toothed; stalkless. **Flower Cluster:** Heads 2–3 cm across, several in a flat-topped cluster; young heads often exuding sticky, white sap; ray florets absent or 24–36, bright yellow, 8–14 mm long; disc florets yellow; bracts linear, very sticky, in 5–6 rows. **Fruit:** Achene whitish, 1.5–4.5 mm long; pappus of 2–8 deciduous awns 2.5–5.5 mm long.

Notes: The genus name honours Russian botanist David Hieronymus Grindel (1776–1836). The species name *squarrosa* means "spreading in all directions, rough-textured."

Sneezeweed, Swamp Sunflower · *Helenium autumnale*

Habitat: Open woods and meadows. **General:** Stems 40–70 cm tall, hairless to slightly hairy; rootstalk short, woody. **Leaves:** Alternate, lance-shaped, 4–15 cm long, 5–40 cm; margins slightly toothed; stalkless with leaf bases continuing down the stem. **Flower Cluster:** Heads 2–3 cm across; ray florets 10–20, 3-lobed, 0.5–1.2 cm long; disc florets numerous, the disc 0.8–2 cm across; bracts bent backward, in 2–3 overlapping rows. **Fruit:** Achene 4–5-angled, 1–2 mm long; pappus of several brown, awn-tipped scales.

Notes: The species name *autumnale* refers to the flowering period.

Common Annual Sunflower · *Helianthus annuus*

Habitat: Dry, sandy areas, roadsides and disturbed sites. **General:** Stem 20–300 cm tall, bristly haired, resembling cultivated sunflower. **Leaves:** Alternate, oval to slightly heart-shaped, 10–40 cm long, 5–40 cm wide, surface sandpapery; margins coarsely toothed; stalks 2–20 cm long. **Flower Cluster:** Heads 7–15 cm across, solitary or in groups of 2–9, on stalks 2–20 cm long; ray florets 17–30, yellow, 2.5–5 cm long; disc florets more than 150, reddish purple to brown; bracts 20–100, oval to lance-shaped, 1.3–2.5 cm long, bristly haired, in several overlapping rows. **Fruit:** Achene 4–5 mm long; pappus of 25 small scales 2–3.5 mm long.

Notes: The genus name *Helianthus* comes from the Greek words *helios*, "sun," and *anthos*, "flower." The species name *annuus* means "annual."

Rough Woodland Sunflower, Rough Sunflower

Helianthus divaricatus

Habitat: Dry, open woodlands. General: Stems 20–150 cm tall, hairless to bristly haired; rhizome woody; roots fibrous. Leaves: Opposite, oval to lance-shaped, 6–15 cm long, 1–8 cm wide, covered in short, bristly hairs; margins smooth; stalks short. Flower Cluster: Heads 2–7.5 cm across, solitary or in a branched cluster of 2–9, on stalks 0.5–9 cm long; ray florets yellow, 8–15, 1.5–3 cm long; disc florets more than 40, yellowish brown, the disc 1–1.5 cm across; bracts 18–25, 0.6–1.2 cm long, bristly haired, in 2–3 overlapping rows. Fruit: Achene 3–4 mm long; pappus of 2 awned scales 2–2.5 mm long.

Notes: The species name *divaricatus* means "spreading at a wide angle."

Jerusalem-artichoke

Helianthus tuberosus

Habitat: Roadsides, waste areas and gardens; native species with a weedy nature. General: Stems 50–300 cm tall, bristly haired; rhizomes producing tubers in late summer. Leaves: Opposite and alternate, oval to lance-shaped, 10–25 cm long, 4–15 cm wide, bristly haired; margins smooth to toothed; stalks 2–8 cm long, often winged. Flower Cluster: Heads 5–10 cm across, in a cluster of 3–15, on stalks 1–15 cm long; ray florets 10–20, yellow, 2.5–4 cm long; disc florets more than 60, yellow, the disc 1.5–2.5 cm across; bracts 22–35, dark green, 8.5–15 mm long, in 2–3 overlapping rows. Fruit: Achene 5–7 mm long; pappus of 2 awned scales 2–3 mm long.

Notes: The species name *tuberosus* means "tuber-bearing." • The tubers were collected by First Nations peoples as a food source, leading to the use of this plant as a food crop in Europe.

Orange Hawkweed, Devil's-paintbrush

Hieracium aurantiacum

Habitat: Waste areas, roadsides and lawns; introduced from Europe. **General:** Flowering stems 20–70 cm tall, covered with stiff, black hairs; exuding milky juice when broken; rhizome creeping; stolons present. **Leaves:** Basal leaves 3–8, club- to spatula-shaped, 5–20 cm long, 1–3.5 cm wide, covered with stiff, black hairs, margins smooth, stalked; stem leaves 1–2, small and scale-like when present. **Flower Cluster:** Heads 2–2.5 cm across, 3–30 in a flat-topped cluster; ray florets 25–120, burnt orange to red, 1–1.4 cm long; disc florets absent; bracts 13–30, 6–8 mm long, bristly haired, in 2–3 overlapping rows. **Fruit:** Achene purplish brown, 1–2 mm long; pappus of 25–30 greyish white capillary bristles 3.5–4 mm long.

Notes: *Hieracium* comes from the word *hierax*, the Greek and Latin word for "hawk." The species name *aurantiacum* means "orangey red," referring to the flower colour.

King Devil · *Hieracium caespitosum*

Habitat: Roadsides and waste areas; introduced from Europe. **General:** Stems 20–90 cm tall, covered with bristly, black hairs; exuding milky juice when broken; rhizome creeping; stolons present. **Leaves:** Basal leaves 3–8, oblong to lance-shaped, 4–25 cm long, 1–3 cm wide, covered with bristly, black hairs, stalked, margins smooth; stems leaves 1–3, small, stalkless. **Flower Cluster:** Heads 8–14 mm across, 5–25 in an umbel-like panicle; ray florets 25–50, yellow; disc florets absent; bracts 12–18, bristly haired, in 2–3 overlapping rows. **Fruit:** Achene 1.5–2 mm long; pappus of 25–30 capillary bristles 4–6 mm long.

Notes: The species name *caespitosum* means "tufted" or "growing in clumps," a reference to this plant's status.

Spotted Cat's-ear
Hypochaeris radicata

Habitat: Roadsides and lawns; introduced from Europe and North Africa. **General:** Flowering stems 15–60 cm tall, branched above; base thickened, woody; roots fibrous. **Leaves:** Basal, oblong to lance-shaped, 3–35 cm long, 0.5–7 cm wide, upper surface hairy; margins toothed to pinnatifid; short-stalked to stalkless. **Flower Cluster:** Heads 2–3 cm across, in clusters of 1–7 on branched, leafless stems; ray florets 20–100, yellow or greyish green, 1–1.5 cm long; disc florets absent; bracts 20–30, 0.3–2 cm long, margins papery, in 3–4 rows. **Fruit:** Achene golden brown, 0.6–1 cm long; pappus of 40–60 white bristles, outer bristles stiff, inner bristles feather-like.

Notes: The genus name comes from the Greek words *hypo*, "beneath," and *choiras*, "pigs," alluding to pigs digging up the roots. The species name *radicata* means "conspicuously rooted."

Elecampane
Inula helenium

Habitat: Roadsides, waste areas and abandoned fields; introduced medicinal plant from Europe. **General:** Stems 50–200 cm tall, leafy; taproot woody. **Leaves:** Basal and alternate, oblong to elliptic, 10–50 cm long, 4.5–20 cm wide, hairy above, woolly below; margins wavy to slightly toothed; lower leaves long-stalked, upper leaves clasping. **Flower Cluster:** Heads 5–8 cm across, on long stalks; ray florets 50–100, yellow, 1–3 cm long; disc florets brownish yellow, 0.9–1.1 cm long; bracts in 2 rows, outer series leaf-like and 1.2–2.5 cm long, inner series narrow and translucent. **Fruit:** Achene 4-angled; pappus of capillary bristles.

Notes: *Inula* is the Latin name for this plant. The species name means "resembling *Helenium*," another yellow-flowered member of this family.

Tansy Ragwort, Stinking Willie
Jacobaea vulgaris

Habitat: Waste ground and roadsides; introduced from Europe. **General:** Stems 30–120 cm tall, branched, often purplish, ill-scented when crushed. **Leaves:** Alternate, 4–23 cm long, 2–11 cm wide, divided 2–3 times into narrow, irregular lobes, underside with cobwebby hairs; margins wavy-toothed; lower leaves long-stalked, upper leaves stalkless. **Flower Cluster:** Heads 2–3 cm across, 20–60 in a flat-topped cluster; ray florets 10–15 (usually 13), yellow, 0.8–1.2 cm long; disc florets 5–80, yellow, 0.7–1 cm long; bracts 13, 3–5 mm long, black-tipped, in a single row. **Fruit:** Achene light brown, 1.5–3 mm long; pappus of numerous soft, white bristles 3–6 mm long.

Notes: The genus name commemorates St. James (Jacobus), one of the 12 Apostles, or refers to the island of St. Iago in the Cape Verde Islands. • Tansy ragwort is toxic to livestock, causing animals to stagger and giving the plant another common name, "staggerwort."

Prickly Lettuce
Lactuca serriola

Habitat: Roadsides, waste areas and gardens; introduced from Europe. **General:** Stems 30–180 cm tall, hollow, somewhat woody, often prickly at the base. **Leaves:** Alternate, bluish green, oblong to lance-shaped, 5–30 cm long, 2.5–10 cm wide, underside midrib with sharp, yellowish spines; margins with deep, backward-pointing lobes, sharply toothed to prickly; short-stalked to stalkless. **Flower Cluster:** Heads 0.3–1 cm across, in a large, pyramidal panicle; ray florets 5–20, yellow; disc florets absent; bracts 5–13, green, 0.9–1.6 cm long, in 3–4 overlapping rows, reflexed in fruit. **Fruit:** Achene olive brown, 2–3 mm long; pappus of white capillary bristles 3–5 mm long.

Notes: The genus name comes from the Latin word *lac*, "milk," a reference to the milky white sap. The species name *serriola* means "in ranks" or "of salads."

Fall Hawkbit
Leontodon autumnalis

Habitat: Roadsides and lawns; introduced from Europe and Asia. **General:** Stems 10–80 cm tall, branched above; stem and leaves exuding milky juice when broken. **Leaves:** Basal, oblong to lance-shaped, 4–35 cm long, 0.5–4 cm wide; margins smooth to deeply toothed or lobed. **Flower Cluster:** Heads 0.7–1.3 cm across, in a loose cluster of 2–5; ray florets 20–30, yellow, 1.3–1.6 cm long; disc florets absent; bracts 18–20, lance-shaped, 1–1.2 cm long, in 2 series. **Fruit:** Achene 4–7 mm long; pappus of yellowish white, feathery bristles 5–8 mm long.

Notes: The genus name comes from the Greek words *leon* and *odons*, meaning "lion's tooth," a reference to the deeply toothed leaves. The species name means "autumn-flowering."

Ox-eye Daisy
Leucanthemum vulgare

Habitat: Roadsides, waste areas and fields; introduced ornamental from Europe. **General:** Stems 30–100 cm tall, hairless; rhizome creeping. **Leaves:** Basal leaves spatula-shaped, 4–15 cm long, 3–5 cm wide, glossy, margins deeply lobed to coarsely toothed, stalks 1–3 cm long; stems leaves alternate, clasping, reduced in size upward; margins irregularly toothed. **Flower Cluster:** Heads 2–6 cm across, showy, solitary on a long stalk; ray florets 15–35, white, 1–2 cm long; disc florets 120–200, yellow, 2–3 mm long; bracts 35–60, green with brown margins, in 3–4 overlapping rows. **Fruit:** Achene greenish brown to black, 1–2 mm long; pappus absent.

Notes: The genus name *Leucanthemum* comes from the Latin words *leucos*, "white," and *antemon*, "flower," referring to the flower colour. The species name *vulgare* means "common."

Pineapple-weed · *Matricaria discoidea*

Habitat: Yards, roadsides and waste areas; native species with a weedy nature. **General:** Stems 5–40 cm tall, hairless, leafy, pineapple-scented when crushed. **Leaves:** Alternate, 1–5 cm long, 0.2–2 cm wide, hairless, fern-like, divided several times into narrow segments 1–2 mm wide; short-stalked to stalkless. **Flower Cluster:** Heads cone-shaped, 0.5–1 cm across, on stalks 0.2–2.5 cm long; ray florets absent; disc florets 125–535, yellowish green, about 1 mm long; bracts 29–47, greenish yellow with translucent, papery margins, in 2–3 overlapping rows. **Fruit:** Achene 1–2 mm long; pappus absent.

Notes: The genus name comes from the Latin word *mater* or *matrix*, a reference to this plant's medicinal value. The species name *discoidea* means "without rays, disc-like." • The common name refers to the pineapple fragrance of the crushed leaves.

Scotch Thistle · *Onopordum acanthium*

Habitat: Waste areas and roadsides; introduced ornamental from the eastern Mediterranean region of Europe. **General:** Stems 50–400 cm tall, branched, winged, spiny, covered in white-woolly hairs. **Leaves:** Alternate, oblong to rectangular, 6–60 cm long, 3–20 cm wide, covered with velvety grey hairs; margins with 8–10 pairs of coarse triangular spiny lobes; leaf bases continuing down stem, forming winged margins to 1.5 cm wide. **Flower Cluster:** Heads globe-shaped, 2.5–5 cm across, solitary or in groups of 2–7; ray florets absent; disc florets numerous, violet to reddish, 2–2.5 cm long; bracts linear to lance-shaped, 2–2.5 mm wide at the base, in 8–10 overlapping rows, covered in short, cobwebby hairs, spine-tipped, the spines 2–6 mm long. **Fruit:** Achene mottled dark brown, 4–5 mm long; pappus of numerous pale yellow to reddish barbed bristles 7–9 mm long.

Notes: The genus name comes from the Greek word *onopordon*, meaning "cotton thistle." The species name means "resembling *Acanthus*," a genus of thistle-like plants.

White Rattlesnake-root, White Lettuce · *Prenanthes alba*

Habitat: Rich woodlands. **General:** Stems 20–175 cm tall, often mottled purplish green and covered in a powdery, white film; stem and leaves exuding milky juice when broken. **Leaves:** Alternate, oval to triangular or heart-shaped, 4–30 cm long, 3–18 cm wide, hairless above, slightly hairy below; margins toothed to shallowly or deeply lobed; lower leaves long-stalked, upper leaves short-stalked to stalkless. **Flower Cluster:** Heads nodding, fragrant, 1.3–1.5 cm long, 10–15 in a panicle; ray florets 7–12, white or pinkish, 0.9–1.5 cm long; disc florets absent; bracts 8, purplish to maroon, 1–1.3 cm long, hairless. **Fruit:** Achene 3.5–6 mm long; pappus of tan to reddish brown capillary bristles 6–7 mm long.

Notes: The genus name comes from the Greek words *prenes*, "drooping," and *anthe*, "blossom," a reference to the nodding flowers. The species name *alba* means "white." • Traditionally, the root was boiled with milk and then taken as a treatment for snakebites, hence the common name "white rattlesnake-root."

Purple Rattlesnake-root, White Lettuce · *Prenanthes racemosa*

Habitat: Streambanks, wet meadows and marshy flats. **General:** Stems 30–175 cm tall, light green to purple; exuding milky juice when broken; roots tuberous. **Leaves:** Alternate, oblong to lance-shaped or spatulate, 4–25 cm long, 1–8 cm wide, leathery; margins smooth to finely toothed; stalkless and clasping the stem. **Flower Cluster:** Heads bell-shaped, 1.1–1.2 cm long, 4–7 mm wide, in an elongated raceme or panicle; ray florets 9–29, pinkish white to lavender, 0.7–1.3 cm long; disc florets absent; bracts 7–14, green to purple, 1–1.2 cm long. **Fruit:** Achene yellowish brown, 5–6 mm long; pappus of 30–50 pale yellow bristles 6–7 mm long.

Notes: The species name *racemosa* refers to the raceme-like flower cluster.

Pinnate Prairie Coneflower, Greyhead · *Ratibida pinnata*

Habitat: Meadows and dry, open wood-lands. **General:** Stem 30–125 cm tall; rhizome short, woody. **Leaves:** Alternate, lance-shaped, 5–40 cm long, 3–15 cm wide, glandular-dotted, pinnately 3–9-lobed, the lobes lance-shaped; lower leaves stalked, upper leaves short-stalked to stalkless. **Flower Cluster:** Heads 1–12, on stalks 3–27 cm long; ray florets 6–15, yellow, 1.6–6 cm long; disc elliptic to ovoid, 1–2.5 cm long; disc florets 100–200, greenish yellow; bracts 10–15, 3–15 mm long, in a single series. **Fruit:** Achene 2–4 mm long; pappus absent.

Notes: The species name *pinnata* means "feather-like," a reference to the leaves. • The common name "coneflower" refers to the cone-shaped disc.

Black-eyed Susan

Rudbeckia hirta

Habitat: Roadsides and waste ground; introduced ornamental from western North America. **General:** Stems 30–100 cm tall, purplish green, bristly haired. **Leaves:** Alternate, lance-shaped, 5–17.5 cm long, 0.5–7 cm wide, surface bristly haired; margins slightly toothed to smooth; stalks long, winged; upper leaves shorter, stalkless. **Flower Cluster:** Heads 5–8 cm across, solitary on a long stalk; ray florets 8–20, orangey yellow, 2–4 cm long; disc florets 250–500, dark brown to purple, the disc 1–2 cm across; bracts leaf-like, reflexed, in 2 rows. **Fruit:** Achene 4-angled, 1.5–3 mm long; pappus absent.

Notes: The genus name commemo-rates Swedish botanist Olaus Rudbeck (1660–1740). The species name *hirta* means "bristly haired."

Sticky Groundsel · *Senecio viscosus*

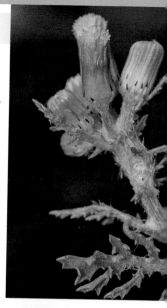

Habitat: Roadsides and waste ground; introduced from Europe. **General:** Stems 10–40 cm tall, hollow, somewhat succulent, sticky-haired throughout, ill-scented. **Leaves:** Alternate, oblong, 3–15 cm long, 1–5 cm wide, covered with sticky hairs, underside often purplish; margins deeply lobed; lower leaves stalked, upper leaves stalkless. **Flower Cluster:** Heads 0.5–1.5 cm wide, on long, hairy stalks; ray florets 11–13, inconspicuous, 1–2 mm long; disc florets yellow, 2–5 mm long; bracts 6–8 mm long, often black-tipped, in a single row. **Fruit:** Achene 2–3 mm long; pappus of numerous soft, white bristles.

Notes: The genus name comes from the Latin *senex,* "old man," a reference to the fluffy, white seed heads. The species name *viscosus* means "sticky" or "viscous," referring to the sticky hairs on the plant's stem and leaves.

Common Groundsel

Senecio vulgaris

Habitat: Gardens, roadsides and waste areas; introduced from Europe by the Pilgrims to treat the early stages of cholera. **General:** Stems 10–60 cm tall, hollow, somewhat succulent, branched, hairless. **Leaves:** Alternate, oblong, 5–15 cm long, 1–4 cm wide, underside often purplish green; margins wavy to deeply lobed; lower leaves stalked, upper leaves stalkless and clasping the stem. **Flower Cluster:** Heads 0.5–1 cm across, 8–20 in a cyme-like cluster; ray florets absent; disc florets 5–80, yellow, 5–8 mm long; bracts 20–23, yellowish green with green or black tips, 4–6 mm long, in a single row. **Fruit:** Achene tan, 2–3 mm long; pappus of numerous soft, white bristles 3–4 mm long.

Notes: The species name *vulgaris* means "common." • This widespread weed is capable of producing over 1700 seeds per plant. With the possibility of 5 generations within a single growing season, each seed has the potential to produce over 1 billion more seeds.

Cup-plant · *Silphium perfoliatum*

Habitat: Moist, open meadows and river bottoms. **General:** Stems 75–350 cm tall, square; roots fibrous. **Leaves:** Opposite or in whorls of 3, triangular to oval or lance-shaped, 2–41 cm long, 0.5–24 cm wide; margins smooth to toothed; stalks fused at the base, forming a cup. **Flower Cluster:** Heads 7.5–10 cm across, in a large panicle; ray florets 17–35, yellow; disc florets 85–200, yellow; bracts 25–37, in 2–3 series. **Fruit:** Achene 8–12 mm

long; pappus of 2 awns 0.5–1.5 mm long.

Notes: The genus name comes from the Greek word *silphion*, an unknown resinous plant

that appears on the coins of Cyrene, an ancient Greek colony in Libya. The species name means "perfoliate," a reference to the fusion of leaf bases.

White Goldenrod · *Solidago bicolor*

Habitat: Dry, open, rocky to sandy woodlands. **General:** Stems 20–100 cm tall, branched only in flower cluster; surface with stiff to woolly hairs. **Leaves:** Basal leaves oblong to oval, 3.5–21 cm long, 1.5–5 cm wide, margins toothed, stalks winged; stem leaves alternate, elliptic, 1.5–5 cm long, 5–15 mm wide, margins smooth, middle to upper leaves stalkless. **Flower Cluster:** Heads 3–5 mm long, 12–270 in a spike-like panicle, on stalks 1.5–2.5 mm long; ray florets 7–9, white, 3.5–4 mm long; disc florets 9–12, pale yellow, 3–4 mm long; bracts in 3–4 rows. **Fruit:** Achene 1.5–2.5 mm long; pappus of white capillary bristles 2.5–3.5 mm long.

Notes: The species name *bicolor* means "2-coloured," a reference to the white ray florets and the pale yellow disc florets.

Blue-stem Goldenrod, Wreath Goldenrod

Solidago caesia

Habitat: Rich, shady woods. General: Stems 20–80 cm tall, erect, arching, purplish blue with a powdery white film; rhizome short, woody. Leaves: Alternate, lance-shaped, 6–12.5 cm long, 1–3 cm wide, slightly hairy above and on the midrib below; margins toothed; stalkless; basal leaves absent at flowering. Flower Cluster: Heads 3–5 mm across, 9–380 in a nodding, spike-like panicle 13–36 cm long; heads; ray florets 1–6, yellow, 2–4 mm long; disc florets 3–8, yellow, 2–3 mm long; bracts 3–5 mm long, hairless, in 3 overlapping rows. Fruit: Achene 1–2 mm long; pappus of white capillary bristles 2–4 mm long.

Notes: The genus name comes from the Latin word *solido*, "to strengthen," a reference to this plant's medicinal properties. The species name *caesia* means "bluish grey," referring to the stem colour.

Canada Goldenrod

Solidago canadensis

Habitat: Open meadows and forests. General: Stems 30–120 cm tall, leafy, densely haired below flower cluster; rhizome creeping. Leaves: Alternate, numerous, oval to lance-shaped, 5–19 cm long, 0.5–3 cm wide, surfaces with short hairs, prominently 3-veined; margins toothed; short-stalked to stalkless. Flower Cluster: Heads 2–5 mm across, 70–1300 in a pyramidal panicle, on stalks 3–4 mm long; ray florets 7–17, yellow, 1–2 mm long; disc florets 2–8, yellow, 2–3 mm long; bracts 2–3 mm long, yellowish green, in 3–4 overlapping rows. Fruit: Achene 1–1.5 mm long; pappus of white capillary bristles 1.5–2 mm long.

Notes: The pollen from goldenrod is heavy and sticky and does not cause hayfever and allergies. These symptoms are caused by ragweed (*Ambrosia* spp.), p. 372, which blooms at the same time as goldenrod.

Zigzag Goldenrod
Solidago flexicaulis

Habitat: Shady woods and thickets.
General: Stems 15–90 cm tall, zig-zagged near the top; rhizome short, woody. **Leaves:** Basal leaves oval, 7–18 cm long, 2.7–8 cm wide, margins toothed, stalks winged; stem leaves alternate, oval to lance-shaped, 3.8–9 cm long, 1–3 cm wide, margins smooth to toothed, middle to upper leaves stalk-less. **Flower Cluster:** Heads 4.5–8 mm long, 25–250 in a panicle 7–31 cm long, on stalks 0.5–5 mm long; ray florets 1–5, yellow, 2.5–4 mm long; disc florets 4–8, yellow, 2–4 mm long; bracts 1–6 mm long, in 3 rows. **Fruit:** Achene 1–2 mm long; pappus of white capillary bristles 3–4.5 mm long.

Notes: The species name *flexicaulis* means "with flexible stems."

Early Goldenrod
Solidago juncea

Habitat: Open, sandy areas and disturbed sites. **General:** Stems 30–120 cm tall, branched, hairless; rhizome creeping. **Leaves:** Alternate, oblong to oval or lance-shaped, 3–30 cm long, 0.8–7 cm wide; margins smooth to finely toothed; stalks winged. **Flower Cluster:** Heads 3–4 mm long, 60–450 in a pyramidal panicle, on stalks 1.5–6 mm long; ray florets 7–12, yellow, 2–2.5 mm long; disc florets 8–15, yellow, 2–3 mm long; bracts in 3–4 rows. **Fruit:** Achene 1–1.5 mm long; pappus of white capillary bristles 2.5–3.5 mm long.

Notes: The species name *juncea* means "like *Juncus*, resembling a rush." • The common name "early golden-rod" refers to the blooming period, which is much earlier than that of other goldenrods.

Grey-stemmed Goldenrod
Solidago nemoralis

Habitat: Dry, open meadows and sites with poor soil. **General:** Stems 20–100 cm tall, greyish green owing to the presence of hairs; rootstalk woody. **Leaves:** Alternate, oblong to lance-shaped, 5–25 cm long, 0.8–4 cm wide, covered with short, spreading hairs; margins slightly toothed; lower leaves stalked, upper leaves stalkless; basal leaves present at flowering. **Flower Cluster:** Heads 3–6 mm across, 10–300 in a compact, cylindric cluster 8–25 cm long, on stalks 2–3.5 mm long; branched clusters often 1-sided; rays florets 5–11, yellow, 2.8–5.5 mm long; disc florets 3–10, yellow, 2.5–5 mm long; bracts in 3 rows. **Fruit:** Achene 1–2 mm long; pappus of white capillary bristles 2–4 mm long.

Notes: The species name *nemoralis* means "of the woods" and "leafy."

Perennial Sowthistle
Sonchus arvensis ssp. *arvensis*

Habitat: Roadsides and waste areas; introduced from the Caucasus region of Asia. **General:** Stems 30–200 cm tall, succulent, hollow; exuding milky juice when broken; rhizome creeping. **Leaves:** Alternate, lance-shaped, 6–40 cm long, 0.2–1.5 cm wide; lower leaves deeply lobed with soft, prickly margins; upper leaves with 4–14 backward-pointing, spine-tipped lobes per side. **Flower Cluster:** Heads 3–5 cm wide, in a flat-topped cluster, stalks yellow-haired; ray florets 30–250, yellow; disc florets absent; bracts 27–50, 1.4–1.7 cm long, dark green with yellow hairs, in 3–5 overlapping rows. **Fruit:** Achene 2–3 mm long; pappus of soft, white bristles 0.8–1 cm long.

Notes: *Sonchus* is the Greek name for sowthistle. The species name *arvensis* means "of cultivated fields," a reference to this plant's weedy nature. • Another perennial sowthistle, *S. a.* ssp. *uliginosus*, may be mistaken for *S. a.* ssp. *arvensis* but is distinguished by smooth, hairless flowerhead stalks and bracts 1–1.5 cm long.

Annual Spiny Sowthistle
Sonchus asper

Habitat: Waste ground, gardens and roadsides; introduced from Europe and Asia. **General:** Stems 20–150 cm tall, hollow, reddish green, leafy, hairless; exuding milky juice when broken. **Leaves:** Alternate, lance-shaped, 6–30 cm long, 1–15 cm wide, glossy and dark green above, pale green below; margins shallowly lobed and spiny-toothed, turning purplish green and prickly with age; stalkless and clasping the stem with large, rounded lobes. **Flower Cluster:** Heads pear-shaped, 0.9–1.2 cm long, 1.5–2.5 cm wide, in a flat-topped cluster; ray florets 30–250, light yellow; disc florets absent; bracts 27–50, thickened at the base, in 3–5 overlapping rows. **Fruit:** Achene reddish brown, 2.5–3 mm long; pappus of numerous white bristles 6–9 mm long.

Notes: *Sonchus* is the ancient Greek name for sowthistle. The species name *asper* means "rough-textured."

Annual Sowthistle
Sonchus oleraceus

Habitat: Roadsides and waste areas; introduced from Europe and Asia. **General:** Stems 50–100 cm tall, hollow, hairless; exuding milky juice when broken. **Leaves:** Basal leaves oblong to lance-shaped, 6–50 cm long, 1–15 cm wide, pinnately lobed, margins spine-toothed, stalks winged; stem leaves alternate, bluish green, hairless, 1–3 lobes per side, margins prickly and toothed, upper leaves stalkless and clasping the stem. **Flower Cluster:** Heads 1.8–2.5 cm across, solitary at end of stem; ray florets 30–250, yellow; disc florets absent; bracts 27–50, in 3–5 rows. **Fruit:** Achene dark brown, 2.5–3 mm long; pappus of white bristles 5–8 mm long.

Notes: The species name *oleraceus* means "used as food" or "cultivated." • Annual sowthistle is recognized as a serious weed in 56 countries.

Heartleaf Aster, Common Blue Wood Aster

Symphyotrichum cordifolium

Habitat: Rich, moist soils, open slopes and streambanks. **General:** Stems 20–120 cm, often reddish, hairless; rhizome branched, woody. **Leaves:** Alternate, oval to elliptic, 1–15 cm long, 1–7.5 cm wide, sparsely haired; base heart-shaped; margins toothed; stalks winged; basal leaves absent at flowering time. **Flower Cluster:** Heads cylindric to bell-shaped, 3–6 mm long, 5–300 in a dense panicle, on stalks 3–20 mm long, bracts linear to oblong or lance-shaped; ray florets 8–20, blue to purple, 5–10 mm long, 1–2 mm wide; disc florets 8–20, creamy white to pale yellow becoming purple with age, 4–5 mm long; floral bracts in 3–6 rows, green, often red-tipped. **Fruit:** Achene dull purple to pale brown, 2–3 mm long; pappus of white to pinkish hairs 2–5 mm long.

Notes: The genus name comes from the Greek words *symphysis*, "junction," and *trichos*, "hair," referring to the pappus bristles, which are fused at the base. The species name *cordifolium* means "heart-leaved."

White Heath Aster

Symphyotrichum ericoides

Habitat: Dry, open meadows and roadsides. **General:** Stems 30–80 cm tall, tufted; branches numerous, rough-haired; rhizome short, creeping. **Leaves:** Alternate, numerous, linear to oblong, 1–6 cm long, 1–2.5 cm wide, surface smooth to bristly haired; margins smooth; stalks short; leaves mostly shed by flowering time. **Flower Cluster:** Heads 0.8–1.2 cm across, on stalks 0.5–1 cm long at branch ends, often 1-sided; heads; ray florets 8–20, white, 0.6–1.2 cm long; disc florets 6–25, yellowish white; bracts oblong to lance-shaped, in 3–4 overlapping rows, outer bracts often spine-tipped. **Fruit:** Achene purplish brown, 1–2 mm long; pappus of white capillary bristles 3–4 mm long.

Notes: The species name *ericoides* means "heather-like," a reference to the numerous small leaves.

Smooth Aster
Symphyotrichum laeve

Habitat: Dry, open areas. **General:** Stems 30–100 cm tall, hairless; rhizome short, woody. **Leaves:** Alternate, numerous, oblong to lance-shaped, 2–20 cm long, 1–4.5 cm wide; margins smooth to toothed; lower leaf stalks winged, upper leaves stalkless and clasping the stem. **Flower Cluster:** Heads 2–3 cm across, in a panicle, often flat-topped, on stalks 0.5–6 cm long; ray florets 15–25, blue or purple, 0.7–1.2 cm long; disc florets 19–43, yellow turning purplish red; bracts sharp-pointed, green with white base, in 4–6 overlapping rows. **Fruit:** Achene purplish brown; pappus of pale brown capillary bristles 5–7 mm long.

Notes: The species name *laeve* means "smooth," a reference to the hairless stems.

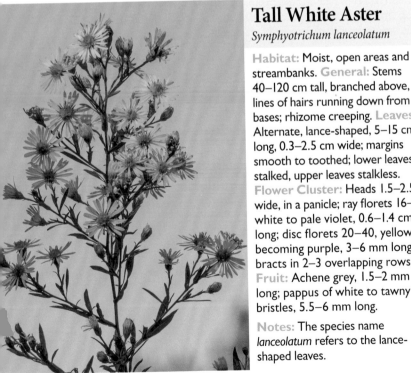

Tall White Aster
Symphyotrichum lanceolatum

Habitat: Moist, open areas and streambanks. **General:** Stems 40–120 cm tall, branched above, lines of hairs running down from leaf bases; rhizome creeping. **Leaves:** Alternate, lance-shaped, 5–15 cm long, 0.3–2.5 cm wide; margins smooth to toothed; lower leaves stalked, upper leaves stalkless. **Flower Cluster:** Heads 1.5–2.5 cm wide, in a panicle; ray florets 16–50, white to pale violet, 0.6–1.4 cm long; disc florets 20–40, yellow becoming purple, 3–6 mm long; bracts in 2–3 overlapping rows. **Fruit:** Achene grey, 1.5–2 mm long; pappus of white to tawny bristles, 5.5–6 mm long.

Notes: The species name *lanceolatum* refers to the lance-shaped leaves.

New England Aster
Symphyotrichum novae-angliae

Habitat: Moist, open areas. **General:** Stems 30–120 cm tall, leafy, somewhat woody; rhizome fleshy with woody, bulb-like segments. **Leaves:** Alternate, lance-shaped, 3–12 cm long, 0.6–2 cm long, often stiff, surfaces rough-textured and soft-haired; margins smooth; stalkless and clasping the stem; lower leaves absent at flowering time. **Flower Cluster:** Heads 2.5–5 cm across, in panicle-like clusters on stalks 0.3–4 cm long; ray florets 40–100, reddish to purplish blue or lavender, 0.9–1.3 cm long; disc florets 50–110, yellow becoming purple, 4–7 mm long; bracts 0.6–1 cm long, often purplish green and covered in sticky hairs, in 3–6 overlapping rows. **Fruit:** Achene purple to brown, 1.5–2.5 mm long, bristly haired; pappus of white or tan bristles, 4.5–6 mm long.

Notes: The species name *novae-angliae* means "of New England," a reference to this plant's range.

Tansy
Tanacetum vulgare

Habitat: Roadsides, waste areas and pastures; introduced medicinal plant from Europe. **General:** Stems 40–180 cm tall, robust, often purplish at the base, aromatic; rhizome woody. **Leaves:** Alternate, oblong to oval or elliptic, 5–20 cm long, 2–10 cm wide, divided

into numerous toothed segments, surface with numerous small glands; stalked. **Flower Cluster:** Heads 20–200, 0.5–1 cm across, button-shaped, in a flat-topped cluster; ray florets absent; disc florets 60–300, yellowish orange; bracts 30–60, papery, margins pale, in 2–5 overlapping rows. **Fruit:** Achene 1–2 mm long; pappus a small, 5-lobed crown or absent.

Notes: *Tanacetum* is the medieval Latin name for tansy. The species name *vulgare* means "common." • Prairie farmers placed these fragrant plants in their grain bins to repel mice and insects.

Dandelion
Taraxacum officinale

Habitat: Waste areas, roadsides and lawns; introduced from Europe as a garden vegetable. **General:** Flowering stems 5–75 cm tall, hollow; exuding milky juice when broken; taproot stout, fleshy. **Leaves:** Basal, in a rosette, oblong to lance-shaped, 5–40 cm long, 1–10 cm wide; margins coarsely toothed to lobed or deeply divided, terminal segment larger than lateral lobes; stalks obscurely winged. **Flower Cluster:** Heads 2–5 cm across, solitary at top of stem; ray florets 100–300, yellow; disc florets absent; bracts green, in 2 overlapping rows, outer row often reflexed. **Fruit:** Achene 3–4 mm long; pappus of numerous feathery bristles 4–6 mm long; fruiting head 3.5–5 cm across.

Notes: The genus name *Taraxacum* is believed to have originated from the Persian words *talkh* and *chakok*, meaning "bitter herb." The species name *officinale* means "medicinal." • The common name "dandelion" comes from the French *dents de lion*, meaning "lion's teeth," a reference to the deeply toothed leaves. • Young dandelion leaves make a tasty addition to salads. The roasted root can be ground and used as a coffee substitute.

Goat's-beard
Tragopogon dubius

Habitat: Roadsides and waste areas; introduced from Eurasia. **General:** Stems 40–100 cm tall, hairless; exuding milky juice when broken; taproot fleshy. **Leaves:** Alternate, grass-like, 10–30 cm long, 0.5–1.5 cm wide, bluish green, hairless; margins smooth; stalkless and clasping the stem. **Flower Cluster:** Heads 3–7 cm across, solitary atop a swollen stalk; ray florets 30–80, yellow; disc florets absent; bracts 10–14 (commonly 13), 2.5–4 cm long, in a single row. **Fruit:** Achene 2.5–3 cm long; pappus of feathery, webbed hairs 2.3–3 cm long; fruiting head 7–10 cm across.

Notes: The genus name comes from the Greek words *tragos*, "goat," and *pogon*, "beard."

Meadow Salsify
Tragopogon pratensis

Habitat: Disturbed sites and roadsides; introduced from Europe. **General:** Stems 15–100 cm tall, hairless; exuding milky juice when broken. **Leaves:** Alternate, grass-like, 5–30 cm long, 1–2 cm wide, bluish green, hairless; margins smooth; stalkless and clasping the stem. **Flower Cluster:** Heads 2.5–5.5 cm across, solitary atop stem; ray florets yellow, 1–2.5 cm long, toothed at tip; disc florets absent; bracts 8, 1.2–2.4 cm long, in a single row. **Fruit:** Achene 1.2–2.4 cm long; pappus of feathery, webbed hairs; fruiting head 5–9 cm across.

Notes: The species name *pratensis* means "growing in meadows." The common name also alludes to this species habitat.

Scentless Chamomile
Tripleurospermum perforata

Habitat: Waste areas, roadsides and disturbed sites; introduced from Europe. **General:** Stems 10–100 cm tall, branched, hairless, odourless when crushed. **Leaves:** Alternate, 2–8 cm long, hairless, divided into numerous narrow or thread-like, branched segments; margins smooth; short-stalked or stalkless; odourless when crushed. **Flower Cluster:** Heads resembling daisies, 3–4.5 cm wide, 10–300 per stem; ray florets 12–25, white, 8–16 mm long; disc florets 300–500, yellow, 2–3 mm long; bracts 28–60, in 2–5 rows. **Fruit:** Achene dark brown, 1–3 mm long, prominently 3-ribbed; pappus absent or, if present, small scales.

Notes: The genus name comes from the Greek words *tri*, "three," *pleuro*, "ribbed," and *sperma*, "seed," a reference to the strongly 3-ribbed achenes. The species name *perforata* means "pierced."

Colt's-foot, Coughwort
Tussilago farfara

Habitat: Moist, open areas; introduced from Europe as a cold remedy. **General:** Flowering stems 5–50 cm tall, woolly haired, appearing in early spring; rhizome creeping. **Leaves:** Basal leaves angular to heart-shaped, 7.5–15 cm long, 10–20 cm wide, green above, white-woolly below, margins irregularly toothed to lobed, stalks long, appearing after flowering stems have emerged; stem leaves alternate, linear to lance-shaped, 0.4–2.5 cm long, 2–8 mm wide. **Flower Cluster:** Heads 2–3.5 cm wide, solitary atop stem; rays florets 100–200, yellow, 0.4–1 cm long; disc florets 30–40, yellow, 1–1.2 cm long; bracts 21, purplish green, 0.8–1.5 cm long, in 1–2 rows. **Fruit:** Achene 3–4 mm long; pappus of 60–100 white capillary bristles 0.8–1.2 cm long.

Notes: The genus name comes from the Latin word *tussis*, "cough," a reference to this plant's medicinal properties. • The common name "coughwort" refers to the folkloric use of the leaves in cough drops and candies.

Yellow Ironweed, Wingstem
Verbesina alternifolia

Habitat: Streambanks and woodlands. **General:** Stems 1–3 m tall, hairless, strongly winged; rhizome thick, creeping. **Leaves:** Alternate (occasionally opposite), elliptic to linear or lance-shaped, 10–25 cm long, 2–8 cm wide, base of stalk often purplish; margins coarsely toothed. **Flower Cluster:** Heads 8–50, 2–5 cm across, in a loose, flat-topped cluster; ray florets 5–8, yellow, 1.5–2.5 cm long, notched at tip; disc florets 40–60, yellow, 4–6 mm long; bracts 8–12, 0.5–1 cm long, in a single series. **Fruit:** Achene dark brown to black, 3–5 mm long; pappus of 2 awns 1.5–2 mm long.

Notes: The genus name comes from this plant's resemblance to species in the genus *Verbena*. The species name means "alternate-leaved." • The common name "wingstem" refers to the prominently winged stems.

Tall Ironweed

Vernonia gigantea

Habitat: Riverbanks and alluvial flats. General: Stems 80–300 cm tall, purplish green, hairless. Leaves: Alternate, lance-shaped, 15–25 cm long, 3–7 cm wide, underside soft-haired; margins finely toothed; stalkless. Flower Cluster: Heads 0.5–1 cm across, in a loose terminal cluster 20–50 cm wide; ray florets absent; disc florets 9–30, deep purple; bracts 30–40, purplish green, 4–5 mm long, in 4–5 overlapping rows. Fruit: Achene 2.5–3.5 mm long; pappus purple, consisting of 2 rows of bristles, outer scales 20–25, less than 1 mm long, inner scales 35–40, 4–6 mm long.

Notes: The genus name commemorates William Vernon, a 17th-century English botanist. The species name *gigantea* means "large" or "gigantic," a reference to the tall stature of this species. • The common name "ironweed" refers to the rigid stem.

Cocklebur

Xanthium strumarium

Habitat: Moist, sandy areas. General: Stems 30–150 cm tall, spotted purplish black, bristly haired. Leaves: Alternate and opposite, oval to triangular, 2–12 cm long, 3–8 cm wide, palmately 3–5-lobed, sandpapery, base heart-shaped; margins wavy to toothed; stalks 2–14 mm long.

Flower Cluster: Heads male or female; male heads 5–8 mm across, ray florets absent, disc florets 15–20, bracts 6–16 in 1–3 rows; female heads 2–5 mm across, ray florets absent, disc florets 2, green, bracts 2 and becoming woody with age. Fruit: Achene 8–15 mm long, in a woody bur 2.2–2.8 cm long, with hooked spines 2–5 mm long; pappus absent.

Notes: The genus name comes from the ancient Greek word *xanthos*, the name of a plant used to produce a yellow dye. The species name *strumarium* means "tumours" or "ulcers," a reference to this plant's medicinal properties and its use to treat ulcers.

Aquatic Plants

Larger duckweed (*Spirodela polyrrhiza*)

AQUATIC PLANTS

This section includes submersed or floating plants. Occasionally, stems, leaves or flowers are raised above the water's surface, but never more than 30 cm.

Key to the Aquatic Plants

Species in this key are identified by leaf arrangement (i.e., floating, alternate, basal, opposite or whorled).

Leaves absent

Lemna minor p. 422

Lemna trisulcata p. 423

Spirodela polyrrhiza p. 423

Leaves alternate

Azolla caroliniana p. 413

Polygonum amphibium var. *stipulaccum,* p. 424

Brasenia schreberi p. 426

Nuphar lutea ssp. *variegata* p. 427

Nymphaea odorata p. 428

Nasturtium officinale p. 429

Leaves alternate

Menyanthes trifoliata p. 431

Leaves alternate

Potamogeton crispus p. 416

Potamogeton gramineus p. 417

Potamogeton richarsonii p. 417

Ranunculus aquatilis p. 428

Utricularia cornuta p. 432

Utricularia macrorhiza p. 432

411

Leaves alternate

*Utricularia
minor*
p. 433

Leaves basal

*Isoetes
echinospora*
p. 414

*Marsilea
quadrifolia*
p. 415

*Eriocaulon
aquaticum*
p. 420

*Hydrocharis
morsus-ranae*
p. 421

*Vallisneria
americana*
p. 422

*Lobelia
dortmanna*
p. 434

Leaves opposite and/or whorled

*Zannichellia
palustris*
p. 418

*Najas
flexilis*
p. 419

*Elodea
canadensis*
p. 421

*Callitriche
palustris*
p. 433

*Ceratophyllum
demersum*
p. 425

*Cabomba
caroliniana*
p. 426

Leaves opposite and/or whorled

*Myriophyllum
sibiricum*
p. 430

*Myriophyllum
spicatum*
p. 430

*Hippuris
vulgaris*
p. 434

*Utricularia
macrorhiza*
p. 432

*Utricularia
minor*
p. 433

MOSQUITO-FERN FAMILY

Azollaceae

The mosquito-fern family is a small family of aquatic ferns commonly found on calm water or stranded on mudflats. These moss-like plants have translucent to brown, unbranched roots. The dichotomously branched stems and leaves are often reddish green. The stalkless, alternate leaves are 2-lobed, the upper leaves above water and the lower ones submersed. Two types of spores are produced in sporocarps, which are borne in pairs at the base of the lateral branches.

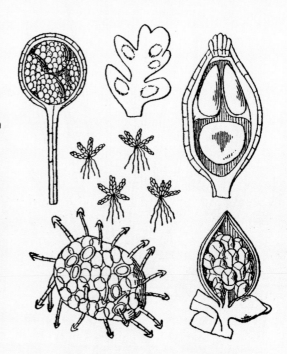

Mosquito-fern

Azolla caroliniana

Habitat: Ponds, marshes and slow-moving water. **General:** Free-floating, often forming mats; stems 0.5–1 cm long, dichotomously branched, reddish green; roots 3–5 cm long. **Leaves:** Alternate, scale-like, 0.5–0.6 mm long, overlapping those above. **Reproductive Structures:** Sporocarps, in pairs on the first leaf of each branch.

Notes: The genus name comes from the Greek words *azo*, "to dry," and *oulla*, "to kill," a reference to the plants dying when they dry out. The species name means "of Carolina." • This aquatic fern may survive under ice for prolonged periods.

413

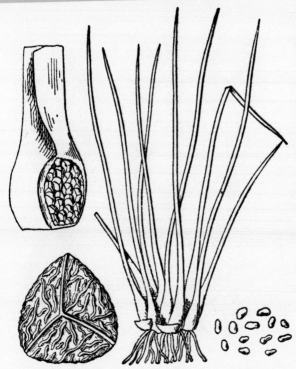

QUILLWORT FAMILY

Isoetaceae

The quillworts are rush-like, aquatic to semi-aquatic plants that reproduce by spores, which are produced in sporangia located in enlarged, hollow leaf bases. Two types of spores are produced: large (megaspores) and small (microspores). The tufted basal leaves resemble those of onions or rushes. The leaves are often 2–4-sided.

Spiny-spored Quillwort

Isoetes echinospora

Habitat: Shallow water of slightly acidic lakes and ponds. **General:** Rootstock 2-lobed, globe- to spindle-shaped, cork-like. **Leaves:** Basal, linear, grass-like, 1–100 cm long. **Reproductive Structures:** Sporangia ovoid to oblong, 3–15 mm wide; megaspores white, grey or black, ridged, spiny; microspores grey to brown, smooth, spiny.

Notes: The genus name comes from the Greek words *isos*, "equal," and *etos*, "year," a reference to the evergreen nature of some species. *Echinospora* means "spiny-spored."

WATERCLOVER FAMILY

Marsileaceae

Members of the waterclover family are aquatic to semi-aquatic plants that reproduce by spores. The rhizomes are long-creeping, producing shoots at the nodes. The leaves are 2-ranked. Sporangia are produced in hard, bony structures called sporocarps that are borne on short stalks at the base of the leaves. The sporangia each contain 1 megaspore and 20–64 microspores.

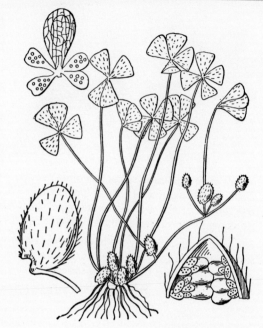

Waterclover, Pepperwort

Marsilea quadrifolia

Habitat: Ponds and slow-moving water. **General:** Resembling clover; roots 1–3 per node. **Leaves:** Basal, stalks 5.4–16.5 cm long, 4 palmately arranged pinnae; pinnae 0.7–2.1 cm long, 0.6–1.9 cm wide, wedge- to heart-shaped, hairless. **Reproductive Structures:** Sporocarps on stalks originating at the leaf base, oval, 4–5.6 mm long, 2.3–2.8 mm wide, sparsely hairy; sori 10–17.

Notes: The species name means "4-leaved." • The common name "pepperwort" refers to the sporocarp, which is about the size of a peppercorn.

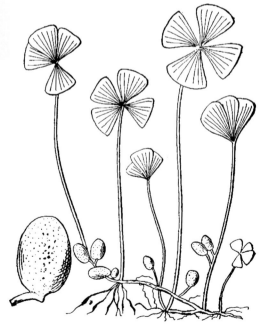

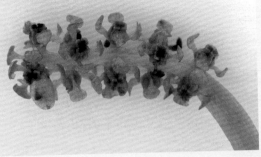

PONDWEED FAMILY
Potamogetonaceae

Pondweeds are aquatic, herbaceous plants with jointed stems and simple, alternate or opposite leaves. Several membranous sheaths are found at the base of each leaf. The whorls of small, green flowers appear in spikes that often rise above the water's surface. The perfect flowers are composed of 4 tepals (petals and sepals that are indistinguishable from one another), 4 stamens and 1 pistil with 4 styles. Unisexual male flowers have a single stamen, whereas female flowers have 1–8 pistils. The fruit is an achene or drupe.

Pondweeds reproduce by seeds, tubers, rhizomes and turions. Turions, often called winter buds, are compact stem tips that overwinter on the bottom of a water body and form new plants the following spring.

The horned pondweed family (Zannichelliaceae) is now included in the pondweed family.

Curled pondweed (*Potamogeton crispus*), an introduced species from Europe, has been found in the Toronto area. This weedy, invasive plant appears in early spring and outcompetes native vegetation.

Curly Pondweed
Potamogeton crispus

Habitat: Ponds and shores of shallow lakes; introduced from Europe and Asia.
General: Aquatic, submersed; stems 20–100 cm long, 1–2 mm thick, flattened; rhizome absent; producing turions in leaf axils in spring; turions 1.5–3 cm long, bur-like, spindle-shaped, consisting of 3–7 small, thickened leaves, germinating in fall and overwintering as small plants. **Leaves:** Alternate, all submersed, 2-ranked, 1–4 per side, 1.2–9 cm long, 0.4–1 cm wide, 3–5-nerved, midvein reddish; margins crinkled to wavy, finely toothed; stalkless; stipules persistent, brown, 0.4–1 cm long, not fibrous or shredding with age.
Flower Cluster: Spike cylindric, 2.5–4 cm long; stalk 2–10 cm long, curved.
Flowers: Green; tepals 4; stamens 4; pistils 4. **Fruit:** Achene 3–5 mm long, red to reddish brown, beak 2–2.5 mm long.

Notes: The species name *crispus* means "curled" or "crinkled," a reference to the leaf margins.

Slender-leaved Pondweed
Potamogeton gramineus

Habitat: Ponds, lakes and streams. **General:** Aquatic, submersed; stems 30–70 cm long, jointed, often spotted, branches numerous; rhizome present. **Leaves:** Alternate; floating leaves elliptic, leathery, 13–19-nerved, 5–13 cm long, 1–4 cm wide, margins smooth, stalks 5–20 cm long; submersed leaves 1–12 cm long, 1–2 cm wide, 3–9-nerved, linear to lance-shaped, stalkless; stipules membranous, 4–10 cm long. **Flower Cluster:** Spike dense, cylindric, 3–5 cm long, underwater; stalk 3.5–4.5 cm long. **Flowers:** Green; tepals 4; stamens 4; pistils 4. **Fruit:** Nutlet drupe-like, 2–3 mm long, green to brown.

Notes: *Potamogeton* comes from the Greek words *potamos*, "river," and *geiton*, "neighbour." The species name *gramineus* means "grass-like."

Clasping-leaved Pondweed
Potamogeton richardsonii

Habitat: Lakes and ponds. **General:** Aquatic, submersed; stems 30–100 cm long, leafy, branched, 1–2.5 mm thick; rhizome unspotted. **Leaves:** Alternate, lance-shaped, 1.5–12 cm long, 0.5–2 cm wide, prominently 3–7-nerved; margins wavy; stalkless and clasping the stem; stipules 1–2 cm long, soon disintegrating into whitish fibres. **Flower Cluster:** Spike 1.5–3 cm long, often rising above the water's surface; stalk club-shaped, 2–10 cm long; flowers in 6–12 whorls. **Flowers:** Green; tepals 4; stamens 4; pistil 1. **Fruit:** Nutlet drupe-like, 1.5–3 mm long, greenish brown.

Notes: The species name commemorates Sir John Richardson (1787–1865), an English surgeon and naturalist on the Franklin expeditions of 1819–22.

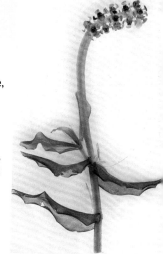

Horned Pondweed

Zannichellia palustris

Habitat: Slow-moving water; ponds and streams. **General:** Aquatic, submersed; stems 5–50 cm long, less than 1 mm thick, leafy, branched. **Leaves:** Opposite, thread-like, 3–8 mm long, less than 1 mm wide, 1-nerved, base membranous and sheathing. **Flower Cluster:** 1–6 flowers per leaf axil (1 male, 1–5 female); bract cup-shaped. **Flowers:** Green, inconspicuous, tepals absent; male flowers with 1 stamen; female flowers with 1–8 pistils. **Fruit:** Achene drupe-like, 2–5 mm long, toothed on 1 side; beak 1–2 mm long.

Notes: The genus name commemorates Venetian botanist Gian Girolamo Zanichelli (1662–1729). The species name *palustris* means "marsh" or "bog," a reference to this plant's habitat.

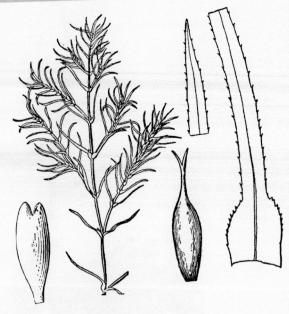

WATER-NYMPH FAMILY

Najadaceae

The water-nymph family is a small group of submersed aquatic plants with simple, opposite or whorled leaves. The small axillary flowers are of 2 types, male and female, and are often surrounded by a small, spathe-like bract. Male flowers are composed of a single stamen and several fused sepals. The female flowers consist of 1 pistil with 3 styles. Sepals are absent in female flowers. The fruit is an achene.

Slender Naiad

Najas flexilis

Habitat: Shallow water in ponds and streams. **General:** Aquatic, submersed; stems profusely branched and tufted, leaves crowded near stem tips. **Leaves:** Opposite, 1–3 cm long, less than 1 mm wide, pale green, collapsing when removed from water, sheathing the stem, widest at the base, tapering to a fine point; margins spiny-toothed, 35–80 teeth per side. **Flower Cluster:** Solitary, in leaf axils. **Flowers:** Green, inconspicuous; male flowers with 1 stamen enclosed in a membranous sheath; female flowers with 1 pistil. **Fruit:** Achene dark brown to yellow, 2–3 mm long.

Notes: The genus name comes from the Greek word *naias*, meaning "water nymph." The species name refers to the flexible stems of this plant.

PIPEWORT FAMILY

Eriocaulaceae

The pipewort family is a group of semi-aquatic plants with grass-like, non-sheathing, basal leaves. The flowers, borne in terminal heads, have 2–3 petals and 2–3 sepals. Male flowers have 4 or 6 stamens, whereas female flowers have a single pistil. The fruit is a capsule.

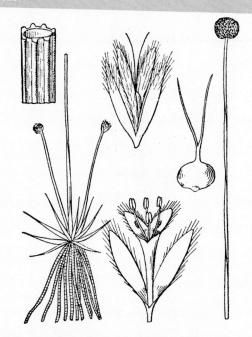

Seven-angled Pipewort
Eriocaulon aquaticum

Habitat: Shallow water along lakeshores. **General:** Aquatic to semi-aquatic; plants male or female; flowering stems 7-angled, 5–20 cm tall (to 100 cm in deep water); roots fleshy, segmented. **Leaves:** Basal, grass-like, 1–10 cm long (to 40 cm, if submersed), 1–2 mm wide, prominently 3–9-cross-veined. **Flower Cluster:** Head terminal, button-like, 4–10 mm across; bracts white and hair-like below head, interspersed with flowers. **Flowers:** Greyish white, 1–3 mm long; male flowers with 2 petals, 2 sepals and 4 stamens; female flowers with 2 petals, 2 sepals and 1 pistil. **Fruit:** Capsule 1-seeded.

Notes: The genus name comes from the Greek words *erion*, "wool," and *kaulos*, "stem," a reference to the white hairs at the base of the flower cluster.

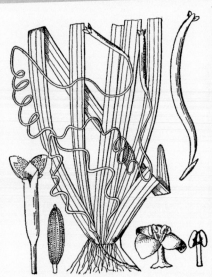

WATERWEED FAMILY
Hydrocharitaceae

Members of this family are submersed or floating aquatic plants. Plants may be male, female or both. The simple leaves may be alternate, opposite, whorled or basal, and sheathing or sheathless. The flowers are of 2 types, male and female. Male flowers are composed of 3 petals, 3 green sepals and 1 to many stamens, whereas female flowers have similar petals and sepals and 2–15 styles. The fruit is a capsule or berry.

The primary form of reproduction is by winter buds, which are clusters of tightly held leaves near the end of the stem. They develop when water temperatures drop and day length shortens, breaking off and falling to the bottom, where they overwinter. When temperatures and day-light increase in the spring, the buds anchor to the bottom and produce new stems.

Several members of this family are noxious weeds. A native species, eelgrass (*Vallisneria americana*), has been designated the worst weed in the world.

Waterweed · *Elodea canadensis*

Habitat: Shallow ponds and lakes.
General: Aquatic, submersed; plants male or female; stems 20–100 cm long, leafy; roots creeping. **Leaves:** Opposite and in whorls of 3–4, 6–20 mm long, 1–5 mm wide, dark green, translucent; margins finely toothed; stalkless.
Flower Cluster: Solitary, in leaf axils; male or female; spathe 2–15 mm long.
Flowers: White, less than 12 mm across; petals 3; sepals 3; male flowers with 7–9 stamens, on short stalks that break off and float to the surface to release pollen; female flowers with 1 pistil reaching the water's surface on thread-like stalks 10–20 cm long. **Fruit:** Capsule egg-shaped, 6–9 mm long; seeds 1–6.

Notes: The genus name comes from the Latin word *elodes*, meaning "bog" or "marsh." The species name means "of Canada."

European Frogbit

Hydrocharis morsus-ranae

Habitat: Edges of lakes and ponds; introduced from the Netherlands and currently spreading throughout the St. Lawrence River basin. **General:** Rooted or free-floating; plants male or female; stems 5–20 cm tall; stolons cord-like, producing turions in autumn (to 150 annually); roots fibrous, 5–30 cm long, becoming entangled in vegetation and helping to stabilize the colony. **Leaves:** Basal, emergent or floating, heart- to kidney-shaped or round, leathery, 1.3–6.3 cm long, 1.2–6 cm wide, hairless, lower surface of floating leaves with spongy tissue confined to midvein region; margins smooth; stalked. **Flower Cluster:** Cyme or solitary; male spathes with 1–5-flowered cyme, stalk 1–4 cm long; female flowers solitary, stalks 1–9 cm long. **Flowers:** White, 8–14 mm across; male or female; male flowers with 3 petals, 3 sepals and 9–12 stamens; female flowers with 3 petals, 3 sepals and 1 pistil; lasting a single day; after pollination, female flowers are pulled underwater and re-emerge as fruit 4–6 weeks later. **Fruit:** Capsule berry-like; seeds 1–1.5 mm long, rarely produced.

Notes: The genus name comes from the Greek words *hydor*, "water," and *charis*, "grace." The species name *morsus-ranae* means "bit" and "frog," hence the common name "frogbit."

Eelgrass

Vallisneria americana

Habitat: Streams, rivers and lakes. **General:** Aquatic, submersed; plants male or female; rhizomes and stolons present. **Leaves:** Basal, linear, grass-like, 10–200 cm long, 0.3–1.5 cm wide; margins smooth to toothed; stalkless. **Flower Cluster:** Cyme long-stalked; male scapes 3–5 cm long, borne underwater; female scapes borne on water's surface. **Flowers:** Greenish white; male or female; 3 petals; 3 sepals; male flowers 1–1.5 mm wide, stamens 2, breaking free underwater and releasing pollen on the water's surface; female flowers 4–6 mm wide, pistil 1. **Fruit:** Capsule berry-like, cylindric to elliptic, 5–10 cm long; ripening underwater as it is pulled downward by a recoiling stalk.

Notes: The genus name commemorates Antonio Vallisneri (1661–1730), an Italian naturalist. The species name means "of America."

ARUM FAMILY · Araceae

Family description on p. 159.

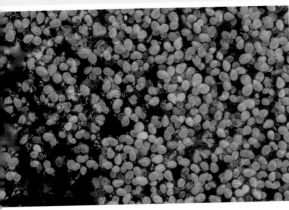

Common Duckweed

Lemna minor

Habitat: Surface of still water; often forming a green carpet. **General:** Aquatic, free-floating; appearing singly or in clusters of 2–8; plant body (thallus) not differentiated into stem or leaves, oval, 2–5 mm long, 1–3 mm wide, faintly 3-nerved, visible under low magnification; root 1, to 1.5 cm long. **Flowers:** Inconspicuous, tiny and rarely seen; male or female; male flowers with 1 stamen; female flowers with 1 pistil. **Fruit:** Capsule less than 1 mm long; seeds 1–7.

Notes: The genus name *Lemna* is believed to have originated from the Greek word *limnos*, "lake," a reference to this plant's habitat. The species name *minor* means "small."

Ivy-leaved Duckweed

Lemna trisulca

Habitat: Still to slow-moving water. **General:** Aquatic, submersed, free-floating; plant body (thallus) not differentiated into stem and leaves, 6–10 mm long, 1–7 mm wide, connected to others and forming

T-shaped colonies; root 1, 5–25 mm long. **Flowers:** Inconspicuous, tiny and rarely seen, green; male or female; male flowers with 1 stamen; female flowers with 1 pistil. **Fruit:** Capsule less than 1 mm long; seeds 1–7.

Notes: The species name *trisulca* means "3-cleft" or "3-forked," probably a reference to the branched nature of the plants.

Larger Duckweed

Spirodela polyrrhiza

Habitat: Still water along lakeshores and ponds. **General:** Aquatic, free-floating; plant body (thallus) round, 0.3–1 cm across, often with a reddish spot, underside reddish purple, 4–15-nerved; roots 7–21, 0.5–3 cm long. **Flowers:** Inconspicuous, tiny and rarely seen, green; male or female; male flowers with 1 stamen; female flowers with 1 pistil. **Fruit:** Capsule 1–1.5 mm long; seeds 1–7.

Notes: The genus name comes from the Greek words *speira*, "a cord," and *delos*, "evident," a reference to the visible roots. The species name *polyrrhiza* means "many roots."

BUCKWHEAT FAMILY · Polygonaceae

Family description on p. 195.

Water Smartweed
Polygonum amphibium var. *stipulaceum*

Habitat: Ponds and quiet water. **General:** Aquatic to semi-aquatic; stems 20–100 cm long; rhizome creeping. **Leaves:** Alternate, 2–23 cm long; submersed leaves collapsing when removed from water; floating leaves oval; above-water leaves lance-shaped, 6–12 cm long; ocrea hairless to densely haired, 0.5–5 cm long; stalks to 7 cm long. **Flower Cluster:** Raceme 1–4 cm long, 1–2 cm thick, on stalks 2–20 cm long, held above the water's surface. **Flowers:** Pink, showy; tepals 4–6; stamens 3–8; pistil 1. **Fruit:** Achene pale brown, lens-shaped, 2–3 mm long.

Notes: The genus name comes from the Greek word *persica*, "peach," referring to the leaves, which resemble those of a peach tree. The species name *amphibia* means "adapted for growing on land or water."

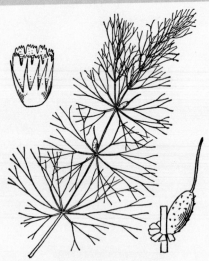

HORNWORT FAMILY
Ceratophyllaceae

Hornworts are free-floating aquatic herbs that rarely have roots. Leaves appear in whorls of 5–12 and are divided 2–3 times into forked, linear segments, which are often spiny-toothed. The small, inconspicuous flowers are of 2 types, male and female, and are borne in leaf axils. Male flowers are composed of 8–12 sepals and 12–16 stamens. Female flowers have a similar number of sepals and 1 pistil. Petals are absent. The fruit is an achene with 2 basal spines.

Reproduction is by seeds and winter buds. Winter buds, produced when day length shortens and water temperatures begin to drop, are clusters of tightly held leaves near the tip of the stem. These buds break off and sink to the bottom, where they overwinter. When water temperatures and daylight increase in the spring, the buds rise and elongate to form new plants.

Coontail

Ceratophyllum demersum

Habitat: Lakes and ponds. **General:** Aquatic, free-floating; stems 30–150 cm long, branched; roots absent. **Leaves:** Whorl of 5–12, forked into 2–4 narrow, toothed segments, the segments 1–4 cm long, less than 1 mm wide; margins spiny-toothed; collapsing when removed from water. **Flower Cluster:** Solitary, in leaf axils; male or female. **Flowers:** Inconspicuous, enclosed by 8–12 transparent bracts; male flowers with 12–16 stamens; female flowers with 1 pistil. **Fruit:** Achene spiny, black, 4–6 mm long.

Notes: The genus name comes from the Greek words *keras*, "leaf," and *phyllon*, "leaf," a reference to the branched leaves. The species name *demersum* means "living underwater." • Leaves crowded near the tip of the stem give the plant the appearance of a raccoon's tail, hence the common name.

WATER-SHIELD FAMILY

Cabombaceae

The water-shield family is a small family of aquatic plants with submersed and floating, alternate or opposite leaves. Flowers are borne singly on long stalks originating in leaf axils. The perfect flowers have 3 petals, 3 sepals, 3–36 stamens and 1–18 pistils. The fruit, a capsule, resembles a leathery achene or follicle. The fruit usually has 1–3 seeds.

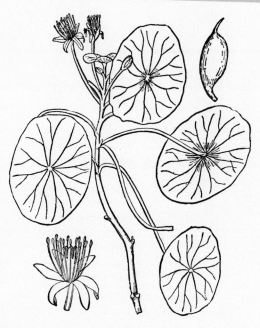

A few members of this family are known to be weedy outside their native range. One species, Carolina fanwort (*Cabomba caroliniana*), has recently been found in the Greater Toronto Area.

Water-shield
Brasenia schreberi

Habitat: Shallow water in lakes and ponds. General: Aquatic; stems to 2 m long, with a slimy coating; rhizome creeping. Leaves: Floating, alternate, long-stalked, elliptic, 3.5–13.5 cm long, 2–8 cm wide, underside with

a slimy coating; stalks long, attached to the centre of the leaf. Flower Cluster: Solitary, on stalks to 15 cm long. Flowers: Dull red, 1.5–3 cm across; petals 3; sepals 3; stamens 18–36; pistils 4–18. Fruit: Capsule leathery, club-shaped, 0.6–1 cm long.

Notes: The genus name commemorates Christoph Brasen (1738–72), a surgeon and missionary at the Nain Mission in Labrador, where he collected and catalogued numerous plant species.

Carolina Fanwort
Cabomba caroliniana

Habitat: Lakes, ponds and streams; introduced from the eastern U.S. General: Aquatic, submersed; stems unbranched, occasionally covered with a slimy film; rhizomes branched, slender. Leaves: Opposite; submersed leaves 1–3.5 cm long, 1.5–5.5 cm wide, divided into numerous linear segments each 1–2 mm wide, stalks 0.5–4 cm long; floating leaves linear to elliptic, shield-like, 0.6–2 cm long, appearing only during flowering. Flower Cluster: Solitary, on stalks 3–10 cm long, in leaf axils. Flowers: White to purplish yellow, 0.6–1.5 cm across; petals 3; sepals 3, petal-like; stamens 6; pistils 2–4. Fruit: Capsule pear-shaped, 4–7 mm long, somewhat leathery; seeds 1–3.

Notes: The genus name is the First Nations name for this plant. • This species, commonly grown in aquariums, has been released into the natural environment and has become weedy in several southern Ontario lakes.

WATER-LILY FAMILY

Nymphaeaceae

Members of this family are aquatic herbs with horizontal rootstalks. The floating, heart-shaped leaves are basal and long-stalked. The solitary, long-stalked flowers are composed of numerous petals, 4–6 sepals, numerous stamens and 1 large stigmatic disc. The compound pistil has 8–30 locules (compartments in which seeds are produced in the ovary). The fruit, a berry, is many-seeded and has a leathery rind.

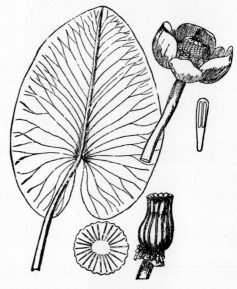

Several species of water-lilies are grown ornamentally. These include giant water-lily (*Victoria amazonica*), which has large, raft-like leaves to 2 m across, water-lotus (*Nelumbo* spp.) and water-lilies (*Nymphaea* spp.).

Yellow Pond-lily

Nuphar lutea ssp. *variegata*

Habitat: Lakes and ponds. **General:** Aquatic; rhizome 2.5–7 cm thick, scaly. **Leaves:** Alternate, floating, 10–35 cm long, 8–25 cm wide, heart-shaped with rounded basal lobes, green, often purple-tinged; margins smooth; stalks long. **Flower Cluster:** Solitary, on a leafless stalk arising from the rhizome. **Flowers:** Yellow, 4–7 cm across; petals several, small, inconspicuous; sepals 6, yellow, petal-like; stamens numerous; pistil 1; stigmatic disc 0.8–2 cm across. **Fruit:** Berry purplish brown, spongy, 2–4 cm long, outer surface leathery.

Notes: The genus name comes from the ancient Persian word *nufar*, which means "yellow water-lily." The species name *lutea* means "yellow."

Fragrant Water-lily
Nymphaea odorata

Habitat: Ponds, lakes and slow-moving streams. **General:** Aquatic; rhizome branched. **Leaves:** Alternate, oval to round, 5–40 cm long, 5–40 cm wide, veins 6–27, upper surface green and shiny, lower surface purple or red; margins smooth; stalks long. **Flower Cluster:** Solitary, in leaf axils. **Flowers:** White or occasionally pinkish, fragrant, 6–19 cm across, floating, opening in the morning and closing at night; petals 14–43; sepals 4, green or reddish; stamens 35–120, yellow; pistil 1, stigmatic disc 0.3–1 cm across. **Fruit:** Berry globe-shaped; seeds numerous, 1.5–2.5 cm long.

Notes: The genus name *Nymphaea* comes from Greek and Roman mythology and refers to the attractive water nymphs who played in the same habitat. The species name *odorata* means "fragrant."

CROWFOOT OR BUTTERCUP FAMILY
Ranunculaceae

Family description on p. 215.

White Water Crowfoot
Ranunculus aquatilis

Habitat: Slow-moving water and ponds. **General:** Aquatic or semi-aquatic; stems 10–100 cm long, branched; rhizome creeping. **Leaves:** Alternate, submersed, kidney-shaped, 0.4–1.1 cm long, 0.7–2.5 cm across, deeply divided into numerous thread-like segments; collapsing when removed from water. **Flower Cluster:** Solitary, on stalks 1–6 cm long, floating on the water's surface. **Flowers:** White, 1–1.5 cm across, usually floating on the water's surface; petals 5; sepals 5, spreading or reflexed; stamens numerous; pistils 15–25. **Fruit:** Achene hairless, 1–2 mm long, in clusters of 15–25, 2–5 mm across.

Notes: The genus name comes from the Greek word *rana*, "frog," a reference to the marshy habitat of many species. The species name *aquatilis* means "growing in water," also referring to this species' habitat.

MUSTARD FAMILY · Brassicaceae

Family description on p. 231.

Watercress
Nasturtium officinale

Habitat: Shallow water in ponds and streams; introduced from Europe as a salad green.
General: Aquatic to semi-aquatic; stems 10–60 cm long, hairless, weak, trailing, often rooting at nodes; roots fibrous, thin. **Leaves:** Alternate, 2–15 cm long, compound with 3–9 leaflets; leaflets oval to round, terminal segment largest; margins smooth; stalked; pungent when crushed. **Flower Cluster:** Raceme 5–20 cm long; flower stalks 0.8–2 cm long. **Flowers:** White, 4–6 mm wide; petals 4; sepals 4; stamens 6; pistil 1. **Fruit:** Silique sickle-shaped, 1–2.5 cm long, 1–3 mm wide; fruiting stalks 0.8–1.5 cm long.

Notes: The genus name comes from the Latin word *nasustortus*, meaning "twisted nose," a reference to the pungent smell of the leaves. The species name *officinale* means "medicinal."

WATER-MILFOIL FAMILY
Haloragaceae

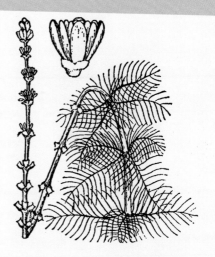

The water-milfoil family consists primarily of aquatic plants with whorled leaves. The leaves are deeply divided into numerous thread-like segments. Male flowers, composed of 4 petals, 3–4 sepals and 2–8 stamens, are found at or near the top of the flower cluster. Female flowers have similar petals and sepals and 2–4 feathery styles. They are found near the bottom of the flower cluster. The petals and sepals often fall off when the flower opens. The fruit is nut-like and splits into 4 seeds.

The primary form of reproduction is by winter buds (turions), clusters of tightly held leaves near the end of the stem. They develop when water temperatures drop and day length shortens, breaking off and falling to the bottom, where they overwinter. When temperatures and daylight increase in the spring, the buds anchor to the bottom and produce new stems.

An introduced species, Eurasian water-milfoil (*Myriophyllum spicatum*) has been designated a noxious weed throughout much of North America.

Northern Water-milfoil
Myriophyllum sibiricum

Habitat: Lakes and streams. **General:** Aquatic, submersed; stems 30–150 cm long, branches often purplish red. **Leaves:** Whorls of 4, 1–3 cm long, feather-like, divided into 12–22 thread-like segments; collapsing when removed from water. **Flower Cluster:** Spike 4–10 cm long, interrupted; flowers in whorls; male flowers at top of spike, female flowers lower on same spike. **Flowers:** Non-showy, 3–5 mm wide; petals 4, falling as the flower opens; sepals 4; male flowers purplish, stamens 8; female flowers with 1 pistil; styles with 2–4 feathery stigmas. **Fruit:** Nutlet globe-shaped, 2.3–3 mm long.

Notes: The genus name *Myriophyllum* comes from the Greek words *murios* and *phyllon*, meaning "innumerable leaves," a reference to the finely divided leaves. The species name means "of Siberia."

Eurasian Water-milfoil
Myriophyllum spicatum

Habitat: Lakes, ponds and quiet waters; introduced from Europe. **General:** Aquatic, rooting in the substrate and free-floating; stems to 12 m long, profusely branched; rhizome creeping; turions developing in autumn. **Leaves:** Whorls of 3–4, 1–3.5 cm long, feather-like, divided into 28–48 thread-like segments; segments 6–12 mm long. **Flower Cluster:** Spike 5–20 cm long, stalk usually pink, raised above the water, becoming thick and succulent to keep spike afloat; flowers in whorls of 4; male flowers at top of spike, female flowers lower on same spike. **Flowers:** Reddish; male flowers with 4 wine red petals, 4 sepals and 8 stamens; female flowers 3–10 whorls per spike, with 4 reddish petals, 4 sepals and 1 pistil; petals of both flower types falling off soon after opening. **Fruit:** Schizocarp globe-shaped, nut-like, breaking into 4 single-seeded sections; seeds 2.5–3 mm across.

Notes: The species name *spicatum* means "spike," a reference to the flower cluster. • This species, designated a noxious weed, was accidentally introduced into North America in shipping ballasts in the late 1800s at Chesapeake Bay in the northeastern U.S. The first Canadian report was from Rondeau Provincial Park in Ontario in 1961.

430

BUCKBEAN FAMILY

Menyanthaceae

The buckbean family is a family of semi-aquatic plants with thick, creeping, scaly rootstocks and alternate, compound leaves with 3 leaflets. The flowers are composed of 5 united petals, 5 united sepals, 5 stamens and 1 pistil. The stamens are attached to the funnel-shaped corolla. The fruit is a capsule with several shiny seeds.

Buckbean

Menyanthes trifoliata

Habitat: Bog, swamps and pond edges. **General:** Flowering stems 10–30 cm tall; rootstock creeping, scaly. **Leaves:** Alternate, crowded near the stem base, compound with 3 leaflets; leaflets oval to elliptic, 3–10 cm long, 1–5 cm wide, margins smooth, stalks long, bases sheathing. **Flower Cluster:** Raceme 10–30 cm long. **Flowers:** White or pinkish, funnel-shaped, 1–1.5 cm long; petals 5, bearded within; sepals 5; stamens 5; pistil 1. **Fruit:** Capsule, ovoid, 0.6–1 cm long; seeds shiny.

Notes: The species name *trifoliata* means "3 leaflets."

BLADDERWORT FAMILY · Lentibulariaceae
Family description on p. 352.

Horned Bladderwort
Utricularia cornuta

Habitat: Wet, sandy shorelines. **General:** Semi-aquatic; flowering stems 10–25 cm tall, brownish; roots finely branched; bladders minute, borne underground; insectivorous. **Leaves:** Alternate, underground and seldom seen, thread-like. **Flower Cluster:** Raceme 1–6-flowered; bracts 1–2 mm long; flower stalks less than 2 mm long. **Flowers:** Yellow, irregular, 2-lipped, 1.8–2.4 cm long; petals 5, spurred, the spur 0.7–1.4 cm long; sepals 5; stamens 2; pistil 1. **Fruit:** Capsule; seeds numerous.

Notes: The species name *cornuta* means "horned." • The common name refers to the horn-like spur and the small bladders that are found on the thread-like leaves.

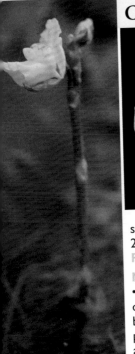

Common Bladderwort · *Utricularia macrorhiza*

Habitat: Lakes, streams and ponds. **General:** Aquatic, submersed, free-floating; stems to 2 m long, leafy; insectivorous. **Leaves:** Alternate or whorled, 1–4 cm long, divided 3–7 times into narrow segments; bladders numerous, clear, 3–5 mm long. **Flower Cluster:** Raceme 1–15-flowered, rising above the water's surface. **Flowers:** Yellow with red or brown stripes, irregular, 2-lipped, 1.4–2 cm long; petals 5; sepals 5; stamens 2; pistil 1. **Fruit:** Capsule; seeds numerous.

Notes: The species name *macrorhiza* means "large-rooted." • The bladders of this plant close when insects and small organisms enter, trapping them. Digestive enzymes and bacteria in the bladders digest the insects, supplying the plant with nutrients. The bladders also help keep the plant afloat and the flowers above the water's surface.

Small Bladderwort · *Utricularia minor*

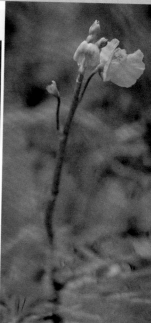

Habitat: Ponds, ditches and slow-moving water. **General:** Aquatic, sub-mersed; stems delicate, thread-like, branched; insectivorous. **Leaves:** Alternate or whorled, dissected into numerous forked segments, each 1–6 mm long; bladders 1–3, 1.5–2 mm long, on the same branch as the leaves. **Flower Cluster:** Raceme 2–6-flowered; stalks erect, curved. **Flowers:** Pale yellow, irregular, 2-lipped, 4–8 mm long; petals 5, spurred, the spur very short; sepals 5; stamens 2; pistil 1. **Fruit:** Capsule many-seeded.

Notes: The genus name comes from the Latin word *utriculus*, "little bottle," referring to the insect-catching bladders. The species name *minor* means "small."

PLANTAIN FAMILY · Plantaginaceae

Family description on p. 353.

Vernal Water-starwort

Callitriche palustris

Habitat: Ponds, ditches and mudflats. **General:** Aquatic, sub-mersed; stems 10–20 cm long. **Leaves:** Opposite; submersed leaves linear, 1–1.5 cm long, 1-nerved; floating leaves spatula-shaped, 3-nerved. **Flower Cluster:** Solitary, in leaf axils; bracts 2, below each flower. **Flowers:** Green, inconspicuous; male or female; male flowers with 1 sta-men; female flowers with 1 pistil. **Fruit:** Nutlet.

Notes: The genus name comes from the Greek words *calos*, "beautiful," and *trichos*, "hair," a reference to the delicate stems.

Mare's-tail
Hippuris vulgaris

Habitat: Ponds, lakes and shallow streams. **General:** Aquatic to semi-aquatic; stems 5–30 cm tall, unbranched; rhizome extensive, creeping. **Leaves:** Whorls of 6–12, abovewater leaves dark green, linear, 1–5 cm long, 1–2 mm wide, stalkless; underwater leaves to 6 cm long, thin, collapsing when taken from water. **Flower Cluster:** Solitary, in leaf axils. **Flowers:** Green, inconspicuous, less than 3 mm long, stalkless; sepals and petals absent; stamen 1; pistil 1. **Fruit:** Nutlet 2–3 mm long.

Notes: The genus name *Hippuris* means "horse-tail" or "mare's-tail."

HAREBELL FAMILY · Campanulaceae

Family description on p. 366.

Dortmann's Lobelia
Lobelia dortmanna

Habitat: Shallow water and shorelines. **General:** Stems hollow, leafless, 20–100 cm tall. **Leaves:** Basal leaves often submersed, tubular, hollow, 2–9 cm long; stem leaves small, thread-like. **Flower Cluster:** Raceme 1–6-flowered, borne above the water's surface. **Flowers:** White to pale violet, tubular, 2-lipped, 1.2–1.7 cm long; petals 5, upper lip 2-lobed, lower lip 3-lobed; sepals 5; stamens 5, fused into a tube; pistil 1. **Fruit:** Capsule; seeds several.

Notes: The genus name commemorates Belgian botanist Matthias de L'Obel (1538–1616).

Grasses, Sedges & Rushes

Hop sedge (*Carex lupulina*)

GRASSES, SEDGES & RUSHES

This section contains 3 large families—grasses, sedges and rushes. All families have narrow leaves with parallel venation and non-showy flowers.

Key to the Grasses, Sedges & Rushes

Species in this key are arranged by stem shape (round, oval or triangular) and number of leaf ranks.

STEMS ROUND

Leaves 2-ranked

Agrostis stolonifera p. 440

Alopecurus aequalis p. 440

Andropogon gerardii p. 441

Avena fatua p. 441

Bromus inermis p. 442

Bromus tectorum p. 442

Leaves 2-ranked

Calamagrostis canadensis p. 443

Cenchrus longispinus p. 443

Dactylis glomerata p. 444

Digitaria sanguinalis p. 444

Echinochloa crus-galli p. 445

Elymus hystrix p. 445

Leaves 2-ranked

Elymus repens p. 446

Festuca rubra p. 446

Glyceria grandis p. 447

Hordeum jubatum p. 447

Lolium persicum p. 448

Oryzopsis asperifolia p. 448

Leaves 2-ranked

Panicum capillare p. 449

Pennisetum glaucum p. 449

Phalaris arundinacea p. 450

Phleum pretense p. 450

Phragmites australis ssp. *australis*, p. 451

Poa annua p. 451

STEMS ROUND

STEMS TRIANGULAR TO OVAL

Leaves 2-ranked

Schizachyrium scoparium var. *scoparium*, p. 452

Setaria viridis p. 452

Spartina pectinata p. 453

Zizania aquatica p. 453

Leaves 3-ranked, flowers in spikelets

Carex aquatilis p. 454

Carex aurea p. 455

Carex bebbii p. 455

Carex brunnescens p. 456

Carex crawei p. 456

Carex crinita p. 457

Leaves 3-ranked, flowers in spikelets

Carex eburnea p. 457

Carex gracillima p. 458

Carex hystericina p. 458

Carex intumescens p. 459

Carex lasiocarpa p. 459

Carex lupulina p. 460

Leaves 3-ranked, flowers in spikelets

Carex pedunculata p. 460

Carex pensylvanica p. 461

Carex plantaginea p. 461

Carex pseudocyperus p. 462

Carex retrorsa p. 462

Carex richardsonii p. 463

Leaves 3-ranked, flowers in spikelets

Carex rosea p. 463

Carex rostrata p. 464

Carex scoparia p. 464

Carex stipata p. 465

Carex tenera p. 465

Carex viridula p. 466

Leaves 3-ranked, flowers in spikelets

*Carex
vulpinoidea*
p. 466

*Cyperus
bipartitus*
p. 467

*Cyperus
esculentus*
p. 467

*Cyperus
odoratus*
p. 468

*Dulichium
arundinaceum*
p. 468

*Eleocharis
acicularis*
p. 469

Leaves 3-ranked, flowers in spikelets

*Eleocharis
palustris*
p. 469

*Eriophorum
gracile*
p. 470

*Rhynchospora
alba*
p. 470

*Schoenoplectus
americanus*
p. 471

*Schoenoplectus
tabernaemontani*
p. 471

*Scirpus
atrovirens*
p. 472

Leaves 3-ranked, flowers in spikelets

*Scirpus
cyperinus*
p. 472

*Scirpus
microcarpus*
p. 473

STEMS ROUND TO OVAL

Leaves 3-ranked, flowers in cymes, heads or corymbs

*Juncus arcticus
var. balticus*
p. 474

*Juncus
bufonius*
p. 474

*Juncus
effusus*
p. 475

*Juncus
gerardi*
p. 475

*Juncus
nodosus*
p. 476

*Juncus
tenuis*
p. 476

Leaves 3-ranked, flowers in cymes, heads or corymbs

*Juncus
torreyi*
p. 477

*Luzula
parviflora*
p. 477

GRASS FAMILY

Poaceae

The grass family, formerly Graminaceae, is one of the largest and most important plant families in the world. Grasses are quite diverse in their growth habit, ranging from small annuals to perennial bamboos with stems over 40 m tall. Their habitats are equally varied and include everything from tropical rainforests to barren arctic tundra.

Members of the grass family have distinctive vegetative and floral structures. The family is often divided into several subfamilies or tribes based largely on the structure and composition of the spikelets.

The roots of grasses may be fibrous or rhizomatous. The round, hollow stems, often called culms, are solid at the nodes. The 2-ranked, sheathing leaves consist of a blade, ligule, auricles and sheath, 4 characteristics that are often used for

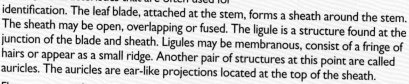

identification. The leaf blade, attached at the stem, forms a sheath around the stem. The sheath may be open, overlapping or fused. The ligule is a structure found at the junction of the blade and sheath. Ligules may be membranous, consist of a fringe of hairs or appear as a small ridge. Another pair of structures at this point are called auricles. The auricles are ear-like projections located at the top of the sheath.

Flowers are produced in small inflorescences called spikelets, which occur in clusters called spikes, racemes or panicles. The spikelets are arranged along a jointed stem called a rachilla. At the base of each spikelet are 2 bracts called glumes. The lower bract is referred to as the first glume and the inner one as the second glume. Below each flower are 2 small bracts. The outer bract, called the lemma, is often boat-shaped. The palea is the inner bract and is often enclosed by the margins of the lemma. Individual flowers, whether perfect or unisexual, are borne inside these bracts. Perfect flowers often have 3 small, translucent scales called lodicules, which are evolutionary remnants of the petals and sepals. Each flower has 3 stamens and 1 pistil with 2 feathery styles. The feathery styles assist in trapping wind-blown pollen. The single-seeded fruit is called a grain or caryopsis.

Several species of grasses, including bamboo (*Bambusa* spp.), are grown for their ornamental value. Economically important grasses are sugar cane (*Saccharum officinale*), corn (*Zea mays*), rice (*Oryza sativa*), wheat (*Triticum aestivum*), barley (*Hordeum vulgare*) and oats (*Avena sativa*).

439

Creeping Bentgrass
Agrostis stolonifera

Habitat: Moist, open areas; introduced from Europe. **General:** Stems 20–100 cm tall, often rooting at lower nodes; nodes 2–5; rhizomes creeping. **Leaves:** Blades flat to rolled, 1–10 cm long, 0.4–1 cm wide, both surfaces roughened; sheaths open, margins overlapping and hairless; ligules lance-shaped, tattered, 2–6 mm long. **Flower Cluster:** Panicle open, pyramid-shaped, 5–30 cm long, 0.4–2.5 cm wide, branches whorled and reddish green; spikelets 2–3.5 mm long; glumes rough on midnerve, 1.5–2.5 mm long; lemmas about ⅔ as long as glumes. **Fruit:** Caryopsis elliptic, smooth, 1–1.3 mm long.

...es: *Agrostis* is the Greek word for "field." The species name *stolonifera* refers ...e creeping nature of this plant's roots.

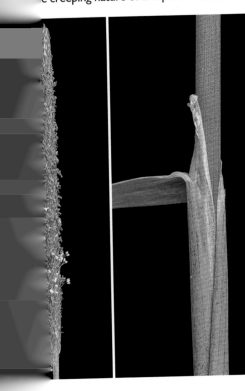

Water Foxtail
Alopecurus aequalis

Habitat: Shallow water or muddy shorelines. **General:** Stems 20–50 cm tall, occasionally rooting from lower nodes. **Leaves:** Blades greyish green, 2–15 cm long, 1–5 mm wide; sheaths open; ligules membranous, 2–5 mm long. **Flower Cluster:** Panicle spike-like, cylindric, 2–7 cm long, 3–6 mm thick; spikelets 2–2.5 mm long, hairless; glumes 2–3 mm long, woolly at the base; lemmas 2.5–4 mm long, awned. **Fruit:** Caryopsis 1–1.8 mm long.

Notes: The genus name comes from the Greek words *alopex* and *oura*, meaning "fox tail," a reference to the slender panicle. The species name means "equal."

Big Bluestem, Turkeyfoot

Andropogon gerardii

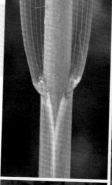

Habitat: Moist, open, sandy soils.
General: Stems 80–200 cm tall, tufted; rhizomes absent or short.
Leaves: Blades 0.5–1 cm wide; margins scabrous; sheaths dull green to bluish, loose-fitting; ligule membranous. **Flower Cluster:** Racemes 2–7, terminal, 4–7 cm long; spikelets in pairs, 0.7–1.1 cm long; first glume as long as spikelet; lemmas 0–2 cm long. **Fruit:** Caryopsis 1–1.3 mm long.

Notes: The genus name comes from the Greek words *andros*, "man," and *pogon*, "beard," a reference to the hairy appearance of the fruiting cluster. The species name commemorates Louis Gerard (1733–1819), the French botanist who first described this species.

Wild Oat · *Avena fatua*

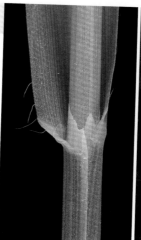

Habitat: Waste areas and roadsides; introduced from Europe. **General:** Stems erect, 40–150 cm tall, nodes 3–5, dark-coloured, tufted, 3–5 stems from the base; roots fibrous. **Leaves:** Blades flat, 20–60 cm long, 0.4–1.8 cm wide, twisted counter-clockwise; sheaths open, slightly hairy, edges transparent; ligules papery, irregularly torn, 2–6 mm long. **Flower Cluster:** Panicle open, pyramidal, nodding, 10–40 cm long, 5–20 cm wide; spikelets 2–2.5 cm long, 2–3-flowered, single at ends of drooping branches; glumes smooth, 1.8–2.8 cm long, 7–11-nerved, fine-lined; lemmas 1.4–2 cm long, 7–9-veined, awns twisted, slightly bent, 2.5–4 ccm long. **Fruit:** Caryopsis elliptic, light yellow to black, 0.6–1.2 cm long, base with numerous brown hairs; awn black, 2–5 cm long, bent at 90° angle.

Notes: The genus name *Avena* is the Greek word for oats. The species name *fatua* means "worthless."

441

Awnless Brome
Bromus inermis

Habitat: Open fields, roadsides and disturbed sites; introduced forage crop from Europe. **General:** Stems 50–150 cm tall, erect, hairless; rhizomes creeping, extensive, dark-coloured, jointed, the internodes covered in large, brown to black, scaly sheaths. **Leaves:** Blades flat, 10–40 cm long, 4–8 mm wide, nearly hairless; sheaths closed, with a small V-shaped notch, a few hairs present; ligules 1–2 mm long, brownish at the base; auricles absent. **Flower Cluster:** Panicle open, nodding, 5–20 cm long, 4–10 cm wide; branches 1–4 per node, 1–10 cm long, with several spikelets; spikelets 1.5–3 cm long, 3–5 mm wide, purplish brown, 7–13-flowered; glumes 0.6–1.1 cm long, prominently 1–5-nerved; lemmas 0.7–1.6 cm long, 5–9-nerved, awns less than 3 mm long. **Fruit:** Caryopsis elliptic, pale yellow to dark brown, 9.5–10.6 mm long, 1.9–2.7 mm wide; awn short, less than 3 mm long.

Notes: The genus name comes from the Greek word *bromos*, an ancient name for oats. The species name *inermis* means "unarmed."

Downy Brome
Bromus tectorum

Habitat: Waste areas and roadsides; introduced from southern Europe. **General:** Stems 5–60 cm tall, tufted, nodes 2–5, often rooting at lower nodes, surface with silky hairs; roots fibrous. **Leaves:** Blades flat, 5–16 cm long, 2–4 mm wide, surface with long, soft, white hairs; sheaths closed, soft-haired; ligules thin, transparent, 2–5 mm long, irregularly toothed; auricles absent. **Flower Cluster:** Panicle nodding, 4–20 cm long, purplish green, surfaces with soft, white hairs, branches drooping; spikelets 2–4 cm long (including awn), 3–10-flowered; glumes 0.6–1.3 cm long, woolly, 1-nerved; lemmas 0.9–1.6 cm long, lance-shaped, woolly, awns 1–2.5 cm long. **Fruit:** Caryopsis elliptic, pale brown with red tinge, 10.5–12.1 mm long, 1.1–1.5 mm wide; awn straight, 1.2–1.7 cm long.

Notes: The species name *tectorum* means "of roofs or houses."

Marsh Reedgrass
Calamagrostis canadensis

Habitat: Marshes, swamps and moist woodlands. **General:** Stems 40–180 cm tall, nodes purplish blue; rhizome creeping. **Leaves:** Blades numerous, 15–30 cm long, 4–8 mm wide, drooping, sheath hairless; ligules membranous, 2–10 mm long, torn. **Flower Cluster:** Panicle dense to open, 8–20 cm long, more than 2 cm wide, somewhat nodding; spikelets 2–6 mm long; glumes 3–4 mm long; lemmas 2–6 mm long, 5-nerved; awns straight. **Fruit:** Caryopsis elliptic, yellowish brown, 1.5–2 mm long.

Notes: The genus name comes from the Greek word *kalamagrostis*, the name for an ancient grass that resembles a reed. The species name means "of Canada."

Longspine Sandbur
Cenchrus longispinus

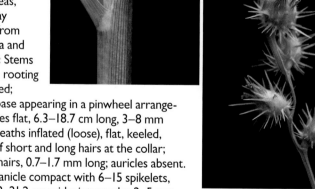

Habitat: Waste areas, roadsides and railway grades; introduced from Europe, South Africa and Australia. **General:** Stems 10–90 cm tall, often rooting at lower nodes, tufted; branches from the base appearing in a pinwheel arrangement. **Leaves:** Blades flat, 6.3–18.7 cm long, 3–8 mm wide, dark green; sheaths inflated (loose), flat, keeled, margins hairy; tuft of short and long hairs at the collar; ligules a ring of fine hairs, 0.7–1.7 mm long; auricles absent. **Flower Cluster:** Panicle compact with 6–15 spikelets, 4.1–10.2 cm long, 1.2–21.2 cm wide; internodes 2–5 mm long; spikelets 1, stalkless, 1–4-flowered; glumes shorter than lemmas; first glume 1.5–3.8 mm long, 1-nerved; second glume 4–6 mm long, 3–5-nerved; lemmas 5–7.6 mm long, 5–7-nerved. **Fruit:** Caryopsis 1–4, oval, reddish brown, 2.2–3.8 mm long, 1.5–2.6 mm wide, in globe-shaped burs; burs 8.3–11.9 mm long, 3.5–6 mm wide, surface with 40 or more spines 3.5–7 mm long and 1 mm wide.

Notes: *Cenchrus* comes from the ancient Greek word *kenchros*, meaning "millet." The species name *longispinus* means "long-spined," a reference to the fruit.

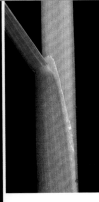

Orchard Grass
Dactylis glomerata

Habitat: Roadsides, open woodlands and meadows; introduced from Europe and North Africa. **General:** Stems coarse, 20–150 cm tall, clump-forming. **Leaves:** Blades flat to strongly V-shaped, 7–45 cm long, 0.2–1.4 cm wide, bluish green, surface and margins rough; sheaths hairless, flattened, partly open; ligules 5–7 mm long, membranous; auricles absent. **Flower Cluster:** Panicle open, 2–30 cm long; spikelets 5–7 mm long, in 1-sided clusters, the clusters 3–6-flowered and flat; first glume 5–6 mm long; second glume 6–7 mm long, sharp-pointed, keeled and bristly haired; lemmas 5–8 mm long, 5-nerved, awns 0.5–1.5 mm long. **Fruit:** Caryopsis elliptic to lance-shaped, tan to light brown, 3–8 mm long.

Notes: The genus name comes from the Greek word *dactylos*, meaning "finger." The species name *glomerata* comes from the Latin word for "clustered," a reference to the clustered spikelets.

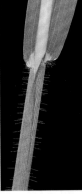

Large Crabgrass
Digitaria sanguinalis

Habitat: Lawns, roadsides and waste areas; introduced forage crop from Europe. **General:** Stems ascending to prostrate, 30–120 cm long, nodes 3–8, often rooting at lower nodes; roots fibrous. **Leaves:** Blades flat, 5–20 cm long, 0.4–1 cm wide; sheaths with overlapping, hairy margins, numerous long, white hairs at junction of blade and sheath; ligules membranous, 1–2 mm long; auricles absent. **Flower Cluster:** Panicle of 3–13 finger-like spikes, 4–18 cm long, whorled at the top of the stem; spikelets 2.5–3.3 mm long, 1-flowered, in pairs on 1 side of stem; first glume minute or absent; second glume 0.8–1.5 mm long, 3-veined; lemmas 2.5–3.3 mm long, 5-nerved, pale brown. **Fruit:** Caryopsis elliptic to lance-shaped, dull olive to brown, 2.7–3.0 mm long, 0.8–0.9 mm wide.

Notes: The species name *sanguinalis* means "blood red."

Barnyard Grass

Echinochloa crus-galli

Habitat: Waste areas, roadsides and disturbed sites; introduced from Europe and India. **General:** Stems erect to prostrate, 20–150 cm long, producing several branches from the base and forming large mats; base reddish purple, lower nodes often rooting when in contact with soil; roots fibrous. **Leaves:** Blades flat or V-shaped, 5–50 cm long, 0.5–2 cm wide, hairless; sheaths open, flattened, margins overlapping, slightly hairy; ligules and auricles absent. **Flower Cluster:** Panicle of racemes, 5–20 cm long, greenish red or purplish; racemes 2–10 cm long; spikelets numerous, 3–4 mm long, in pairs, in 2–4 rows; glumes 1–1.6 mm long, prominently 3-nerved; lemmas 3–4 mm long, prominently 5–7-nerved, awns 0–5 cm long. **Fruit:** Caryopsis elliptic (rounded on one side, flat on the other), whitish to greyish brown, 3.5–4.2 mm long, 1.6–2.0 mm wide, surface with numerous short, stiff hairs; awn 1–2 cm long.

Notes: The genus name comes from the Greek word *echinos*, "hedgehog," and *chloa*, "grass," a reference to the bristly awns. The species name *crus-galli* means "cock's spur." • Barnyard grass is listed as the third worst weed in the world.

Bottlebrush Grass

Elymus hystrix

Habitat: Dry, open woodlands and thickets. **General:** Stems 60–150 cm tall, erect; nodes 4–8, often whitish. **Leaves:** Blades 10–30 cm long, 0.7–1.6 cm across; sheaths greyish green, hairless; ligules short, papery, turning brown with age; auricles absent. **Flower Cluster:** Panicle spike-like, 12–25 cm long, internodes 0.5–1 cm long; spikelets in groups of 2–3 at each node, 2–4-flowered; glumes absent or reduced to a bristle 1–1.6 cm long; lemmas 0.9–1.5 cm long, 5-nerved, awns 1–4 cm long. **Fruit:** Caryopsis straw-coloured, 1–2 mm long.

Notes: The genus name *Elymus* is Greek for millet. The species name *hystrix* means "porcupine-like," a reference to the flower cluster.

Quackgrass · *Elymus repens*

Habitat: Open fields, roadsides and waste areas; introduced from Europe.
General: Stems 30–120 cm tall, nodes 3–5 per stem; rhizomes extensive, creeping, yellowish white, 1–4 mm thick, nodes with several brown, scaly sheaths. **Leaves:** Blades flat, 6–20 cm long, 0.4–1 cm wide, sparsely hairy above, hairless below; sheaths open, margins overlapping and soft-haired; ligules less than 1 mm long, papery; auricles 1–3 mm long, pointed. **Flower Cluster:** Spike 5–25 cm long; spikelets numerous, 1–2 cm long, 3–6 mm wide, 2–9-flowered, set edgewise to stem, short-awned; glumes 0.7–1.2 cm long, prominently 3–7-nerved; lemmas 0.8–1.3 cm long, 5-veined, awns 0–1 cm long. **Fruit:** Caryopsis elliptic, pale yellow to brown, 9–11 mm long, 0.8–1.7 mm wide; awn 1–10 mm long.

Notes: The species name *repens* means "creeping," a reference to the aggressive, weedy nature of the rhizomes.

Red Fescue
Festuca rubra

Habitat: Open woodlands and meadows.
General: Stems 10–90 cm tall, 1–3 nodes per stem, loosely tufted, often bent near the reddish purple base.
Leaves: Blades dark green, soft, 3–40 cm long, 0.5–3 mm wide (thread-like), V-shaped to inward-rolled; sheaths finely haired; ligules about 0.2 mm long, membranous. **Flower Cluster:** Panicle erect to nodding, 3–20 cm long; spikelets reddish purple, 0.7–1.2 cm long, 3–9-flowered; first glume 2–4 mm long; second glume 3–5 mm long; lemmas 4–8 mm long, margins papery, awns 0.5–3 mm long. **Fruit:** Caryopsis.

Notes: *Festuca* is the Latin word for "grass stalk." The species name *rubra* means "red," a reference to the stem colour.

Tall Manna Grass
Glyceria grandis

Habitat: Marshes and shore-lines of sloughs and lakes.
General: Stems 90–160 cm tall, tufted, often rooting at lower nodes; rhizomes present. **Leaves:** Blades flat, 15–40 cm long, 0.6–1.5 cm wide, margins rough; sheaths closed, round or compressed; ligules papery, 3–9 mm long.
Flower Cluster: Panicle open, 20–40 cm long, branches numerous and arching; spikelets 5–6 mm long, 4–7-flowered; first glume 1.2–1.8 mm long; second glume 2–3 mm long; lemmas 2–2.5 mm long, membranous, purple, 7-veined. **Fruit:** Caryopsis 1–1.5 mm long.

Notes: The genus name *Glyceria* comes from the Greek word *glukeros*, "sweet," a reference to the taste of the seeds of some species. The species name means "grand" or "large."

Foxtail Barley
Hordeum jubatum

Habitat: Waste areas and roadsides; native to western North America.
General: Stems 20–100 cm tall, bluish green, tufted; nodes brown, often swollen.
Leaves: Blades flat or V-shaped, 5–15 cm long, 2–4 mm wide, greyish green, sandpapery; sheaths split, margins overlapping and soft-haired; ligules less than 1 mm long, transparent; auricles very small or absent. **Flower Cluster:** Raceme nodding, 5–12 cm long; spikelets 3 per node (1 fertile, 2 sterile); fertile spikelets 4–7 mm long, glumes 4–7 mm long, awns 3.5–8 cm long, lemmas 4–7 mm long and 5-veined, awns 3.5–8 cm long; sterile spikelets with glume awns 3.5–8 cm long, lemma absent (if present, with awns 3.5–8 cm long); all awns purplish green fading to straw-coloured; cluster breaking into several pieces at maturity. **Fruit:** Caryopsis elliptic, yellowish brown, 3–4 mm long, surface with sharp, backward-pointing barbs; awns 4–8, 1.5–6 cm long.

Notes: *Hordeum* is the Latin name for barley (*Hordeum vulgare*). The species name *jubatum* means "maned," a reference to the flower cluster.

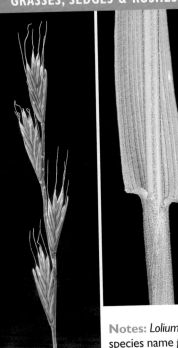

Persian Darnel · *Lolium persicum*

Habitat: Waste areas and roadsides; introduced from southwestern Asia. **General:**
Stems 15–90 cm tall, bright green, nodes
3–4, often branching from lower nodes,
base usually reddish purple. **Leaves:** Blades
flat, twisted, 3–20 cm long, 2–6 mm wide,
upper surface rough; sheaths open, round,
prominently veined, margins overlapping
and hairless; ligules 1–2 mm long, papery;
auricles present. **Flower Cluster:** Panicle
narrow, 3–10 cm long; spikelets 1–2 cm
long, 1.5–7 mm wide, 4–14-flowered,
set edgewise to stem; first glume absent;
second glume 7.5–23 mm long, 5–9-nerved;
lemmas 0.9–1 cm long, 5-nerved, awns
0.5–1.5 cm long. **Fruit:** Caryopsis light
brown, 0.8–1 cm long; awn slightly bent,
0.5–1.2 cm long.

Notes: *Lolium* is the ancient Latin name for ryegrass. The
species name *persicum* means "of or from Persia."

White-seeded Mountain Grass

Oryzopsis asperifolia

Habitat: Open woodlands.
General: Stems 20–70 cm tall,
tufted, erect to spreading, nearly
leafless (2–3 leaves reduced to
bladeless sheaths). **Leaves:**
Blades of lower leaves 15–40 cm
long, 0.3–1 cm wide, flat or rolled,
rough-textured; upper blades less
than 5 cm long, 3–8 mm wide;
sheaths open, margins overlapping; ligules less than 2 mm
long; auricles absent. **Flower
Cluster:** Panicle narrow, 5–8 cm
long, branches erect; spikelets
1-flowered, 6–8 mm long; glumes and lemmas 6–8 mm
long; lemmas with awns 5–10 mm long. **Fruit:** Caryopsis
4–6.5 mm long.

Notes: The genus name *Oryzopsis* comes from the Greek
words *oruza* and *opsis*, a reference to the rice-like grains.
The species name *asperifolia* means "rough-leaved."

Witch Grass
Panicum capillare

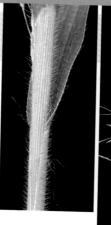

Habitat: Roadsides, railway grades and waste areas; introduced from Europe. **General:** Stems 20–90 cm tall, simple or branched, becoming stiff and bristly with age; nodes bearded. **Leaves:** Blades flat, 7–30 cm long, 0.5–1.5 cm wide, with prominent white midrib, surfaces densely haired; sheaths open, strongly ribbed, margins overlapping with hairs 2–3 mm long; ligules a fringe of hairs 1–2 mm long; auricles absent. **Flower Cluster:** Panicle open, 15–50 cm long, profusely branched, breaking off at maturity and becoming a tumbleweed; spikelets 2–3 mm long; first glume 1–2 mm long, 3-veined; second glume 2–3 mm long, 5–7-veined; lemmas 2–3 mm long, 5–7-nerved. **Fruit:** Caryopsis elliptic, dull brown to dark grey, 1–1.5 mm long, less than 1 mm wide.

Notes: The genus name *Panicum* is the ancient Latin name for millet. The species name means "hair-like" or "slender," a reference to the thread-like flower stalks.

Yellow Foxtail
Pennisetum glaucum

Habitat: Waste areas and roadsides; introduced from Europe or Asia. **General:** Stems 40–130 cm tall, solitary or tufted, often reddish at the base; nodes bearded. **Leaves:** Blades flat or V-shaped, 3–30 cm long, 0.2–1.2 cm wide, loosely twisted, hairless except for long, silky, white hairs, 0.3–1 cm long, at the base; sheaths open, hairless, flattened by third leaf stage, distinctly nerved, margins overlapping; ligules a fringe of hairs 2–3 mm long; auricles absent. **Flower Cluster:** Panicle cylindric, 5–10 cm long, 0.7–1.2 cm wide; spikelets 2–3 mm long, 1-flowered; bristles 6–8 in a whorl, yellowish brown, 4–9 mm long; glumes 1–3 mm long; lemmas 1.5–3.5 mm long. **Fruit:** Caryopsis elliptic, dark greyish brown, 2.3–2.5 mm long, 1.5–1.6 mm wide.

Notes: The genus name comes from the Latin words *penna*, "feather," and *seta*, " bristle." The species name means "whitish."

Reed Canary Grass
Phalaris arundinacea

Habitat: Wet meadows, riverbanks and ditches. **General:** Stems 60–200 cm tall, hairless, leaves 5–10 per stem; rhizomes dark, scaly. **Leaves:** Blades flat, 10–30 cm long, 0.6–1.5 cm wide; sheaths open, margins overlapping; ligules 0.6–1 cm long; auricles absent. **Flower Cluster:** Panicle narrow, dense, 6–40 cm long, 1–4 cm wide, lower branches 3–5 cm long at flowering time; spikelets 4–7.5 mm long, pale green, purple-tinged, 3-flowered; glumes 3.5–7.5 mm long; lemmas 1–5 mm long. **Fruit:** Caryopsis elliptic, 2–3 mm long.

Notes: *Phalaris* is the Greek word for "shining." The species name *arundinacea* means "reed-like," a reference to the habit of this species.

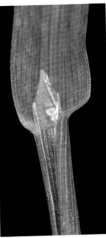

Timothy
Phleum pratense

Habitat: Fields, roadsides and disturbed sites; introduced forage crop from Europe. **General:** Stems 40–150 cm tall, smooth, nodes 3–6; base white, swollen, bulb-like. **Leaves:** Blades flat, 10–30 cm long, 5–8 mm wide, upper blade shortest, erect with a clasping sheath; ligules membranous, 1–6 mm long; auricles absent. **Flower Cluster:** Panicle dense, cylindric, spike-like, 4–15 cm long, 0.5–2 cm across; spikelets numerous, 3–4 mm long, 1-flowered, flattened, stiff-haired; glumes 3–3.8 mm long, 3-veined; lemmas 1.4–2.1 mm long, papery, 5–7-veined, awns 1–2 mm long. **Fruit:** Caryopsis dull to light brown, oval to oblong, 1.2–1.5 mm long.

Notes: The genus name comes from the Greek word *phleos*, the name for a kind of reed. The species name *pratense* means "of meadows," a reference to this plant's native habitat.

Reed
Phragmites australis ssp. *australis*

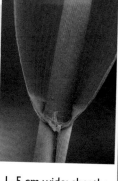

Habitat: Swamps, marshes, lakeshores and ditches; introduced subspecies from Europe and Asia. **General:** Reed-like, somewhat woody; stems 1.5–6 m tall, to 1 cm wide, basal internodes yellow to brown; rhizomes creeping, 0.5–3 cm thick. **Leaves:** Blades flat, 15–70 cm long, 1–5 cm wide; sheaths loose and overlapping, margins with minute, soft hairs; ligules about 0.2 mm long, yellowish purple membrane topped with short, white hairs about 0.5 mm long and firm, white hairs 5–15 mm long. **Flower Cluster:** Panicle large, purplish green, 10–50 cm long, 6–15 cm wide, profusely branched, silky-haired; spikelets 1–1.7 cm long, 3–11-flowered; first glume 2.5–7 mm long; second glume 5–12 mm long, 3–5-veined; lemmas 8–15 mm long, 3-nerved. **Fruit:** Caryopsis dark brown, elliptic, 1.2–1.5 mm long; awn 6–7 mm long; rarely produced.

Notes: The genus name comes from the Greek word *phragma*, meaning "a fence, screen or hedge." The species name *australis* means "southern." • Both native and introduced *Phragmites australis* subspecies are found in Ontario.

Annual Bluegrass
Poa annua

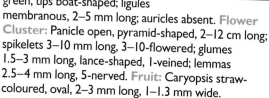

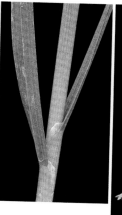

Habitat: Lawns, waste areas and roadsides; introduced from southern Europe. **General:** Stems 3–30 cm long, bright green; nodes 2–4, often rooting at lower nodes and forming large mats. **Leaves:** Blades flat, 1–14 cm long, 1–5 mm wide, surface with soft hairs; sheaths flattened, loose, hairless, light green, tips boat-shaped; ligules membranous, 2–5 mm long; auricles absent. **Flower Cluster:** Panicle open, pyramid-shaped, 2–12 cm long; spikelets 3–10 mm long, 3–10-flowered; glumes 1.5–3 mm long, lance-shaped, 1-veined; lemmas 2.5–4 mm long, 5-nerved. **Fruit:** Caryopsis straw-coloured, oval, 2–3 mm long, 1–1.3 mm wide.

Notes: The genus name *Poa* is the Greek word for grass. The species name *annua* refers to the annual life cycle of this plant, though annual bluegrass may complete its life cycle in as few as 6 weeks.

Little Bluestem

Schizachyrium scoparium var. *scoparium*

Habitat: Dry to moist prairies.
General: Stems 50–150 cm tall, tufted;
rhizomes absent. **Leaves:** Blades
15–30 cm long, 3–6 cm wide; sheaths
keeled; ligules membranous, 1–3 mm
long; auricles absent. **Flower Cluster:**
Raceme 3–6 cm long, with 5–20 pairs
of spikelets; sterile spikelets 1.5–7 mm
long, glumes with awn 7–14 mm long;
fertile spikelets 2-flowered (1 fertile),
6–8 mm long, first glume 5–10 mm long,
lemmas with awns 8–15 mm long. **Fruit:**
Caryopsis 1.5–4 mm long.

Notes: The genus name comes from the Greek word *schizo*,
meaning "to split apart." The species name comes from the Latin
word *scopa*, "broom," a reference to the flower cluster.

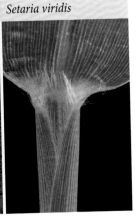

Green Foxtail

Setaria viridis

Habitat: Waste areas, roadsides and disturbed sites; introduced from Europe.
General: Stems 30–100 cm tall, tufted, surface with numerous upward-pointing
hairs, nodes 3–5, hairless; roots fibrous. **Leaves:** Blades flat, 3–25 cm long,
0.5–1.5 cm wide, usually light green; sheaths open, slightly flattened, margins overlapping and hairy; ligules a fringe of hairs 1.5–2 mm long; auricles absent. **Flower
Cluster:** Panicle spike-like, cylindric, 1–11 cm long, 0.5–2.3 cm wide, purplish
green; spikelets numerous, 1-flowered, 1.5–2 mm long, each with 1–3 bristles, the
bristles purplish green, 0.6–1.2 cm long; glumes 1–2 mm long, 1–3-veined; lemmas
1–2 mm long, 5–7-veined. **Fruit:** Caryopsis elliptic, pale green to greyish brown,
1.5–1.9 mm long, about 1 mm wide.

Notes: The species name *viridis* means "green."

Slough Grass, Tall Cord Grass · *Spartina pectinata*

Habitat: Shallow marshes, wet meadows and muddy shorelines. **General:** Stems 70–250 cm tall; rhizome scaly. **Leaves:** Blades 30–90 cm long, 0.3–1.5 cm wide, margins rough-toothed; sheaths green to yellowish green, finely ribbed, hairless; ligules a fringe of hairs 1–3 mm long; auricles absent. **Flower Cluster:** Panicle of 4–30 spikes, 1-sided, 1.5–15 cm long; spikes ascending, 3–10 cm long, 10–80-flowered; spikelets 8–17 mm long; first glume 5–10 mm long, awn 1–5 mm long; second glume 8–17 mm long, awn 2–10 mm long; lemmas 7–12 mm long, keeled, 3-nerved. **Fruit:** Caryopsis rarely produced.

Notes: The genus name comes from the Greek word *spartine*, "cord," a reference to the cord-like rhizomes. The species name *pectinata* means "comb-like," referring to the flower cluster.

Annual Wild Rice
Zizania aquatica

Habitat: Shallow water along lakes, rivers and marshes. **General:** Stems 90–300 cm tall, reed-like, spongy; roots fibrous, yellowish orange. **Leaves:** Blades 90–130 cm long, 1–4 cm wide, surface rough, margins toothed; sheaths open, hairless; ligules membranous, 0.5–2 cm long; auricles absent. **Flower Cluster:** Panicle open, 30–50 cm long, branches 15–20 cm long; male spikelets borne at base of panicle, 7–9 mm long, florets purple; female spikelets 1.8–2.2 cm long, terminal, florets purplish green; glumes absent; lemmas 1–3 cm long, awns 1–4 cm long. **Fruit:** Caryopsis oval, yellow to reddish becoming purplish black at maturity, 1.2–1.9 cm long, 0.8–1 mm wide.

Notes: The genus name comes from *zizanion*, the Greek word for wild grains. The species name *aquatica* means "growing in or near water." • This native grass was collected by First Nations peoples and early settlers. Today, wild rice is cultivated and is available at most supermarkets.

SEDGE FAMILY
Cyperaceae

The sedges are a large family of grass-like plants commonly found in moist to wet habitats. In several genera, the stems are 3-sided and solid. The leaves may be well developed or reduced to small scales. When present, the simple, basal or alternate leaves arise from closed sheaths and are 3-ranked. Leaf blades may be flat, oval or triangular. Flowers are produced in spikelets that form racemes, spikes, panicles or umbel-like heads. The flowers may be male, female or bisexual. Petals and sepals, when present, are often reduced to bristle-like hairs. Perfect flowers have 1–3 stamens and 1 pistil. The fruit is an achene.

Carex, the largest genus in the family, has over 1500 species. Members have unisexual flowers and triangular stems. Male and female flowers are borne in separate spikes. Male flowers consist of a scale-like bract and 3 stamens. Female flowers have a small bract with a hollow, sac-like structure called the perigynium. The ovary is contained in the perigynium, with the style extending through an opening at the top.

Economically important plants in this family include papyrus (*Cyperus papyrus*), used to make paper and mats, and Chinese water-chestnut (*Eleocharis dulcis*), grown for its edible tubers.

Water Sedge
Carex aquatilis

Habitat: Marshes, swamps and lakeshores. **General:** Stems 20–150 cm tall, triangular, reddish at the base; rootstock stolon-like. **Leaves:** Blades bluish green, 3–25 cm long, 2–5 mm wide, margins flat to rolled. **Flower Cluster:** Spikes male or female; lowest bract longer than flower cluster; male spikes 1–3, terminal, 1.5–3 cm long; female spikes 2–6, oblong, 1–4 cm long; perigynia numerous, flask-shaped, 2–4 mm long, short-beaked, scales purplish black. **Fruit:** Achene lens-shaped, 2–4 mm long.

Notes: The species name *aquatilis* means "growing in or near water" and refers to this plant's habitat.

Golden Sedge
Carex aurea

Habitat: Moist, open areas. **General:** Stems 5–30 cm tall, triangular, loosely tufted; stolons present. **Leaves:** Blades 3–20 cm long, 1–4 mm wide, exceeding flower cluster. **Flower Cluster:** Panicle 5–10 cm long; male spikes terminal, 3–10 mm long; female spikes 3–5, lower in the cluster, 5–15 mm long; perigynia ovoid, 2–3 mm long, beak-less, scales pale reddish brown and toothed; perigynia turning yellowish orange and somewhat fleshy at maturity. **Fruit:** Achene rounded, 1–2 mm long.

Notes: The species name *aurea* means "golden," a reference to the colour of the mature fruit.

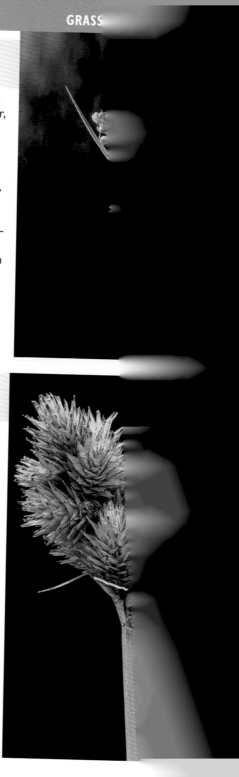

Bebb's Sedge
Carex bebbii

Habitat: Wetlands and marshy meadows. **General:** Stems 20–80 cm tall, triangular, tufted, rough-textured below flower cluster. **Leaves:** Blades flat, light green to yellowish green, 3–4 per stem, 11–25 cm long, 2–4 mm wide. **Flower Cluster:** Panicle head-like 1–3 cm long, 0.5–1.4 cm wide, brown, with 3–12 spikes; spikes 0.4–1 cm long, female spikes above male, lower spikes with green, bristle-like bract at the base; perigynia brownish, 3–3.5 mm long, more than 1 mm wide, scales reddish brown. **Fruit:** Achene ovate to elliptic, 1–1.5 mm long.

Notes: The species name commemorates Michael Bebb (1833–95), an American naturalist who specialized in willows.

Brownish Sedge
Carex brunnescens

Habitat: Wet woods and bogs.
General: Stems 10–90 cm tall,
triangular, thread-like. **Leaves:**
Blades flat or channelled, green
to yellowish green, 10–25 cm long,
1–3 mm wide. **Flower Cluster:**
Panicle 3–7 cm long with 5–10
globe-shaped spikes bract bristle-
like, located below lowest spike;
spikes 0.5–2.5 cm apart, 5–10
female flowers above several male
flowers; perigynia yellow to brownish, 2–2.5 mm long; scales silvery
brown, midrib green. **Fruit:** Achene
oval to round, 1–1.5 mm across.

Notes: The species name
brunnescens means "deep brown"
and refers to the colour of the
mature spikes.

Crawe's Sedge
Carex crawei

Habitat: Dry to moist, open areas
on limestone pavements. **General:**
Stems 2–40 cm tall; rhizomes long,
creeping. **Leaves:** Blades light
green, often recurved, 0.6–9.7 cm
long, 1.5–4.4 mm wide. **Flower
Cluster:** Spikes terminal, on flowering stalk 0.9–9.1 cm long; bracts
shorter than stem, longest bract
2–6.7 mm long; spikes 0.5–2.7 cm
long, 3–6.3 mm wide, terminal
spike 1.1–2.4 cm long; perigynia
yellowish green, 2.2–3.4 mm long,
beak less than 0.3 mm long; scales
1.2–2.9 mm long. **Fruit:** Achene
1.4–1.9 mm long.

Notes: The species name commemorates Ithamar Bingham Crawe
(1792–1847), an American botanist
who discovered this species in the
Lake Champlain area in 1846.

Fringed Sedge
Carex crinita

Habitat: Swamps, marshes and wet meadows. **General:** Stems 40–135 cm tall, strongly triangular, tufted, rough-textured. **Leaves:** Blades 14–50 cm long, 0.3–1.1 cm wide; sheaths reddish brown to brown **Flower Cluster:** Panicle of spikes; lowest bract longer than the flower cluster, 0.3–1.1 cm long; male spikes 1–3, erect, 1–5 cm long; female spikes 2–5, pendent, 3.5–11.5 cm long, 4–6.8 mm wide, on stalks 1.4–6.9 cm long; perigynia spreading, inflated, 1.8–4 mm long; scales pale to copper brown, 3.4–11 mm long, awn 1–10 mm long. **Fruit:** Achene oblong to oval, 2–4 mm long.

Notes: The species name *crinita* means "with long hairs." • The seeds of this sedge are often eaten by waterfowl.

Bristle-leaf Sedge
Carex eburnea

Habitat: Coniferous to mixed-wood forests. **General:** Stems 7–31 cm tall, tufted. **Leaves:** Blades bristle-like, 3–21 cm long, 0.2–1 mm wide. **Flower Cluster:** Panicle of spikes; bracts 2.5–8 mm long; male spikes often inconspicuous, 3–10 mm long, 0.5–1.5 mm wide, scales 2.6–4 mm long; female spikes 3–7 mm long, 1.5–6 mm wide; perigynia light green to dark brown, 1.5–2.2 mm long, beaked; scales white, papery, midvein green or brown, 1–2 mm long. **Fruit:** Achene, elliptic to oblong.

Notes: The species name *eburnea* means "of or pertaining to elephants" or "made with ivory."

Graceful Sedge
Carex gracillima

Habitat: Moist to dry, deciduous and mixedwood forests. **General:** Stems 20–90 cm tall, often dark maroon at the base. **Leaves:** Basal leaves bladeless, often maroon; stem leaves flat, 3–9 mm wide. **Flower Cluster:** Panicle of spikes; bracts leaf-like with blades 3–4 mm wide; terminal spike 1–6 cm long, male flowers near the base, female flowers at tip; lower 3–5 spikes female, cylindric, 3–6 cm long, nodding onstalks 1–4 cm long; perigynia green, often red-dotted, 8–12-veined, 2–3.7 mm long; scales with papery white margins, short-awned. **Fruit:** Achene 2–4 mm long.

Notes: The species name *gracillima* means "graceful."

Porcupine Sedge
Carex hystericina

Habitat: Swamps and wet meadows. **General:** Stems 20–100 cm tall, triangular, tufted, rough on angles; rhizomes short. **Leaves:** Basal, blades pale green, flat to W-shaped, 2.5–8.5 mm wide; lower sheaths reddish purple. **Flower Cluster:** Panicle 2.5–20 cm long; lowest bract 4–30 cm long; male spikes terminal; female spikes 2–4, usually erect; perigynia spreading, strongly veined, 4.5–7.3 mm long; scales 2.3–6.5 mm long. **Fruit:** Achene pale brown.

Notes: The species name *hystericina* means "like a porcupine," a reference to the spreading perigynia.

Bladder Sedge, Inflated Sedge

Carex intumescens

Habitat: Dry to moist forests and wet meadows. **General:** Stems 30–80 cm tall, occasionally tufted; rhizome short. **Leaves:** Blades alternate, 6–12 per stem, 8–27 cm long, 3.5–8 mm wide; basal sheaths purplish red; ligules rounded, 1–8 mm long. **Flower Cluster:** Cluster of spikes 2–15 cm long; bracts leafy, 6–21 mm long, 2–6 mm wide; male spikes terminal, 1–5 cm long; female spikes 1–4, 1–2.7 cm long, 1–12-flowered; perigynia reflexed, strongly veined, 10–16.5 mm long; scales 4–9.5 mm long, awns to 6.5 mm long. **Fruit:** Achene elliptic, 3.5–5.7 mm long.

Notes: The species name *intumescens* means "swollen."

Hairy-fruited Sedge, Wiregrass

Carex lasiocarpa

Habitat: Bogs, marshes and wet meadows. **General:** Stems 40–120 cm tall, triangular, dark purplish red at the base; rhizomes long-creeping, forming large colonies. **Leaves:** Blades very long, 1–2 mm wide, wiry, greyish green, somewhat V-shaped, tips thread-like; basal sheaths reddish purple. **Flower Cluster:** Panicle 6–20 cm long; lowest bracts overtopping the panicle; male spikes terminal, 1–3, 2–6 cm long; female spikes 1–3, cylindric, 1–4 cm long; perigynia 3–4.5 mm long, densely haired; scales reddish brown to purple, lance-shaped, sharp-tipped. **Fruit:** Achene 3–5 mm long.

Notes: The species name *lasiocarpa* means "woolly fruited."

Hop Sedge
Carex lupulina

Habitat: Swamps and wet meadows.
General: Stems 20–130 cm tall, triangular, loosely tufted; rhizome short.
Leaves: Alternate, 4–8 per stem, blades 15–64 cm long, 0.4–1.5 cm wide, base reddish brown; ligules triangular, 3.5–18 mm long. **Flower Cluster:** Panicle 4–40 cm long; bracts leaf-like, 13–55 cm long, 0.3–1.1 cm wide; male spikes terminal, 1.5–8.5 cm long; female spikes 2–5, 4–80-flowered, 1.5–6.5 cm long, 1.3–3 cm wide; stalks of lower spikes 0.5–20 cm long; perigynia strongly veined, 1.1–1.9 cm long, shiny; scales 0.6–1.5 cm long with awns 3–6 mm long. **Fruit:** Achene 3–4 mm long.

Notes: The species name *lupulina* means "hops-like."

Early Flowering Sedge, Long-stalked Sedge
Carex pedunculata

Habitat: Moist to dry woodland openings. **General:** Stems 9–28 cm tall, triangular, tufted; rhizome short.
Leaves: Basal, blades dark green, 5–25 cm long, 1.4–4 mm wide; sheaths reddish to purplish brown at the base
Flower Cluster: Panicle of spikes; male spikes terminal, 7.5–9.8 mm long, often with 2–5 female flowers at the base; female spikes 2–5, often with 1–2 male flowers at the apex; stalks of lower spikes to 13 cm long, upper spikes with stalks 2–6 cm long; perigynia 3.7–6 mm long; scales tan to dark reddish brown. **Fruit:** Achene 2–3 mm long.

Notes: The species name *pedunculata* means "with a distinct stalk," a reference to the flower stems.

Pennsylvanian Sedge
Carex pensylvanica

Habitat: Dry, sandy to gravelly soils under hardwood trees. General: Stems 10–45 cm tall, triangular, tufted, reddish brown, fibrous remnants of old leaves at the base; rhizomes stolon-like. Leaves: Blades stiff, erect, 5–20 cm long, 1–3.6 mm wide; somewhat rough on both surfaces. Flower Cluster: Panicle 1.5–5 cm long; bract leaf-like, below lowest spike, tinged red or brown near the base; male spikes terminal, club-shaped, 0.5–2.4 cm long; female spikes 1–3, 0.5–1 cm long; perigynia globe-shaped, 3–4 mm long; scales green with white margins. Fruit: Achene dark brown, oblong to ovoid, 1.5–2.5 mm long.

Notes: The species name means "of Pennsylvania," the location where this species was first documented.

Plantain-leaved Sedge
Carex plantaginea

Habitat: Rich deciduous to mixedwood forests. General: Stems 24–25 cm tall, triangular, tufted. Leaves: Basal, evergreen, blades 14–42 cm long, 0.8–3.2 cm wide, prominently 3-veined; sheaths purple at the base; older leaves shrivelled or dead at tips. Flower Cluster: Panicle of 3–5 spikes; bracts 0.8–2 cm long; male spikes 1, 0.8–2 cm long, 2–3.5 mm wide; female spikes 0.8–3 cm long, 4–7 mm wide, 4–15-flowered; perigynia 3.7–4.9 mm long; scales 3–6 mm long, midrib green to purple. Fruit: Achene oval, 2.2–2.7 mm long.

Notes: The species name *plantaginea* refers to the leaves, which resemble those of plantains (*Plantago* spp.).

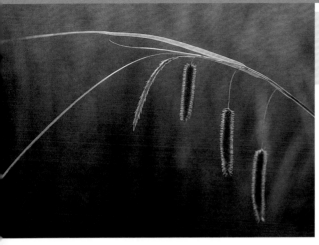

Cyperus-like Sedge
Carex pseudocyperus

Habitat: Swamps, marshes and ditches. **General:** Stems 30–100 cm tall, triangular, tufted, surface rough, often forming large clumps; rhizomes less than 10 cm long. **Leaves:** Blades 10–45 cm long, 0.4–1.3 cm wide; sheaths pale brown; ligules conspicuous. **Flower Cluster:** Panicle 10–20 cm long; bracts 12–55 cm long, longer than flower cluster; male spikes terminal, 2–5 cm long; female spikes 3–7, 3–10 cm long, on long, drooping stalks; perigynia 5–6 mm long, strongly veined; scales greenish red, 2–9 mm long, awns 3–4 mm long. **Fruit:** Achene triangular, 1–2 mm long.

Notes: The species name *pseudocyperus* means "false *Cyperus*," a reference to this plant's resemblance to species in the genus *Cyperus*.

Retrorse Sedge
Carex retrorsa

Habitat: Swamps, marshes and wet meadows. **General:** Stems 10–105 cm tall, triangular, tufted; rhizome very short or absent. **Leaves:** Blades flat to W-shaped, 0.3–1 cm wide; sheaths dark reddish brown. **Flower Cluster:** Panicle of spikes, 3–35 cm long; lowest bract 19–70 cm long, much longer than flower cluster; male spikes 1–3, terminal, slightly above female spikes; female spikes 2–6, cylindric, 20–150-flowered; perigynia 6–13-nerved, 0.6–1 cm long; scales 2.4–4.5 mm long. **Fruit:** Achene pale brown.

Notes: The species name *retrorsa* means "bent backward or downward."

Richardson's Sedge
Carex richardsonii

Habitat: Open woodlands, alvars and rocky outcrops. **General:** Stems 14–32 cm tall; basal sheaths dark purplish red to reddish brown; rhizomes long. **Leaves:** Basal, pale green, 10–20 cm long, 1–3.3 mm wide, thick, stiff; turning dark red in autumn. **Flower Cluster:** Spikes on stalks to 7 cm long; bracts long-sheathed; male spikes terminal, 1.7–2.3 cm long; female spikes 2–4, from stem nodes; perigynia oblong to ovoid, 2.4–2.9 mm long, with white hairs; scales reddish brown. **Fruit:** Achene elliptic, about 1.8 mm long.

Notes: The species name commemorates Sir John Richardson (1787–1865), a Scottish naturalist who collected numerous arctic plant species.

Stellate Sedge
Carex rosea

Habitat: Dry to moist, deciduous forests. **General:** Stems 20–90 cm tall, triangular; rhizomes very short. **Leaves:** Blades flat, green, 20–50 cm long, 1–3 mm wide; sheaths tight. **Flower Cluster:** Panicle 2–7 cm long, 5–8 mm wide; bracts 2–10 cm long; spikes 4–8, each 7–14-flowered; male flowers borne above female flowers; perigynia green, 2.6–4 mm long, spongy at the base; scales papery, oval to round, 1.5–2 mm long. **Fruit:** Achene oval to oblong, 1.6–2.2 mm long.

Notes: The species name *rosea* means "rosy."

Awned Sedge
Carex rostrata

Habitat: Marshes, swamps and ditches. **General:** Stems 10–100 cm tall, bluntly triangular, spongy at the base; rhizome long. **Leaves:** Blades pale green, 20–50 cm long, 2–8 mm wide, 4–10 per stem; sheaths brown, tinged pink or red. **Flower Cluster:** Panicle 10–30 cm long; lowest bract 18–45 cm long; male spikes 2–4, terminal, 1.5–5 cm long; female spikes 2–4, 4–10 cm long, 20–150-flowered; perigynia light brown, 4–8 mm long, prominently beaked, in 8–12 rows; scales purplish brown, 2.5–4.5 mm long. **Fruit:** Achene brown, 2–3 mm long.

Notes: The species name *rostrata* comes from a Latin word meaning "beak," a reference to the prominent beak of the perigynia.

Broom Sedge
Carex scoparia

Habitat: Dry to wet, open areas. **General:** Stems 20–100 cm tall, tufted; vegetative stems with leaves clustered at the tip. **Leaves:** Alternate, 3–5 per flowering stem, 10–32 cm long, 1.4–3.5 mm wide. **Flower Cluster:** Spike cluster dense to open, 1.5–6 cm long; lowest bract with bristle tip; spikes 3–10, 7–16 mm long; perigynia pale green or golden brown, 4–6.8 mm long; scales brownish with a green or gold midvein, 3–4 mm long. **Fruit:** Achene oval to elliptic, 1.3–1.7 mm long.

Notes: The species name *scoparia* means "broom-like."

Awl-fruited Sedge
Carex stipata

Habitat: Wet meadows and shore-lines. **General:** Stems 40–100 cm tall, triangular, yellowish green, winged on angles. **Leaves:** Blades arched, 10–60 cm long, 0.4–1 cm across, longer than flower cluster, often with a puckered band; sheaths loose-fitting, hairless; ligules 0.8–1.2 cm long. **Flower Cluster:** Panicle 3–10 cm long, oval with 5–15 spikes; bract bristle-like, absent or to 5 cm long, below lowest spikelet; male flowers borne above female flowers in the same spike; perigynia tapered, yellowish green, 4–6 mm long, beak toothed; scales pale green. **Fruit:** Achene oval, 1–3 mm long.

Notes: The species name *stipata* means "stalked," a reference to the short stalk of the perigynia.

Straw Sedge
Carex tenera

Habitat: Dry to moist, open woodlands and meadows. **General:** Stems 20–90 cm tall, often nodding, triangular, tufted. **Leaves:** Blades 3–5 per stem, 15–35 cm long, 1.3–2.5 mm wide; ligules 1–2 mm long. **Flower Cluster:** Panicle of spikes, open, often nodding, separated; lower bracts scale-like to 4 cm long; spikes 3–8, loosely clustered, 2.5–5 cm long, 0.7–1 cm wide; perigynia brown, 5–7-nerved, 2.8–4 mm long, winged; scales whitish to pale brown, 2.3–3.3 mm long. **Fruit:** Achene oval, 1.3–1.7 mm long.

Notes: The species name *tenera* means "tender."

Green Sedge
Carex viridula

Habitat: Wet meadows and boggy shorelines. **General:** Stems 10–85 cm tall, triangular, yellowish green, tufted, fibrous at the base. **Leaves:** Blades erect, yellowish green, 5–30 cm long, 1–6 mm wide, often exceeding flowering stem; ligules absent on upper leaves. **Flower Cluster:** Panicle 1–5 cm long; lowest bract leaf-like, 2–18 cm long, often exceeding flower cluster; male spikes terminal, 0.5–2.5 cm long; female spikes 2–8, 0.5–1.8 cm long; perigynia yellowish green to brown, 2–4 mm long; scales pale brown to reddish brown with papery margins, 1.4–3.2 mm long. **Fruit:** Achene 2–3 mm long.

Notes: The species name *viridula* means "greenish," a reference to the yellowish green colour of the plant.

Yellow-fruited Sedge, Fox Sedge
Carex vulpinoidea

Habitat: Wet meadows, marshes and roadside ditches. **General:** Stems 20–100 cm tall, triangular, surface rough. **Leaves:** Blades 20–120 cm long, 1–5 mm wide; sheaths spotted red or pale brown; ligules 1–2 mm long. **Flower Cluster:** Panicle spike-like, 7–10 cm long, 0.5–1.5 cm wide, consisting of numerous compact spikelets; bracts bristle-like; spikelets with a few male flowers above and several female flowers below; perigynia green to pale brown, 2–3.2 mm long, long-beaked; scales pale brown, awns 1–3 mm long. **Fruit:** Achene reddish brown, 1.2–1.4 mm long.

Notes: The species name comes from the Latin word *vulpes*, meaning "fox."

Shining Flatsedge
Cyperus bipartitus

Habitat: Emergent shorelines and muddy, disturbed sites. **General:** Stems 3–25 cm tall, triangular, tufted, hairless. **Leaves:** Alternate, blades 1–3 per stem, V-shaped, 1–8 cm long, 1–2 mm wide; sheaths closed; ligules absent. **Flower Cluster:** Spikes 1, flattened, 0.7–1.4 cm long, 0.9–1.4 cm wide; bracts 2–3, V-shaped to flat, 1–12 cm long, 0.5–1.5 mm wide; rays 1–4, 0.5–3 cm long; spikelets 3–5, 0.8–1.8 cm long; scales 10–26, light to dark brown, 1.9–2.7 mm long. **Flowers:** Greenish brown, less than 2 mm across; petals and sepals absent; stamens 3; pistil 1. **Fruit:** Achene black, 1–1.3 mm long.

Notes: The species name *bipartitus* means "2-parted."

Yellow Nut Sedge, Chufa
Cyperus esculentus

Habitat: Moist, open areas, disturbed sites and landscapes; native species with a weedy nature. **General:** Stems 15–90 cm tall, triangular, often reddish brown at the base; roots creeping, producing tubers at root tips; tubers black, 0.8–1.9 cm long, woody. **Leaves:** Alternate, 3–7 per stem, 3-ranked, flat to V-shaped, 20–90 cm long, 4–9 mm wide; sheaths closed, triangular. **Flower Cluster:** Panicle of spikes, umbrella-shaped; bracts 3–9, leaf-like, 5–30 cm long, 0.5–4 mm wide; rays 4–10, 2–12 cm long; spikelets 10–20, 1–3 cm long, 1.2–3.5 cm wide, many-flowered; scales 6–34, 1.8–2.7 mm long. **Flowers:** Yellowish brown, less than 2 mm across; petals and sepals absent; stamens 3; pistil 1. **Fruit:** Achene oval, 3-sided, whitish brown, 1.2–1.8 mm long, seldom produced.

Notes: The species name *esculentus* means "edible," a reference to the tubers. • This species is designated a noxious weed throughout most of eastern North America. Its range is limited by winter soil temperatures; the tubers die at temperatures lower than −6°C.

Fragrant Galingale, Rusty Flatsedge
Cyperus odoratus

Habitat: Disturbed, muddy places and emergent shorelines. **General:** Stems 10–50 cm tall, triangular. **Leaves:** Alternate, 3-ranked, blades V- or W-shaped, 5–30 cm long, 0.4–1.2 cm wide; sheaths closed. **Flower Cluster:** Panicle of spikes, umbrella-shaped; bracts 4–10, horizontal to ascending, 10–55 cm long, 1–14 mm wide; rays 2–12, 2–13 cm long; spikelets 20–60, oblong to linear, 0.5–1.5 cm long; scales 8–12, midrib green, reddish brown laterally, 2.2–2.8 mm long. **Flowers:** Yellowish green, less than 2 mm across; petals and sepals absent; stamens 1–3; pistil 1; stigmas 2–3. **Fruit:** Achene reddish brown to black, 1.2–1.5 mm long, triangular.

Notes: The species name *odoratus* means "fragrant" or "scented."

Three-way Sedge
Dulichium arundinaceum

Habitat: Shallow water along shorelines. **General:** Stems 20–100 cm tall, oval in cross-section, unbranched, jointed; internodes 2–5 cm long, hollow; rhizomes 2–5 mm thick. **Leaves:** Alternate, 3–15 per stem, 4–15 cm long, 3-ranked; lower sheaths 2–14, bladeless. **Flower Cluster:** Cluster 6–30 cm long; spikes 1–17, each 2–5 cm long, composed of 3–10 spikelets; bracts 5–20, leaf-like; spikelets 4–8-flowered. **Flowers:** Green, 1.4–2.4 mm across; perianth bristles 6–9, 4–7.5 mm long; stamens 3; pistil 1, style 1. **Fruit:** Achene stalked, 2–4 mm long.

Notes: The genus name *Dulichium* is the Latin name for a kind of unknown sedge. The species name *arundinaceum* means "reed-like."

Needle Spikerush
Eleocharis acicularis

Habitat: Emergent shorelines and muddy, disturbed sites. **General:** Stems 1–60 cm tall, oval in cross-section; rhizomes creeping. **Leaves:** Reduced to bladeless sheaths, reddish green to brown, near the stem base. **Flower Cluster:** Spike oval to lance-shaped, 2–8 mm long, 1–2 mm wide; scales 4–25, reddish to purplish brown, 1.5–2.5 mm long. **Flowers:** Pale brown; petals and sepals reduced to 0–4 whitish brown bristles; stamens 3; pistil 1. **Fruit:** Achene angular, 0.7–1.1 mm long.

Notes: The species name *acicularis* means "needle-like," a reference to the stem. • The rhizomes often produce large mats, which often become detached and form floating masses.

Marsh Spikerush
Eleocharis palustris

Habitat: Shallow water, lakeshores and mudflats. **General:** Stems 10–100 cm tall, oval to somewhat flattened, often spongy; rhizomes creeping, reddish brown. **Leaves:** Reduced to bladeless sheaths, reddish green to brown, near the stem base. **Flower Cluster:** Spike oval to lance-shaped, 0.5–2.5 cm long, 3–7 mm wide; scales 1–3, reddish brown, 3–4 mm long, below the spike. **Flowers:** Brown, 2–3 mm across; petals and sepals consisting of 4 bristles; stamens 1–3; pistil 1. **Fruit:** Achene lens-shaped, 1–1.5 mm long.

Notes: The genus name comes from the Greek words *helos* and *charis*, meaning "marsh grace." The species name *palustris* means "of the marsh."

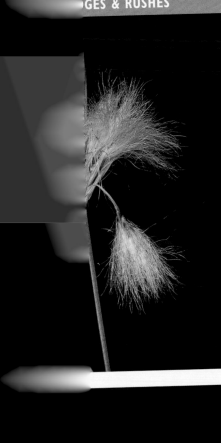

Slender Cottongrass
Eriophorum gracile

Habitat: Bogs, muskeg and swamps.
General: Stems 20–40 cm tall,
tufted. **Leaves:** Alternate, lower
stem leaves channelled, 1–2 mm
wide, tips rounded, upper leaves
reduced; basal leaves similar to lower
stem leaves, often absent at time of
flowering. **Flower Cluster:** Panicle
of 2–5 spreading or nodding spikelets, each spikelet 0.7–1 cm long;
bract greyish black, about 1 cm long.
Flowers: Green, 2–3 mm across;
petals 3; sepals 3; petals and sepals
divided into 4–6 white, silky bristles,
each 1–1.5 cm long; stamens 1–3;
pistil 1. **Fruit:** Achene 1–3 mm long
with numerous white bristles.

Notes: The common name refers
to the cottony, white hairs of the
flowering and fruiting clusters.

White Beak Sedge
Rhynchospora alba

Habitat: Acidic bogs, fens, floating
peat mats and rocky shorelines.
General: Stems 6–75 cm tall,
tufted, triangular to oval in cross-
section, leafy. **Leaves:** Alternate,
linear to thread-like, 0.5–1.5 mm
wide. **Flower Cluster:** Spikes
head-like, terminal, 1.5–2.5 cm
wide; bracts 1–6, white; spikelets
3.5–5.5 mm long; flowers in scale
axils; scales 3–4 mm long, abruptly
pointed. **Flowers:** White to green;
perianth of 10–12 bristles; stamens 3;
pistil 1. **Fruit:** Achene, often 1 per
spikelet, 2.5–3 mm long, beaked.

Notes: The genus name comes
from the Greek words *rhynchos*,
"snout," and *spora*, "seed," a reference to the beak-like tip on the
achene.

Chairmaker's Bulrush
Schoenoplectus americanus

Habitat: Marshes and swamps. **General:**
Stems 40–250 cm tall, triangular, sides
deeply concave; rhizomes 2–5 mm wide.
Leaves: Basal, 1–3 per stem, 20–125 cm
long, 2–8 mm wide, V-shaped near the
base, flattened to triangular elsewhere.
Flower Cluster: Panicle head-like; lowest
bract 1–6 cm long, resembling a leaf; spikelets 2–20, 5–15 mm long, 3–5 mm wide;
scales orange to reddish or purplish brown,
2.7–4 mm long, often awn-tipped. **Flowers:** Yellowish brown; petals and sepals
5–6, reduced to bristles; stamens 3; pistil 1.
Fruit: Achene brown, 1.8–2.8 mm long,
flattened to triangular in cross-section.

Notes: The genus name comes from the
Greek words *schoinos*, "rush" or "reed," and
plectos, "twisted" or "woven," a reference
to the use of this plant in making baskets
and other woven items.

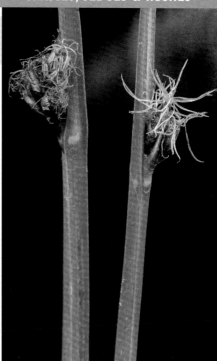

Common Great Bulrush, Softstem Bulrush
Schoenoplectus tabernaemontani

Habitat: Shallow water along lakes and ponds.
General: Semi-aquatic; stems 0.5–3 m tall,
spongy, pale green, circular in cross-section;
rhizomes creeping, 0.3–1 cm thick, edible.
Leaves: Basal, 3–4 per stem, reduced to bladeless sheaths, 2–200 cm long, 1–4 mm wide.
Flower Cluster: Panicle 2–4 times branched,
spreading to drooping, 5–25 cm long, branches
1–6 cm long; bracts C-shaped, 1–7 cm long;
spikelets 15–200, brown, 3–17 mm long,
2.5–4 mm wide, in dense clusters of 2–7; scales
dark orangey brown, 2–3.5 mm long. **Flowers:**
Brown, 1–3 mm across; petals and sepals
reduced to 6 bristles; stamens 2–3; pistil 1.
Fruit: Achene greyish brown, 1.5–3 mm long.

Notes: The species name commemorates Jacobus Theodorus Tabernaemontanus
(1525–97), a German physician and botanical artist. Often called the "father of
German botany," he is best known for his 1588 published work *Neuwe Kreuterbuch*,
which contains over 2300 woodcut illustrations.

Dark Green Bulrush, Black Bulrush
Scirpus atrovirens

Habitat: Moist meadows and marshes. **General:** Stems 40–150 cm tall, tufted; rhizomes short, fibrous. **Leaves:** Alternate, 6–11 per stem, blades 20–54 cm long, 0.7–1.7 cm wide; sheaths light brown. **Flower Cluster:** Panicle of spikelets; rays spreading, often bearing bulblets; bracts green, margins speckled reddish brown; spikelets 2–5 mm long, in clusters of 4–110; scales dark brown, 1.2–2.1 mm long. **Flowers:** Dark green; petals and sepals reduced to 5–6 bristles; stamens 1–3; pistil 1. **Fruit:** Achene light brown, 1–1.5 mm long, flattened to triangular in cross-section.

Notes: The species name *atrovirens* means "dark green," a reference to the dark-coloured spikelets.

Woolgrass Bulrush
Scirpus cyperinus

Habitat: Marshes and moist meadows. **General:** Stems 20–120 cm tall, forming tussocks; rhizomes short, branching, fibrous. **Leaves:** Alternate, 5–10 per stem; blades 22–80 cm long, 0.3–1 cm wide; sheaths green to reddish brown. **Flower Cluster:** Panicle of 2–15 cymes; rays ascending to spreading; bracts reddish brown to black at the base; spikelets oval, 3.5–8 mm long, 2.5–3.5 mm wide; scales reddish brown to black, 1–2.2 mm long. **Flowers:** Greenish brown; petals and sepals reduced to 6 bristles; stamens 1–3; pistil 1. **Fruit:** Achene white to pale brown, less than 1 mm long.

Notes: The species name means "like *Cyperus*," a reference to the flower cluster resembling those of *Cyperus* species.

Small-fruited Bulrush

Scirpus microcarpus

Habitat: Marshes, swamps and ditches. **General:** Stems 30–100 cm tall, leafy, triangular; rhizomes thick, reddish. **Leaves:** Alternate, 4–11 per stem; blades 23–60 cm long, 0.5–2 cm wide, margins sharp and cutting; sheaths often red-tinged. **Flower Cluster:** Panicle umbel-like, 5–20 cm long; rays spreading to drooping, 3–15 cm long; bracts 3–4, leaf-like; spikelets greenish brown, 2–8 mm long, in dense clusters of 3–18; scales green to black, 1–3.5 mm long. **Flowers:** Greenish brown, 1–2 mm across; petals and sepals reduced to 3–6 bristles; stamens 2–3; pistil 1. **Fruit:** Achene whitish brown, 1–1.5 mm long.

Notes: The species name *microcarpus* means "small-fruited."

RUSH FAMILY · Juncaceae

The rush family is a small family of rhizomatous, grass-like plants commonly found in moist to wet sites. The linear leaves are primarily basal, but a few alternate stem leaves may be present. Occasionally, the leaves are reduced to bladeless sheaths. The sheaths are usually open. Flowers are borne in terminal or axillary heads, panicles or corymbs. These non-showy, green or purplish brown flowers have 3 petals, 3 sepals, 3 or 6 stamens and 1 pistil. The petals and sepals are similar in shape, size and appearance, and appear in 2 distinct whorls.

A few species of *Juncus* and *Luzula* are grown for their ornamental value.

473

Wire Rush, Arctic Rush
Juncus arcticus var. *balticus*

Habitat: Wet meadows, low-lying areas and roadside ditches. **General:** Stems 20–60 cm tall, dark green, round in cross-section; rhizomes long, creeping. **Leaves:** Basal, reduced to brown, bladeless sheaths, 8–15 cm long, loose-fitting; auricles absent. **Flower Cluster:** Cyme open to head-like, 2–4 cm long, 6 to many-flowered, on side of stem. **Flowers:** Purplish to chestnut brown, 4–7 mm across; petals 3; sepals 3; stamens 6, yellow; pistil 1, style pink. **Fruit:** Capsule 4–5 mm long.

Notes: The species name means "of the arctic." The variety name *balticus* refers to the Baltic region of northern Europe. • This species is found throughout the Northern Hemisphere.

Toad Rush
Juncus bufonius

Habitat: Moist meadows, lakeshores and roadside ditches; often weedy. **General:** Stems 5–40 cm tall, tufted. **Leaves:** Basal and alternate, blades 3–13 cm long, less than 1 mm wide; auricles very small or absent. **Flower Cluster:** Cyme open, 2–20 cm long; lowest bract shorter than flower cluster. **Flowers:** Greenish, 4–7 mm across; petals 3; sepals 3; stamens 3–6; pistil 1. **Fruit:** Capsule tan to reddish brown, 2.7–4 mm long; seeds yellowish brown.

Notes: The species name comes from the Greek word *bufo*, " toad," a reference to this plant's moist habitat.

Soft Rush
Juncus effusus

Habitat: Swamps, marshes and wet meadows. **General:** Stems 40–130 cm tall, oval in cross-section; rhizomes short, branched, forming dense clumps. **Leaves:** Blades absent. **Flower Cluster:** Panicle, many-flowered, on side of stem; bracts erect, longer than flower cluster. **Flowers:** Brown, 1.9–3.5 mm across; petals 3; sepals 3; stamens 3; pistil 1. **Fruit:** Capsule greenish brown, 1.5–3.2 mm long.

Notes: The species name *effusus* means "scattered," "spread out" or "loose," a reference to the open flower cluster.

Blackgrass Rush
Juncus gerardii

Habitat: Wet meadows and marshes. **General:** Stems 20–90 cm tall; rhizomes creeping. **Leaves:** Basal, 2–4, blades flat, 10–40 cm long, less than 1 mm wide; auricles less than 1 mm long.

Flower Cluster: Panicle 10–80-flowered, 2–16 cm long; primary bract shorter than flower cluster. **Flowers:** Dark brown, 2.6–3.2 mm across; sepals 3; petals 3; stamens 6; pistil 1. **Fruit:** Capsule chestnut brown, 2.5–3.2 mm long; seeds dark brown.

Notes: The species name commemorates Louis Gerard (1733–1819), the French botanist who first described this species.

Knotted Rush
Juncus nodosus

Habitat: Wet, muddy or sandy shorelines, swamps and wet meadows. **General:** Stems 15–60 cm tall, oval in cross-section; rhizomes with swollen nodes. **Leaves:** Basal leaves 2; stem leaves alternate, 2–4, green to pinkish, blades oval in cross-section, 6–30 cm long, 1–2 mm wide; auricles less than 2 mm long. **Flower Cluster:** Clusters of heads, 3–15 per stem, 1–3 cm across; heads globe-shaped, 0.6–1.2 cm across, 6–30-flowered; bracts erect. **Flowers:** Greenish brown, 3–6 mm across; petals 3; sepals 3; stamens 3 or 6; pistil 1. **Fruit:** Capsule chestnut brown, 3.2–5 mm long.

Notes: The species name *nodosus* means "jointed" or "with knotty nodes," a reference to the swollen nodes of the rhizomes.

Path Rush
Juncus tenuis

Habitat: Exposed, sandy or clayey, disturbed sites, often along animal or human trails. **General:** Stems 15–50 cm tall, tufted, 2–20 per plant; rhizomes branched. **Leaves:** Basal, 1–3, blades flat, 3–12 cm long, 1–2 mm wide; auricles 2–5 mm long, membranous. **Flower Cluster:** Panicle 1–5 cm long, 5–40-flowered; lowest bract longer than flower cluster. **Flowers:** Green, 2–4 mm across; petals 3; sepals 3; stamens 6; pistil 1. **Fruit:** Capsule light brown, 3.8–4.7 mm long.

Notes: The species name *tenuis* means "slender" or "thin," a reference to the stems.

Torrey's Rush

Juncus torreyi

Habitat: Swamps, sandy shorelines and wet meadows. **General:** Stems 40–100 cm tall, 3–5 mm wide, oval in cross-section; rhizomes 1–3 mm diameter, nodes swollen. **Leaves:** Basal leaves 1–3; stem leaves alternate, 2–5, blades reddish green to pinkish, 13–30 cm long, 1–5 mm wide; auricles 1–4 mm long. **Flower Cluster:** Cluster of heads, 1–23 per stem, 2–5.5 cm across; heads globe-shaped, 1–1.5 cm across, 25–100-flowered; bracts bristle-like, erect. **Flowers:** Green to straw-coloured or reddish; petals 3; sepals 3; stamens 6; pistil 1. **Fruit:** Capsule pale to chestnut brown, 4.3–5.7 mm long.

Notes: The species name commemorates American botanist John Torrey (1796–1873).

Small-flowered Woodrush

Luzula parviflora

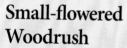

Habitat: Moist, open forests and marshes. **General:** Stems 30–60 cm tall; rootstock spreading. **Leaves:** Grass-like, primarily basal; stem leaves 3–6, alternate, 12–17 cm long, 0.5–1 cm wide; sheaths with soft hairs at throat. **Flower Cluster:** Cyme 5–12 cm long, branches drooping; bracts to 5 cm long. **Flowers:** Brown, 2–3 mm long; petals 3; sepals 3; stamens 6; pistil 1. **Fruit:** Capsule purplish brown, 2–2.7 mm long.

Notes: The genus name possibly originates from the Italian word *lucciola*, "to shine or sparkle," or from the Latin *gramen luzulae* or *luxulae*, "light," a reference to the leaf hairs, which appear shiny when covered with dew.

Ferns & Fern Allies

Sensitive fern (*Onoclea sensibilis*)

FERNS & FERN ALLIES

Ferns and fern allies are herbaceous to semi-woody plants that reproduce by spores. The spores germinate on moist soil and produce a small plant called a prothallus. The prothallus produces sex cells, which after fertilization produce a new plant.

Key to the Ferns & Fern Allies

Species in this key are identified by leaf arrangement and the location of the spore-producing structures.

Sporangia and spores on underside of fronds

| *Asplenium rhizophyllum* p. 491 | *Asplenium trichomanes-racemosum*, p. 492 | *Adiantum pedatum* p. 493 | *Cryptogramma stelleri* p. 493 | *Pellaea glabella* p. 494 | *Athyrium filix-femina* p. 495 |

Sporangia and spores on underside of fronds

| *Cystopteris bulbifera* p. 495 | *Cystopteris fragilis* p. 496 | *Dryopteris carthusiana* p. 496 | *Gymnocarpium dryopteris* p. 497 | *Polystichum acrostichoides* p. 498 | *Polypodium virginianum* p. 499 |

Sporangia and spores on underside of fronds

| *Phegopteris connectilis* p. 500 | *Pteridium aquilinum* p. 501 |

Sporangia and spores on a fertile branch

| *Botrychium lunaria* p. 487 | *Botrychium multifidum* p. 488 | *Botrychium virginianum* p. 488 | *Osmunda cinnamomea* p. 489 | *Osmunda claytoniana* p. 490 | *Osmunda regalis* var. *spectabilis* p. 490 |

LEAVES BASAL

Sporangia and spores on a fertile branch

Matteuccia
struthiopteris
p. 497

Onoclea
sensibilis
p. 498

LEAVES WHORLED

Sporangia and spores in cone-like structures

Equisetum
arvense
p. 481

Equisetum
fluviatile
p. 482

Equisetum
hyemale
p. 482

Equisetum
palustre
p. 483

Equisetum
scirpoides
p. 483

Lycopodium
annotinum
p. 484

Sporangia and spores in cone-like structures

Lycopodium
complanatum
p. 485

Lycopodium
dendroideum
p. 485

Selaginella
rupestris
p. 486

HORSETAIL FAMILY
Equisetaceae

The horsetail family contains a single genus, *Equisetum*. These plants are perennial herbs with jointed, branched stems and creeping rootstalks. Reproduction is by spores, which are produced in terminal, cone-like structures called strobili. The scale-like leaves are whorled and appear at stem nodes.

Equisetum is the most ancient genus of living vascular plants. Ancestors of the horsetails were huge, tree-like plants that dominated the forests during the age of the dinosaurs. Their fossilized remains have been found throughout the world in coal formations.

Horsetails are toxic to livestock, especially horses.

Common Horsetail
Equisetum arvense

Habitat: Moist woods, roadsides, riverbanks and wet meadows.
General: Stems 10 to 50 cm tall, hollow, jointed; reproductive stems unbranched, brown to pinkish, appearing in early May; vegetative stems green; branches in whorls, 3–5-sided, appearing in late May; rhizome creeping. **Leaves:** Whorls of 8–12, scale-like, brown; fertile stem leaves 4–9 mm long; vegetative stem leaves 1–3.5 mm long. **Reproductive Structures:** Strobili terminal, cone-like, 2–3 cm long, withering soon after spores have been shed.

Notes: The genus name comes from the Latins words *equus*, "horse," and *seta*, "bristle." The species name *arvense* means "of cultivated fields." • Also called "scouring-rushes," horsetails have a high silica content. Early settlers scoured their pots and pans with these plants and also used them for sandpaper.

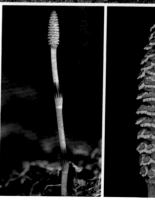

Water Scouring-rush
Equisetum fluviatile

Habitat: Marshes, bogs and roadside ditches. **General:** Semi-aquatic; emergent stems 35–115 cm tall, 3–8 mm thick, jointed, hollow, ridges 10–30; branches short when present; leaves 4–6 at each node; rhizome creeping, reddish brown. **Leaves:** Whorls of 16–20, dark brown, scale-like, 6–10 mm long. **Reproductive Structures:** Strobili terminal, cone-like, 1–2 cm long, stalked, falling off after spores have been shed.

Notes: The species name *fluviatile* means "of or from the river" or "growing in flowing water." • The tubers of this plant are reported to be edible and nutritious.

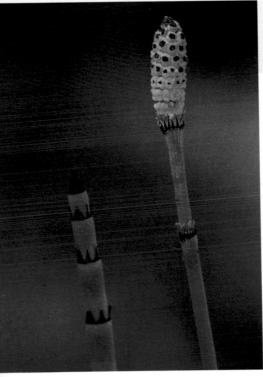

Common Scouring-rush
Equisetum hyemale

Habitat: Moist riverbanks and roadsides. **General:** Stems 18–220 cm tall, evergreen, hollow, jointed, ridges 14–50, green when young, turning ashen with age; sheaths with 2 broad, black bands; rhizome creeping, forming large colonies. **Leaves:** Whorls of 14–50, scale-like, 2–4 mm long, dark brown with papery margins, deciduous. **Reproductive Structures:** Strobili terminal, cone-like, 1–2.5 cm long, pointed.

Notes: The species name *hyemale* means "winter-flowering," a reference to the evergreen stems, which often produce strobili in late winter to early spring.

Marsh Scouring-rush
Equisetum palustre

Habitat: Wet woodlands and marshes.
General: Stems 20–80 cm tall, jointed, hollow, ridges 7–10; branches numerous, whorled, 5–6-sided; rhizome creeping, black.
Leaves: Whorls of 6–8, scale-like, 2–5 mm long, black with pale margins.
Reproductive Structures: Strobili 1–2.5 cm long, long-stalked, terminal, falling off after spores have been shed.

Notes: The species name *palustre* means "marsh-loving."

Dwarf Scouring-rush
Equisetum scirpoides

Habitat: Moss and leaf litter in moist woods. **General:** Evergreen; stems wiry, 2.5–28 cm long, often zigzagged, usually 6-sided; rhizome thread-like, creeping.
Leaves: Whorls of 3–4; scale-like, triangular, 3–4 mm long, brown with white margins. **Reproductive Structures:** Strobili black, 2–5 mm long, sharply pointed.

Notes: The species name *scirpoides* means "like *Scirpus*," a genus of aquatic, grass-like plants in the sedge family (Cyperaceae).

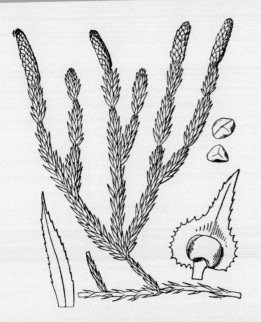

CLUBMOSS FAMILY
Lycopodiaceae

Clubmosses are low, ever-green plants with semi-woody stems. They resemble moss, hence the common name. The small, scale-like leaves are simple and appear in ranks of 4–16. Clubmosses reproduce by spores and not seeds. The spores are produced in terminal, club-shaped structures called strobili. These perennial plants are often found in moist forests and alpine meadows

Ancestors of the clubmosses were large, tree-like plants that grew to 50 m tall. These plants were the dominant woody species of the Carboniferous period. Their fossilized remains have been found in coal formations throughout the world.

Running Clubmoss
Lycopodium annotinum

Habitat: Moist, coniferous forests and rocky outcrops. **General:** Stems prostrate, to 2 m long; branches erect, 5–30 cm tall, forked once or twice. **Leaves:** In ranks of 6 or more, scale-like, 3–8 mm long, sharply pointed. **Reproductive Structures:** Strobili yellowish green, at the top of the stem, stalkless, 1.5–3 cm long; spores yellow.

Notes: The species name *annotinum* means "one year old."

Ground-cedar, Northern Running-pine
Lycopodium complanatum

Habitat: Moist, wooded areas. **General:** Stems 5–30 cm tall, resembling a young cedar tree; branches flattened, forking into numerous branchlets; horizontal stems woody, often just beneath the ground's surface. **Leaves:** In ranks of 4, scale-like, overlapping, 1–4 mm long, tips pointed. **Reproductive Structures:** Strobili 8–33 mm long, on forked stems 5–15 cm high.

Notes: The species name *complanatum* means "flat on the ground."

Ground-pine, Tree Clubmoss
Lycopodium dendroideum

Habitat: Moist, wooded areas. **General:** Stems 5–30 cm tall, resembling a young spruce tree; branches numerous; horizontal stems deep below the surface. **Leaves:** In ranks of 5–8, linear to needle-like, 2.5–5.5 mm long, sharp-tipped. **Reproductive Structures:** Strobili 1.5–3 cm long, 4–5 mm thick, short-stalked.

Notes: The genus name comes from the Greek words *lycos,* "wolf," and *pous,* "foot." The species name *dendroideum* means "tree-like."

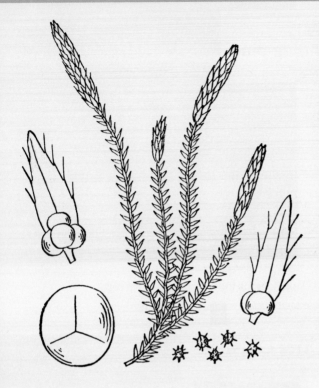

LITTLE CLUBMOSS FAMILY
Selaginellaceae

The little clubmosses are low-growing, perennial plants that resemble mosses, hence the common name. The small, scale-like leaves are spirally arranged or appear in ranks of 4. Like all fern allies, little club-mosses reproduce by spores, which are produced in the leaf-like axils of the strobilus.

Rock Spikemoss
Selaginella rupestris

Habitat: Limestone outcrops and sandy soils. **General:** Stems prostrate, forming cushion-like mats, irregularly branched. **Leaves:** Alternate but appearing in whorls of 4 or 6, crowded, linear to lance-shaped, 2.5–4 mm long, bristle-tipped, base wedge-shaped; margins with long hairs. **Reproductive Structures:** Strobilus solitary, 0.5–3.5 cm long; sporophylls triangular to oval; margins hairy to slightly toothed.

Notes: The species name *rupestris* means "growing among the rocks."

486

ADDER'S-TONGUE FAMILY

Ophioglossaceae

Members of the adder's-tongue family are often called "grape ferns," a name that refers to the terminal, grape-like clusters of sporangia. The sporangia are globe-shaped structures that produce spores. Grape ferns have fleshy roots and consist of a single leaf (frond) and clusters of sporangia. The frond is usually divided once or twice into lobed segments. Unlike true ferns, these plants do not produce fiddleheads.

Common Moonwort

Botrychium lunaria

Habitat: Meadows and open woods. **General:** Stem 3–30 cm tall, fleshy, pale green. **Leaves:** One, attached near midstem, 5–20 cm long; pinnae 3–9 pairs, fan-shaped. **Reproductive Structures:** Sporophyll panicle-like; sporangia in a grape-like cluster rising above the vegetative leaf.

Notes: The species name *lunaria* comes from the Latin word *luna*, which means "moon."

Leather Grape Fern
Botrychium multifidum

Habitat: Moist, sandy areas. **General:** Stem 10–50 cm tall, fleshy. **Leaves:** One, evergreen, leathery, shiny, green, triangular, prominent midrib; pinnae 4–10 pairs, each divided 2–4 times into narrow segments, attached near the plant base. **Reproductive Structures:** Sporophyll panicle-like; sporangia in a grape-like cluster rising above the vegetative leaf.

Notes: The species name *multifidum* refers to the leaf, which is divided into numerous segments.

Rattlesnake Fern, Grape Fern
Botrychium virginianum

Habitat: Mixedwood forests. **General:** Stem 20–80 cm tall, erect. **Leaves:** One, attached near midstem, triangular, 20–50 cm long, 10–36 cm wide, divided 3–4 times into narrow pinnae; pinnae 5–12 pairs, with prominent midrib and toothed margins. **Reproductive Structures:** Sporophyll panicle-like, 5–25 cm long, 2–3 times pinnately branched; sporangia round, in grape-like clusters rising above the vegetative leaf.

Notes: The genus name *Botrychium* comes from the Greek word *botrus*, "cluster of grapes," a reference to the fruiting structure. The species name means "of Virginia."

ROYAL FERN FAMILY

Osmundaceae

Members of the royal fern family have creeping underground stems and black, fibrous roots. The fronds may be of 1 or 2 types, vegetative or reproductive, and may be 1–2-pinnate with the pinnae either vegetative or reproductive. The veins of the leaves are dichotomously branched and run to the margins. Sori are absent. Spores are produced in sporangia borne on modified leaf segments.

Cinnamon Fern

Osmunda cinnamomea

Habitat: Moist areas with acidic soils. **General:** Fronds 60–120 cm tall; rhizomes creeping. **Leaves:** Basal; fronds of 2 types; sterile fronds pinnate with 15–25 pairs of pinnae, the pinnae 7–15 cm long, 1–3 cm wide; fertile fronds borne in the crown of sterile fronds, brown, withering soon, the pinnae 2–4 cm long, 0.8–1.5 cm wide, green becoming cinnamon-coloured with age. **Reproductive Structures:** Sporangia cinnamon-brown, about 0.5 mm wide.

Notes: The species name means "cinnamon-coloured."

Interrupted Fern
Osmunda claytoniana

Habitat: Moist, open areas. **General:** Fronds 50–180 cm tall; rhizome creeping.
Leaves: Basal; fronds of 2 types; sterile fronds 15–25 cm wide at the middle,
pinnae 15–20 pairs, usually alternate, 5–14 cm long, 1–3 cm wide, divided
further into 10–20 pinnules, each to 18 mm long and 6–8 mm wide; fertile
fronds with 1–5 pairs of reduced pinnae in the middle of the frond, fertile pinnae
1–6 cm long, 1–2 cm wide. **Reproductive Structures:** Sporangia green
becoming dark brown with age.

Notes: The species name commemorates John Clayton (1694–1773), an early
American botanist. • The reduced sterile pinnae give rise to the common name.

Royal Fern
Osmunda regalis var. *spectabilis*

Habitat: Moist, open areas and
marsh edges. **General:** Fronds
5–180 cm tall; rhizome creeping.
Leaves: Basal, blades 2-pinnate;
sterile leaves oval, 75–100 cm long,
20–55 cm wide; pinnae 5–7 pairs,
lance-shaped; pinnules alternate,
14–20, oblong, 1–7 cm long,
0.5–2.3 cm wide, margins finely
toothed; fertile branch at tip of sterile blade, pinnae numerous, oblong,
6–11 mm long, 2–3 mm wide; stipes
pinkish, shorter than fronds. **Reproductive Structures:** Sporangia
about 0.6 mm wide, green becoming
red to rusty brown with age.

Notes: The genus name refers to
Osmunder the Waterman, the Saxon
equivalent for the Norse god Thor.
The species name *regalis* means
"regal" or "royal."

SPLEENWORT FAMILY
Aspleniaceae

The spleenwort family is a small family of terrestrial ferns. The rhizomes are short and bear dark, lattice-like scales. Fronds are of 1 type and are simple to 4-pinnate. The sori, borne on the veins, are linear to crescent-shaped. The indusium originates on 1 side of the sori.

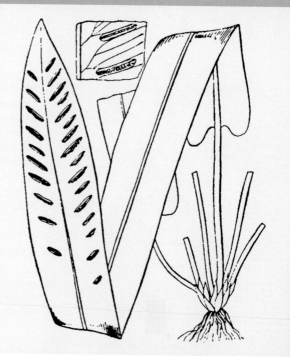

Walking Fern
Asplenium rhizophyllum

Habitat: Moss-covered limestone boulders and ledges in shady woods. **General:** Rhizome erect, 1–3 mm diameter, unbranched; scales dark brown; roots few. **Leaves:** Blades narrowly triangular to linear or lance-shaped, 1–30 cm long, 0.5–5 cm wide, leathery, base heart-shaped, margins smooth to wavy, tip

round or long-tapered (if tapered, often rooting at the tip); stipe reddish brown at the base, 0.5–12 cm long. **Reproductive Structures:** Sori numerous, scattered irregularly, often joined at vein intersections.

Notes: The genus name comes from the Greek words *a*, "without," and *spleen*, "spleen." The species name comes from the words *rhizal*, "root," and *phyllon*, "leaf," for the rooting leaves. • The common name "walking fern" refers to the frond tips, which take root and produce new plants.

491

Green Spleenwort
Asplenium trichomanes-ramosum

Habitat: Cracks and crevices of rock faces. **General:** Rhizome creeping. **Leaves:** Stipe reddish brown at the base; scales dark reddish brown to black; fronds delicate, 5–15 cm long, base wedge-shaped, 9–21 pairs of opposite to alternate pinnae; pinnae 3–8 mm long, 2–4 mm wide; margins toothed. **Reproductive Structures:** Sori 2–4 per pinna; indusium delicate, transparent.

Notes: The species name comes from trichomanes, an ancient Greek word for a type of fern, and ramosum, meaning "branched."

MAIDENHAIR FERN FAMILY
Pteridaceae

Members of the maidenhair fern family are often found growing on rock or soil. Rhizomes are short to long-creeping with hairs and scales. Fronds may be of 2 types and simple to 6-pinnate. The stipe (stalk of the frond) has persistent scales. Sori are borne on the veins and often form a continuous band. The indusium, when present, is formed by the reflexed margin of the leaf and is often referred to as a false indusium.

Maidenhair Fern
Adiantum pedatum

Habitat: Moist woods and shaded, rocky slopes. **General:** Rhizome black, 2–5 mm thick, scales deep yellow. **Leaves:** Basal, arching; fronds fan-shaped, 15–30 cm long, 15–35 cm wide, divided into 2 branches each with 3–5 pinnae; pinnae 15–25 cm long, each with 15–20 pairs of thin pinnules; stipe reddish brown to black, stiff. **Reproductive Structures:** Sori oblong to crescent-shaped, covered by reflexed pinnule margin; indusium 1–3 mm across.

Notes: The genus name *Adiantum* comes from the Greek words *a*, "without," and *dianinein*, "to wet," referring to the way the fronds repel water. The species name *pedatum* means "foot" and refers to the bird's-foot shape of the fronds.

Slender Cliffbrake
Cryptogramma stelleri

Habitat: Limestone rocks in cool, shaded areas. **General:** Rhizome slender. **Leaves:** Basal; stipe yellowish and scaly at the base; fronds of 2 types; sterile fronds 9–15 cm long, 5–6 pairs of pinnae, alternate, segments 1–3; fertile fronds 10–20 cm tall, pinnae divided 2–3 times. **Reproductive Structures:** Sori located on vein tips, partially protected by rolled pinnae margins; indusium absent.

Notes: The genus name comes from the Greek words *cryptos*, "hidden," and *gramme*, "line," a reference to the lines of sori being hidden by the curled pinnae margins.

Smooth Cliffbrake
Pellaea glabella

Habitat: Dry, limestone rock faces. **General:** Tufted; rhizome short, scaly. **Leaves:** Basal; stipe reddish brown, hairless; fronds evergreen, 0.5–20 cm tall; pinnae 0.5–3 cm long, in 5–10 pairs, lower pinnae may be lobed or divided, margins rolled. **Reproductive Structures:** Sori protected by rolled pinnae margins; indusium absent.

Notes: The genus name comes from the Greek word *pellos,* "dark," a reference to the colour of the leaves. The species name *glabella* means "smooth."

WOOD FERN FAMILY
Dryopteridaceae

The wood ferns are large ferns with creeping, scaly rhizomes. Fronds may be of 2 types and simple to 5-pinnate. The stipe (stalk of the frond) is scaly at the base. Sori are borne on or at the tips of veins, but never on the leaf margins, and may be round, oblong or elongated. The indusium, when present, is linear or sickle- or kidney-shaped.

Lady Fern
Athyrium filix-femina

Habitat: Moist, shady woodlands. **General:** Rhizome thick, creeping; scales numerous, brown, 3–10 mm long, 1–2 mm wide.
Leaves: Blades elliptic to oblong or lance-shaped, 2-pinnate, 18–30 cm long, 5–50 cm wide; pinnae linear to oblong, stalkless; pinnules pinnately divided, margins toothed; stipe straw-coloured, 7–60 cm long, dark reddish brown at the base, scales brown, 7–20 mm long, 1–5 mm wide. **Reproductive Structures:** Sori straight, hooked at one end or horseshoe-shaped; indusium toothed.

Notes: The genus name possibly comes from the Greek word *athyros*, "doorless," a reference to the late-opening indusium. The species name *filix-femina* means "lady fern."

Bulblet Bladder Fern
Cystopteris bulbifera

Habitat: Moss-covered limestone boulders and ledges. **General:** Rhizomes 2–2.5 mm thick; scales few, dark brown, old leaf bases present.
Leaves: Blades and stipe 30–80 cm long, broadly to narrowly triangular, 6–9 cm wide, bipinnate; pinnae 20–40 pairs, the lowest being the largest; pinnules 8–10 pairs, oblong, 6–15 mm long, margins toothed; rachis often producing bulblets; stipe reddish when young, turning green with age, shorter than blade. **Reproductive Structures:** Sori produced on late summer leaves; indusium cup-shaped, brown.

Notes: The genus name comes from the Greek word *kystos*, "bladder," and *pteris*, "fern," a reference to the bladder-like indusium. The species name means "bulb-bearing," alluding to the bulblets produced by this fern.

Fragile Bladder Fern, Brittle Fern

Cystopteris fragilis

Habitat: Moist slopes and rocky ledges. **General:** Rhizome creeping, light brown, 1.5–2.5 mm thick, unbranched, scales 2.5–4.5 mm long. **Leaves:** Stipe straw-coloured, scaly at the base, less than 1 mm thick; fronds tufted, 10–30 cm long, 2–8 cm wide, light green, divided 2–3 times with 9–15 pairs of subopposite to alternate pinnae, margins toothed. **Reproductive Structures:** Sori round, borne on veins; indusium hood-like.

Notes: The species name *fragilis* means "fragile" or "brittle," a reference to the thin leaf stalks.

Spinulose Wood Fern

Dryopteris carthusiana

Habitat: Swampy woods and stream-banks; scales cinnamon to pale brown. **General:** Rhizome thick, old leaf bases present. **Leaves:** Blade and stipe 15–75 cm long, 10–30 cm wide, dying in winter; stipe 4–20 cm long, scales tan-coloured; blades light green, oval to lance-shaped, 2–3 times pinnate; pinnae 10–15 pairs, triangular to lance-shaped; pinnules with 3–5 pairs of veins, margins spiny-toothed. **Reproductive Structures:** Sori borne between veins and margin; indusium kidney-shaped, less than 1 mm long.

Notes: The species name *carthusiana* commemorates the monks of the Carthusian Order of Grande Chartreuse in Grenoble, France.

Oak Fern

Gymnocarpium dryopteris

Habitat: Moist woodlands. General:
Rhizome black, 1–1.5 mm thick, scaly;
scales pale brown, 1–4 mm long.
Leaves: Blade and stipe 12–42 cm
long; blade broadly triangular, 5–18 cm
long, 5–25 cm wide, yellowish green,
bipinnate; pinnae 2–12 cm long; pinnules oblong, 0.8–1.8 cm long, margins
smooth to toothed; stipe 9–28 cm
long, scales 2–6 mm long. Reproductive Structures: Sori round, borne
along underside veins; indusium absent.

Notes: The genus name *Gymnocarpium*
means "bearing naked fruit," and refers
to the absence of the indusium, a pro-
tective covering for the sporangia.

Ostrich Fern

Matteuccia struthiopteris

Habitat: Moist, wooded areas.
General: Rhizome thick; scales
thin, brown. Leaves: Fronds of
2 types; sterile fronds 4–6,
50–175 cm tall, 15–35 cm wide,
forming a vase-like arrangement,
pinnae 20–30 pairs, the largest
6.5–13.5 cm long; fertile fronds
stiff, 20–60 cm tall, 4–12 cm
wide, olive green turning dark
brown at maturity, pinnae
30–45 pairs, 3–5.6 cm long,
2–7 mm wide; stipe 11–24 cm
long. Reproductive Struc-
tures: Sori protected by rolled
margins of frond; indusium
hood-like, transparent, lacerated;
spores released in early spring
from previous season's fronds.

Notes: The genus name commemorates Italian physicist
Carlo Matteucci (1800–68). The species name comes from the Greek words
struthokamelos, "ostrich," and *pteris*, "fern." • The common name alludes to the resemblance of the fronds to ostrich feathers.

Sensitive Fern

Onoclea sensibilis

Habitat: Marshes, swamps and wet woodlands. **General:** Rhizome 5–7 mm in diameter; scales broad, fibrous. **Leaves:** Fronds of 2 types; sterile fronds yellowish green, triangular, 13–34 cm long, 15–30 cm wide, stipe black and 22–28 cm long, pinnae 5–11 pairs, lance-shaped, 9–18 cm long, margins smooth to wavy; fertile fronds oblong, 7–17 cm long, 1–4 cm wide, green becoming black at maturity, stipe 19–40 cm long, pinnae linear, 5–11 pairs, 2.5–5 cm long, segment margins rolled and forming a bead-like structure 2–4 mm across. **Reproductive Structures:** Sori borne on veins in bead-like structure; spores released in early spring from previous season's fronds.

Notes: The genus name comes from the Greek words *onos*, "vessel," and *kleio*, "to close," a reference to the inrolled pinnule margins. • The common name refers to the susceptibility of the leaves to frost.

Christmas Fern

Polystichum acrostichoides

Habitat: Shady woods and rocky slopes. **General:** Rhizome 6–9 mm thick, scaly at tip. **Leaves:** Fronds of 2 types; sterile fronds linear to lance-shaped, arched, 30–80 cm long, 5–12 cm wide; pinnae alternate, 20–35 pairs, 2–6 cm long, margins spiny-toothed, stipe 7–20 cm long and densely scaled; fertile fronds with smaller pinnae at tip, larger sterile pinnae at the base. **Reproductive Structures:** Sori borne in 2 rows on either side of the vein, appearing to completely cover the lower surface of the pinna; indusium round.

Notes: The genus name comes from the Greek words *poly*, "many," and *stichos*, "rows," a reference to the multiple rows of sori. The species name means "like *Acrostichum*," a closely related genus of fern.

POLYPODY FAMILY

Polypodiaceae

Ferns in Polypodiaceae have short- to long-creeping, scaly rhizomes. The fronds may be of 2 types, with either simple or pinnate blades. The stipe (stalk of the frond) usually lacks scales. The round to oblong or elongated sori are borne on the veins. The indusium is absent.

This family formerly included the spleenworts (Aspleniaceae), maidenhair ferns (Pteridaceae), wood ferns (Dryopteridaceae) and marsh ferns (Thelypteridaceae).

Rock Polypody

Polypodium virginianum

Habitat: Moist, rocky outcrops. **General:** Rhizome 1.5–6 mm thick, whitish; scales brown, 2.5–3 mm long, fibrous; bittersweet-tasting. **Leaves:** Evergreen, leathery; blades oblong to linear, 8.5–40 cm long, 3–7 cm wide; pinnae oblong, 10–20 pairs, 3–7 mm wide, margins smooth; stipe 3–15 cm long, 1–2 mm thick, scaly at the base. **Reproductive Structures:** Sori round, on edges of pinnae; indusium absent.

Notes: The genus name comes from the Greek words *poly*, "many," and *podos*, "foot." The species name means "of Virginia." • The rhizomes have a licorice flavour and contain compounds that are reported to be about 300 times sweeter than sugar.

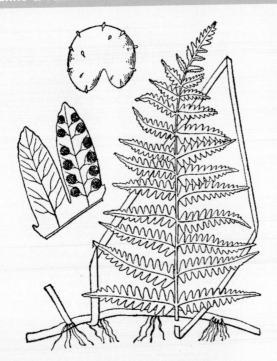

MARSH FERN FAMILY
Thelypteridaceae

Ferns in this family have creeping stems that are scaly at the tip. The fronds are pinnate to bipinnate with veins running to the margins. The sori are often located midway between the veins and the leaf margins. The indusium, when present, is kidney-shaped.

Northern Beech Fern
Phegopteris connectilis

Habitat: Moist, acidic soils.
General: Rhizomes 1–3 mm thick, densely hairy, scaly.
Leaves: Fronds 15–60 cm tall; blades narrowly triangular, 12–25 cm long; pinnae 6–12 cm long, 1–3.3 cm wide, the lowermost separate, the upper connected to the rachis; stipe straw-coloured, 15–36 cm long, scales brown. **Reproductive Structures:** Sori located on veins; indusium absent.

Notes: The genus name comes from the Greek words *phegos*, "oak" or "beech," and *pteris*, "fern." The species name *connectilis*, "connected," refers to the upper pinnae, which are fused to the rachis.

BRACKEN FERN FAMILY

Dennstaedtiaceae

The bracken fern family is a small family of terrestrial ferns. Rhizomes are short- to long-creeping and bear hairs or occasionally scales. Fronds are of a single type, 1 to several times pinnate. The sori are borne at or near the blade margins on vein tips. The indusium may be fused with the curled blade margin.

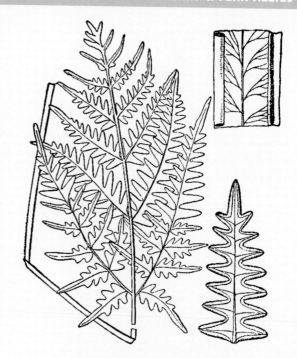

Eastern Bracken Fern

Pteridium aquilinum

Habitat: Moist woodlands. **General:** Rhizome black, scaly, 1.25–2.5 cm in diameter. **Leaves:** Fronds coarse, somewhat triangular, 45–150 cm tall, 20–60 cm across, 2–3-pinnate; pinnae triangular, 25–40 pairs, opposite, margins rolled; stipe 15–100 cm long, scales rusty brown. **Reproductive Structures:** Sori linear, partially covered by rolled pinnae margins; indusium flap-like, hairy.

Notes: The genus name comes from the Greek word *pteris*, meaning "fern." The species name *aquilinum* means "eagle-like." • Bracken fern, whether fresh or dried, is poisonous to livestock. It is believed that ingestion of the plant causes vitamin B deficiency. • The consumption of fiddleheads of this species should be discouraged because they have been implicated in several cases of cancer.

Glossary

Page numbers refer to the diagrams in which the word is illustrated.

achene: a dry, one-seeded fruit that does not open when ripe

annual: a plant that completes its life cycle in a single growing season

anther: the pollen-producing sac of a stamen

adventitious: roots that originate from a stem or leaf

alvar: an open, meadow-like habitat with a thin layer of soil covering limestone

areola (*pl.* areolae): in cacti, a pad-like bud from which spines grow

auricle: a small, ear-like appendage; in grasses, auricles are found at the junction of the blade and the sheath

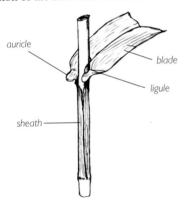

auricle
blade
ligule
sheath

awn: a stiff bristle

axil: the angle between a leaf and the stem

axillary: arising from an axil

basal: from the base; often refers to leaves (p. 10)

biennial: a plant that takes 2 years to complete its life cycle

bog: a peatmoss-filled depression that receives water and nutrients from rainfall; soil tends to be strongly acidic

bract: a leaf-like structure found below a flower or flower cluster

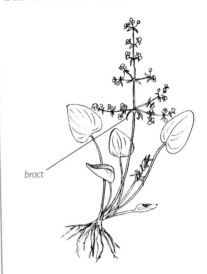

bract

calcareous: soil containing high amounts of calcium carbonate; soil pH is 7.3–8.3.

calyx (*pl.* calyces): a collective term for the sepals of a flower

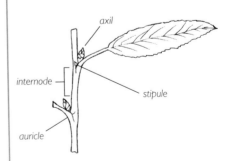

axil
internode
stipule
auricle

capitulum (*pl.* capitula): a head

capsule: a dry fruit that splits open at maturity

capsule

carpel: the basic unit of a pistil; a pea pod has a single carpel, whereas an orange has several carpels (segments)

caryopsis: the fruit or grain of a grass

caryopsis

catkin: a cylindrical cluster of numerous small flowers, usually of a single sex and lacking petals; typically found in willows, alders and birches

cleft: deeply lobed

collar: in grasses, the part of the leaf where the blade and the sheath are connected

column: a structure formed by the fusion of the stamens to the style and stigma

compound leaf: a leaf composed of 2 or more leaflets (p. 10)

coniferous: cone-bearing

corolla: a collective term for the petals of a flower

corona: an appendage derived from the petals and stamen filaments

corymb: a flat-topped flower cluster with flower stalks of different lengths

cupule: a cup-like structure formed by floral bracts surrounding the ovary or ovaries; found in plants of the oak family (Fagaceae)

cyathium: a specialized flower cluster in which male and female flowers appear to resemble a single flower; found in plants of the spurge family (Euphorbiaceae)

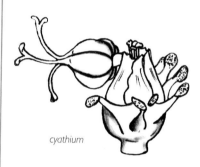

cyathium

cyme: a flower cluster with the oldest flowers at the top and younger flowers below

deciduous: having structures (e.g., leaves, petals) that fall off at maturity or in autumn

decompound: several times compound; usually refers to leaves

decussate: opposite pairs of leaves arranged at right angles to those above and below

dichotomous: branching into 2 equal parts (e.g., leaf veins)

dicotyledon: a plant whose seeds produce 2 cotyledons (embryonic leaves); includes most families of flowering plants

dioecious: producing male and female flowers on separate plants

disc floret: a small flower, usually tubular; found in plants of the aster family (Asteraceae)

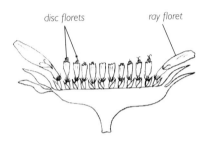

disc florets   *ray floret*

drupe: a fleshy fruit, usually 1-seeded (e.g., a cherry)

drupelet: one part of an aggregate fruit (e.g., a raspberry)

emergent: a plant whose stem rises out of the water

equitant: with leaves folded lengthwise and appearing inserted edgewise on the stem (e.g., irises)

exserted: sticking out; often refers to stamens protruding from the corolla

family: a group of organisms with similar characteristics

fascicle: a bundle or cluster of structures; often used in reference to pine needles

fen: a peatland in which groundwater discharges to the surface; soils are less acidic than those of a **bog**

frond: the leaf of a fern

fruit: a mature ovary, often containing seeds

genus (*pl. genera*): a group of species with similar characteristics; in taxonomy, the rank below family

gland: a structure that produces nectar or another sticky substance

glume: in grasses, a small bract at the base of a spikelet

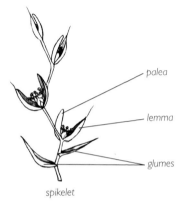

palea

lemma

glumes

spikelet

gynostegium: a structure formed from the fusion of the stamens to the stigma; found in plants of the dogbane family (Apocynaceae)

haustorium (*pl.* haustoria): a highly modified stem or root that enables a parasitic plant to attach itself to a host

herb: a plant with no woody stems

host: an organism on which another organism lives as a parasite

hypanthium: a structure formed by the fusion of the bases of the sepals, petals and stamens

indusium: in some ferns, a flap-like structure that protects the sori

inflorescence: a flower cluster

internode: the section of stem between 2 leaves (p. 502)

involucre: a bract or group of bracts below a flower cluster

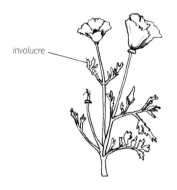

involucre

legume: a dry, pod-like fruit derived from a single carpel; found in plants of the pea family (Fabaceae)

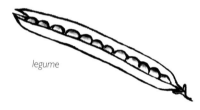

legume

lemma: in grasses, one of 2 small bracts that enclose the flower (p. 504)

lenticel: a specialized opening in the bark of trees that allows for gas exchange

ligule: in grasses, a small projection from the top of the sheath (p. 502)

locule: a chamber inside an ovary

loment: a dry, pod-like fruit derived from a single carpel and breaking into 1-seeded segments at maturity; found in plants of the pea family (Fabaceae)

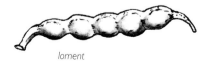

loment

margin: the edge of a leaf or petal

marl: loosely arranged accumulations of shells and calcareous sand, silt and clay; soil is often white owing to the high calcium carbonate content

mericarp: a single-carpel unit of a schizocarp that breaks away at the axis of the ovary

monocot: a plant whose seeds produce a single cotyledon (embryonic leaf); includes lilies, irises, orchids and grasses, among others

muskeg: undrained land, often with peatmoss; synonymous with **bog**

mycorrhizal fungi: soil fungi that form a symbiotic relationship with the roots of most plant species; important in the uptake of nutrients for vascular plants

nerve: a vein of a leaf, petal, sepal or other structure

node: the point at which a leaf is attached to the stem

nodule: a swelling on the root of a leguminous plant; may contain nitrogen-fixing bacteria

noxious: a plant capable of causing damage to agricultural crops

nut: a hard, dry, 1-seeded fruit that does not open at maturity

nutlet: a small nut

ocrea: a sheath formed by the fusion of 2 stipules; found at leaf nodes

ocrea

ovary: the part of the pistil containing the ovules (undeveloped seeds) (p. 10)

ovate: egg-shaped and broader at the base; usually refers to leaves

palmate: having 3 or more lobes or leaflets arising from a single point, like the fingers on a hand; usually refers to leaves (p. 10)

panicle: a branching flower cluster that is often pyramid-shaped

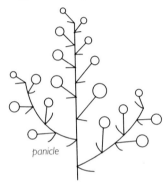

panicle

pappus: hairs or bristles attached to a seed; found in plants of the aster family (Asteraceae)

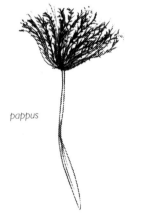

pappus

parasite: a plant that obtains food and nutrients from another living plant

pepo: a type of berry with a thick rind and soft interior; found in the gourd family (Cucurbitaceae)

perennial: a plant, or part of a plant, that lives for more than 2 growing seasons

perfect: a flower that has both stamens and pistils

perfoliate: having fused leaf bases

perianth: a collective term for petals and sepals

perigynium (*pl.* perigynia): a hollow, sac-like bract that encloses the female flower in sedges (*Carex* in Cyperaceae)

petal: one of a number of modified flower leaves, often brightly coloured to attract pollinators, that surround the reproductive parts of a flower (p. 10)

petiole: a leaf stalk

pinna (*pl.* pinnae): the primary division of a compound leaf

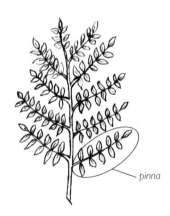

pinna

pinnate: a compound leaf with leaflets arranged on both sides of the stalk (p. 10)

pistil: the female part of a flower; includes the stigma, style and ovary

pistillate: a flower having female parts only

pod: a dry fruit that releases its seeds when mature

pollinia: a pollen mass often shed as a unit; found in plants of the dogbane family (Apocynaceae) and orchid family (Orchidaceae)

poultice: a moist mass of herbs applied to the body as an external medicine

prickle: a spiny structure on the surface of a plant

prothallus: in ferns and fern allies, a small structure that produces male and female sex organs

raceme: an unbranched cluster of stalked flowers on an elongated central stalk, blooming from the bottom up

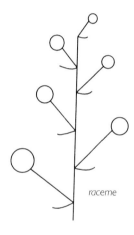

raceme

rachis: the axis of a compound leaf or the stem of a elongated flower cluster

ray floret: a small flower, usually strap-like, found in plants of the aster family (Asteraceae) (p. 504)

recurved: bent or curled backward

reflexed: abruptly bent backward

resupinate: turned upside down during development; found in plants of the orchid family (Orchidaceae)

revolute: with margins curled under

rhizome: an elongated, underground stem

rhizome

rosette: a cluster of basal leaves

samara: a dry, usually 1-seeded fruit with wing-like structures (e.g., the fruit of an ash or maple)

saprophyte: a plant that obtains food and nutrients from decaying material

scape: a leafless stem that produces flowers

schizocarp: a fruit that breaks into sections containing 1 or more seeds

schizocarp

scurfy: covered with scales

semi-aquatic: a plant capable of living in water and on wet shorelines

sepal: the outermost part of a flower; usually green and leaf-like (p. 10)

sessile: stalkless; often refers to leaves

sheath: the base of a leaf that surrounds the stem

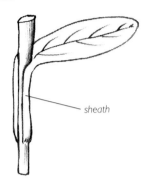

sheath

shrub: a woody plant having many stems arising from a root

silicle: a dry, pod-like fruit, more than twice as long as wide, derived from a 2-unit carpel; found in plants of the mustard family (Brassicaceae)

silique: a dry, pod-like fruit, twice as long as wide or less, derived from a 2-unit carpel; found in plants of the mustard family (Brassicaceae)

simple leaf: a leaf with a single blade (p. 10)

sorus (*pl.* sori): a cluster of sporangia on the underside of a fern leaf

spadix: a thick, fleshy spike bearing many small flowers; often associated with a **spathe**

spathe: a large bract that encloses a flower cluster

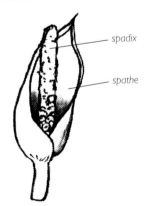

spadix

spathe

spatulate: spoon-shaped

species: a group of similar plants capable of interbreeding to produce offspring like themselves

spike: an unbranched flower cluster in which the flowers are attached directly to the main stem

spike

spikelet: a small spike; a floral unit of grasses and sedges (p. 504)

sporangia: a spore-producing sac

spore: a reproductive structure in ferns and fern allies

sporocarp: a structure containing sporangia; often found in aquatic ferns

sporophyll: a leaf bearing 1 or more sporangia

spur: a slender projection from the corolla or calyx

stamen: the male part of a flower (p. 10)

staminate: a flower having male sex organs only

staminode: a sterile stamen

sterile: without functional reproductive structures

stigma: the receptive tip of the pistil on which pollen grains land (p. 10)

stipe: the stalk of a fern leaf

stipule: a membranous structure found at the base of a leaf (p. 502)

stolon: a horizontal, aboveground stem that roots at the nodes to produce new plants

strobilus (*pl.* strobili): in ferns, a cone-like structure bearing sporangia

strobilus

style: the elongated part of the pistil between the stigma and the ovary (p. 10)

stylopodium: an enlarged structure borne atop the ovary; found in plants of the carrot family (Apiaceae)

subopposite: leaves that are paired but offset so that they are not truly opposite

subshrub: a weakly woody shrub or a herbaceous plant with a woody base

subspecies: in taxonomy, the rank below species; genetically distinct groups within a single species

subtend: to be directly below another structure (e.g., a bract at the base of a flower)

sucker: a stem that grows from the underground root of a plant

swamp: a wooded wetland that is often flooded for part of the year

taproot: the main root of a plant (e.g., a carrot)

tendril: a clasping or twining part of a leaf

tepals: sepals and petals of similar size, appearance and colour

terminal: the end of a stem or leaf

thallus (*pl.* thalli): a plant with undifferentiated stems and leaves (e.g., duckweeds)

tiller: in grasses, a stem or branch that originates at the base of the main stem

tree: a woody plant, typically over 3 m tall, with a single stem arising from the root

turion: a scaly shoot that develops from an underground bud

umbel: a flower arrangement in which the flower stalks originate from a single point

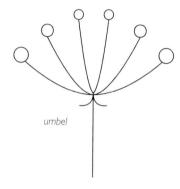

umbel

umbellet: a small umbel; one of the clusters in a compound umbel

unisexual: a flower with functional stamens or pistils, but not both

winter annual: a plant that germinates and produces a leafy rosette in fall, completing its life cycle the following growing season

References

Alex, J.F. 1992. *Ontario Weeds: Descriptions, Illustrations and Keys to their Identification.* Publication 505 Agdex 640, Ontario Ministry of Agriculture and Food, Toronto, ON.

Bannville, Diana. 1990. *Vascular Plants of Metropolitan Toronto, ON.* 2nd ed. Toronto Field Naturalists, Toronto, ON.

Best, K.F., J. Looman and J. Baden Campbell. 1971. *Prairie grasses identified and described by vegetative characters.* Publication 1413, Canadian Department of Agriculture, Ottawa, ON.

Cavers, P.B. 1995. *The Biology of Canadian Weeds.* Contributions 62–83. Agriculture Institute of Canada, Ottawa, ON.

Coombes, Allan J. 1994. *Dictionary of Plant Names.* Timber Press, Portland, OR.

Crockett, Lawrence J. 1977. *Wildly Successful Plants: A Handbook of North American Weeds.* MacMillan Publishing, New York, NY.

Dickinson, Richard, and France Royer. 1999. *Weeds of Canada and the Northern United States.* University of Alberta Press, Edmonton, AB.

Fernald, M.L. 1950. *Gray's Manual of Botany.* 8th (Centennial) ed. American Book Company, New York, NY.

Frankton, Clarence, and Gerald A. Mulligan. 1970. *Weeds of Canada.* Publication 948, Canada Department of Agriculture, Ottawa, ON.

Gregory, Dan, and Roderick MacKenzie. 1986. *Toronto's Backyard: A Guide to Selected Nature Walks.* Douglas and McIntyre, Vancouver, BC.

Gleason, Henry A. 1963. *The New Britton and Brown Illustrated Flora of the Northeastern United States and Adjacent Canada.* Hafner Publishing, New York and London.

Holm, LeRoy G., Donald L. Plucknett, Juan V. Pancho and James P. Herberger. 1977. *The World's Worst Weeds: Distribution and Biology.* University Press of Hawaii, Honolulu, HI.

Hosie, R.C. 1979. *Native Trees of Canada.* Fitzhenry and Whiteside, Toronto, ON.

Kershaw, Linda. 2001. *Trees of Ontario.* Lone Pine Publishing, Edmonton, AB.

Lauriault, Jean. 1989. *Identification Guide to the Trees of Canada.* Fitzhenry and Whiteside, Markham, ON.

Legasy, Karen. 1995. *Forest Plants of Northeastern Ontario.* Lone Pine Publishing, Edmonton, AB.

Little, Elbert L. 1998. *National Audubon Society Field Guide to North American Trees: Eastern Region.* Chanticleer Press, New York, NY.

Looman, J., and K.F. Best. 1987. *Budd's Flora of the Canadian Prairie Provinces.* Agriculture Canada, Ottawa, ON.

Moss, E.H. 1983. *Flora of Alberta.* 2nd ed. Reviewed by J.G. Packer. University of Toronto Press, Toronto, ON.

Mulligan, Gerald A. 1976. *Common Weeds of Canada.* McClelland and Stewart, Toronto, ON.

————. 1979. *The Biology of Canadian Weeds.* Contributions 1–32. Publication 1693. Communications Branch, Agriculture Canada, Ottawa, ON.

————. 1984. *The Biology of Canadian Weeds.* Contributions 33–61. Publication 1765. Communications Branch, Agriculture Canada, Ottawa, ON.

Mulligan, Gerald A., and Derek B. Munro. 1990. *Poisonous Plants of Canada.* Agriculture Canada, Canadian Government Publishing Centre, Ottawa, ON.

Neal, Bill. 1992. *Gardener's Latin: A Lexicon*. Algonquin Books, Chapel Hill, NC.

Newmaster, Steven G., Allan G. Harris and Linda J. Kershaw. *Wetland Plants of Ontario*. Lone Pine Publishing, Edmonton, AB.

Niering, William, and Nancy C. Olmstead. 1998. *National Audubon Society Field Guide to North American Wildflowers: Eastern Region*. Chanticleer Press, New York, NY.

Porsild, A.E., and W.J. Cody. 1980. *Vascular Plants of Continental Northwest Territories, Canada*. National Museums of Canada, Ottawa, ON.

Scoggan, H.J. 1978–79. *The Flora of Canada*. 4 vols. Canadian Museum of Nature, National Museums of Canada, Ottawa, ON.

Smith, James Payne. 1977. *Vascular Plant Families*. Mad River Press. Eureka, CA.

Soper, James H., and Margaret L. Heimburger. 1994. *Shrubs of Ontario*. Royal Ontario Museum, Toronto, ON.

Taylor, T.M.C. 1971. *The Ferns and Fern-allies of British Columbia, Handbook No. 12*. British Columbia Provincial Museum, Victoria, BC.

Tomikel, John. 1976. *Edible Wild Plants of Eastern United States and Canada*. Allecheny Press, California, PA.

Turner, Nancy J., and Adam F. Szczawinski. 1978. *Wild Coffee and Tea Substitutes of Canada*. National Museum of Natural Sciences, Ottawa, ON.

—————. 1979. *Edible Wild Fruits and Nuts of Canada*. National Museum of Natural Sciences, National Museums of Canada, Ottawa, ON.

Walters, Dirk R., and David J. Keil. 1996. *Vascular Plant Taxonomy*. 4th ed. Kendall Hunt Publishing, Dubuque, IA.

Online Sources

Angiosperm Phylogeny Website
www.mobot.org/MOBOT/Research/APweb/welcome.html

Canadian Botanical Conservation Network: Invasive Plants
archive.rbg.ca/cbcn/en/projects/invasives/invade1.html

Global Invasive Species Database
www.issg.org/database/welcome

GrassBase: The Online World Grass Flora Descriptions, Royal Botanic Gardens, Kew
http://www.kew.org/data/grasses-db.html

Flora of North America: eFloras.org
efloras.org/index.aspx

Illinois Wildflowers
www.illinoiswildflowers.info/index.htm

Integrated Taxonomic Information System
www.itis.gov

Missouriplants.com
www.missouriplants.com/index.html

Robert W. Freckmann Herbarium, University of Wisconsin
wisplants.uwsp.edu

U.S. Geological Survey, Northern Prairie Wildlife Research Center: Midwestern Wetland Flora
www.npwrc.usgs.gov/resource/plants/floramw

Index

Entries in **bold** typeface refer to the primary species accounts.

INDEX

One-flowered clintonia (*Moneses uniflora*)

About the Author & Photographer

Richard Dickinson and France Royer have been working together since 1989. Richard graduated with a BSc from the University of Alberta in 1985 and now resides in Toronto. France lives in Edmonton and is a self-taught photographer. They enjoy travelling and exploring the diversity of plant habitats throughout North America.

Since 1996, they have written *Wildflowers of Edmonton and Central Alberta*, *Wildflowers of Calgary and Southern Alberta*, and *Weeds of Canada and the Northern United States*. Other field guides are in the planning stage.